MDPI
St. Alban-Anlage 66
4052 Basel
Switzerland
Tel. +41 61 683 77 34
Fax +41 61 302 89 18
www.mdpi.com

Remote Sensing Editorial Office
E-mail: remotesensing@mdpi.com
www.mdpi.com/journal/remotesensing

Remote Sensing of Land Surface Phenology

www.ingramcontent.com/pod-product-compliance
Ingram Content Group UK Ltd.
Pitfield, Milton Keynes, MK11 3LW, UK
UKHW060458050826
14089UKWH00044B/123

Remote Sensing of Land Surface Phenology

Editors

Xuanlong Ma
Jiaxin Jin
Xiaolin Zhu
Yuke Zhou
Qiaoyun Xie

MDPI • Basel • Beijing • Wuhan • Barcelona • Belgrade • Manchester • Tokyo • Cluj • Tianjin

MDPI

Editors

Xuanlong Ma
College of Earth and Environmental Sciences,
Lanzhou University,
Lanzhou 730000, China

Jiaxin Jin
College of Hydrology and Water Resources,
Hohai University,
Nanjing 210024, China

Xiaolin Zhu
Department of Land Surveying and Geo-Informatics,
The Hong Kong Polytechnic University,
Hong Kong 999077, China

Yuke Zhou
Key Laboratory of Ecosystem Network Observation and Modeling, Institute of Geographic Sciences and Natural Resources Research,
Chinese Academy of Sciences,
Beijing 100101, China

Qiaoyun Xie
Faculty of Science,
University of Technology Sydney,
Ultimo, NSW 2007, Australia

Editorial Office
MDPI
St. Alban-Anlage 66
4052 Basel, Switzerland

This is a reprint of articles from the Special Issue published online in the open access journal *Remote Sensing* (ISSN 2072-4292) (available at: https://www.mdpi.com/journal/remotesensing/special_issues/land_phenology).

For citation purposes, cite each article independently as indicated on the article page online and as indicated below:

LastName, A.A.; LastName, B.B.; LastName, C.C. Article Title. *Journal Name* **Year**, *Volume Number*, Page Range.

ISBN 978-3-0365-5325-2 (Hbk)
ISBN 978-3-0365-5326-9 (PDF)

Contents

About the Editors

Xuanlong Ma

Xuanlong Ma obtained his Ph.D. in ecological remote sensing from the University of Chinese Academy of Sciences. He is now a professor in the College of Earth and Environmental Sciences, Lanzhou University. His areas of expertise are biophysical remote sensing and terrestrial ecology, and his main research interest is in integrating measurements from spaceborne sensors, field instruments, and flux towers to study and analyze vegetation phenological and functional responses to hydroclimatic impacts, with the overarching aim of assessing ecosystem resilience. He is an Editorial Board Member of Remote Sensing and the recipient of the 2021 Kamide Lecture Award from the Asia–Oceania Geoscience Society.

Jiaxin Jin

Jiaxin Jin obtained his Ph.D. in Cartography and Geographical Information Systems from Nanjing University. He is now a professor in the College of Hydrology and Water Resources, Hohai University. His areas of expertise are ecological remote sensing and climate change. His main research interest is in detection and simulation of plant phenology, coupled carbon–water cycles of terrestrial ecosystems in response to climate change, and land-use/cover changes and its eco-hydrological effects using multi-source remotely sensed and field measurements. He is the Review Editor of Frontiers in Plant Science.

Xiaolin Zhu

Dr Zhu received his BSc in 2007 and MSc in 2010, both from Beijing Normal University. He received his PhD in geography at Ohio State University in 2014. He is currently an associate professor in the Department of Land Surveying and Geo-Informatics, The Hong Kong Polytechnic University. His research interests are remote sensing methods and applications, including satellite image processing, data fusion, time series data analysis, night-time light remote sensing, image classification, change detection, environmental and urban remote sensing. He received the Robert N. Colwell Memorial Fellowship Award from the American Society of Photogrammetry and Remote Sensing and the Li Xiaowen Remote Sensing Young Scholar Award. He was awarded China's Excellent Young Scientists Fund in 2020. He is currently an Editorial Board Member for Science of Remote Sensing, Remote Sensing, National Remote Sensing Bulletin, and Big Earth Data.

Yuke Zhou

Yuke Zhou received his Ph.D. in Cartography and Geographic Information Systems from the Institute of Geographical Sciences and Natural Resources Research (IGSNRR), Chinese Academy of Sciences in 2013, where he currently works as an Associate Researcher. His areas of expertise consist of spatial data mining, high-performance GIS computing, and applications of GIS and RS in ecosystem change. His main research interest is in exploring vegetation dynamics (e.g., plant phenological metrics, interannual variability of vegetation indices) and the interactions with vegetation productivity and climatic change.

Qiaoyun Xie

Dr. Qiaoyun Xie is currently a Chancellor's Postdoctoral Research Fellow/Lecturer at the University of Technology Sydney in Australia, and a Visiting Scholar at UC Berkeley in the US. Her main research interest is measuring climate change impacts on vegetation phenology using remote sensing technologies and ground observations. Her expertise spans applied plant biology and plant phenology fields, with a particular focus on quantifying vegetation growth and their shifting seasonality with climate variability at continental and global scales. These provide essential insights for research and management of ecosystem resilience to climate change, bushfire fuel accumulation, crop yields, airborne allergens, native vegetation conditions, and agricultural management. She improved the LSP algorithm locally and developed Australia's national land surface phenology dataset at 500 m resolution.

remote sensing

Editorial

Remote Sensing of Land Surface Phenology: Editorial

Xuanlong Ma [1,*], Jiaxin Jin [2], Xiaolin Zhu [3], Yuke Zhou [4] and Qiaoyun Xie [5]

1 College of Earth and Environmental Sciences, Lanzhou University, Lanzhou 730000, China
2 College of Hydrology and Water Resources, Hohai University, Nanjing 210024, China
3 Department of Land Surveying and Geo-Informatics, The Hong Kong Polytechnic University, Hong Kong SAR, China
4 Key Laboratory of Ecosystem Network Observation and Modelling, Institute of Geographic Sciences and Natural Resources Research, Chinese Academy of Sciences, Beijing 100101, China
5 School of Life Sciences, Faculty of Science, University of Technology Sydney, Sydney, NSW 2007, Australia
* Correspondence: xlma@lzu.edu.cn

1. Background

Land surface phenology (LSP) is an important research field in terrestrial remote sensing and has become an indispensable approach in global change research, as evidenced by many important scientific findings supported by LSP in recent decades. LSP involves the use of remote sensing to monitor seasonal dynamics in vegetated land surfaces and to retrieve phenological metrics (transition dates, rate of change, annual integrals, etc.). LSP is an essential indicator of global change and has played a pivotal role in shaping our understanding about how terrestrial ecosystems are responding to climate change and human activities. Both regional and global LSP products have been routinely generated and played prominent roles in modeling crop yield, ecological surveillance, identifying invasive species, modeling the terrestrial biospheric processes, and assessing global change impacts on urban and natural ecosystems.

Recent advances in field and spaceborne sensor technologies, as well as data fusion techniques, have enabled novel LSP retrieval algorithms that refine LSP retrievals at even higher spatiotemporal resolutions, providing new insights into ecosystem dynamics. Meanwhile, rigorous assessment of the uncertainties in LSP retrievals is undergoing, and efforts to reduce these uncertainties are also forming an active research field. In addition, open-source software and hardware are being developed and have greatly facilitated the use of LSP metrics by scientists beyond the remote-sensing community. As such, we organized this Special Issue to cover the latest developments in sensor technologies, LSP retrieval algorithms and validation strategies, and the use of LSP products in a variety of fields. The objective of this Editorial is to offer the readers an overview of the latest developments in the LSP field and facilitate the distribution of the scientific knowledge from this Special Issue.

Citation: Ma, X.; Jin, J.; Zhu, X.; Zhou, Y.; Xie, Q. Remote Sensing of Land Surface Phenology: Editorial. *Remote Sens.* **2022**, *14*, 4310. https://doi.org/10.3390/rs14174310

Received: 25 August 2022
Accepted: 29 August 2022
Published: 1 September 2022

2. Papers in the Special Issue

The 15 papers published in this Special Issue represent diverse themes in the LSP research field (see Table 1). Figure 1 presents the major keywords contained in the abstracts of the papers. Although natural ecosystems were mostly studied [1–3], urban [4,5] and agricultural ecosystems [6] were also considered in the as an important field of LSP applications. High-altitude and high-latitude ecosystems gain particular attention in this Special Issue, likely due to the sensitivity of these ecosystems to climate change [7–12]. Most studies have a temporal scale greater than a decade, with a few having used NOAA/AVHRR data of longer than three decades [10,13]. Additionally, it can be seen that the use of cloud-based remote-sensing big data analytics facilities such as Google Earth Engine (GEE) have also been adopted by several studies (e.g., [7–9]). While a majority of the papers focused on scientific applications, some studies also looked at the theoretical aspect of LSP such as the scaling effect [13]. Data-wise, most studies used vegetation indices due to their long-term continuity; a few papers also exploited the potential of emerging proxies such as solar-induced chlorophyll fluorescence (SIF) [14]. Lastly, half of the studies published in this Special Issue used some type of ground phenology data, including phenocam, traditional phenology observations, and eddy–covariance flux towers.

Table 1. Publication summary of the Special Issue.

Publication	Topic	Satellite Data	Inclusion of Ground Phenology Data	Target Ecosystems	Temporal Scale	Analytic Platform
Kim et al. [4]	Impact of urbanization on phenology	MODIS EVI	Yes (phenocam)	Urban, rural, and natural	2016	Local
Wang et al. [8]	Mechanism and impact of climatic and soil factors on the phenology of alpine ecosystems	MODIS NDVI	Yes (phenology stations)	Alpine meadow and alpine steppe	2001–2018	GEE
Ma et al. [7]	Phenological trends of GPP dynamics in the Arctic	MODIS GPP	Yes (Fluxnet)	Arctic ecosystems	2001–2019	GEE
Zhang et al. [6]	Crop phenology and yield prediction	MODIS NDVI, EVI, and LAI	No	Maize	2010–2015	Local
Ji et al. [5]	Urban heat island effect on spring phenology	MODIS EVI, LST, Phenology	No	Urban, rural	2006–2018	Local
Guo et al. [9]	Mountain phenology response to meteorological drivers	MODIS NDVI	No	Mountainous ecosystems	2001–2019	Local
Chen et al. [13]	Scaling effect of LSP over complex terrain	MODIS NDVI, GIMMS3g NDVI	Yes (phenology stations)	Grassland, cropland, and forests	1982–2020	Local
Yang et al. [10]	Turning points of grassland autumn phenology	GIMMS3g NDVI	No	Alpine meadow, forests, and shrublands	1982–2015	Local
Guo et al. [15]	Snow phenology and its environmental drivers	MODIS Snow Cover, NDVI	No	Forest, cropland	2001–2018	GEE
Medeiros et al. [3]	Caatinga phenology and environmental drivers	MODIS EVI	No	Caatinga	2000–2019	GEE
Wang et al. [14]	Comparison of LSP from SIF and EVI	MODIS EVI, GOSIF (Reconstructed OCO-2 SIF)	No	Terrestrial ecosystems in China	2003–2016	Local
Costa et al. [2]	Phenology of GPP and WUE	MODIS GPP	Yes (Fluxnet)	Tropical forest, caatinga, and cerrado	2009–2016	Local
Liu et al. [11]	Phenology responses to snow seasonality	MODIS Snow Cover	No	Mountainous ecosystems	2002–2020	Local
Cui et al. [12]	Phenology response to soil moisture and temperature	MODIS NDVI	Yes (phenology stations)	Mountainous ecosystems	2001–2020	Local
Costa et al. [1]	Phenology of ecosystem productivity in dry tropical forest	MODIS GPP, MODIS NDVI and EVI	Yes (Fluxnet)	Caatinga (dry tropical forest)	2014–2015	Local

Figure 1. Cloud of the words contained in the abstracts of the papers published in the Special Issue "Remote Sensing of Land Surface Phenology". The bigger the size, the higher the frequency of each word.

3. Outlook to the Future

As elaborated in the Call of Papers of this Special Issue, our hope was to not only summarize the ongoing diverse LSP developments but also boost discussions on prospects in LSP research. With the 15 selected papers, we believe that this aim should have been accomplished. Based upon the new knowledge we gained from this Special Issue, we hereby offer our perspective on future research directions in the LSP field:

First, improving the quality of the input satellite observations remains important as it lays the foundation for any downstream phenology retrieval and applications. From the sensor perspective, the improvement in sensor performance including spatial, spectral, and temporal resolutions is the direction of the development of new-generation remote-sensing data sources. From the methodological perspective, data fusion can effectively regulate the inherent contradiction between "spatial resolution and temporal resolution" of satellite remote-sensing observation, and realize the simultaneous improvement in spatial resolution and temporal resolution. In addition, with the development of remote-sensing sensor technology, more spaceborne platforms are becoming available, such as geostationary satellites that can observe the full disk of the Earth every 10–15 min, leading to much-improved temporal resolution that is capable of the near-real-time monitoring of vegetation dynamics [16–18]. Meanwhile, dozens or even hundreds of micro-/nanosatellites can form a constellation to significantly increase the spatial coverage and spatial–temporal resolution through a multisatellite synergetic approach. A relatively successful case is the PlanetScope constellation, which has launched more than 200 micro-/nanosatellites so far, providing seamless multispectral observation data at a global scale with a daily temporal resolution and 3 m spatial resolution, significantly improving the capability of resolving the fine details of global phenology change [19,20].

Second, the LSP retrieval workflow needs to be refined including the development of a more generalized algorithm. An often-encountered challenge in remote-sensing phenology applications is the fact that different data sets do not share the same processing algorithm, which makes it difficult to analyze and compare the results from different studies. Therefore, it is an important future direction for remote-sensing phenology to develop a versatile algorithm that can adapt to different circumstances.

Third, like many other remote-sensing subjects, validation is the essential component in any satellite phenology product development. A key issue here is the scale mismatch challenge [21,22]. Scale matching is not only the matching of spatial scales but also the matching between ground-based phenology metrics (e.g., budburst, flowering, leaf-coloring, etc.) and satellite-based metrics (e.g., SOS, EOS, POS, etc.) [23,24]. Therefore, it is critical to advance the theory and method that can resolve scale mismatch issues so that ground and satellite observations can be used in a more tandem manner [25,26]. The use of UAV observations and tower-mounted cameras can, to a certain extent, remediate the scale mismatch issue [27,28]. Meanwhile, considering the complexity of scale effects, computer simulations based on 3D radiative-transfer modelling can be used as a powerful tool to explore the scale effects or mixed image effects in vegetation phenology remote-sensing monitoring [29]. In addition, for low- and medium-resolution remote-sensing phenology products (e.g., MODIS/VIIRS), it is difficult even for UAVs or phenocams to provide validation data at the comparable pixel scale, in which case indirect "validation" can be performed using higher-spatial-resolution satellite data [30].

Author Contributions: Conceptualization, all authors; writing—original draft preparation, all authors; writing—review and editing, all authors. All authors have read and agreed to the published version of the manuscript.

Funding: This study is supported by National Natural Science Foundation of China (No. 42171305, Principal Investigator: X. Ma; No. 41971374, PI: J. Jin); Natural Science Foundation of Gansu Province, China (No. 21JR7RA499, PI: X. Ma); the Open Fund of State Key Laboratory of Remote Sensing Science (NO. OFSLRSS202229, PI: X. Ma); the Open Fund of the State Key Laboratory of Resources and Environmental Information System (No. 4 of the 2022 Round, PI: X. Ma); the Open Fund of Key laboratory of Geographic Information Science (Ministry of Education), East China Normal University (KLGIS2022A02, PI: X. Ma); Fundamental Research Funds for the Central Universities (No. lzujbky-2021-ct11, PI: X. Ma). Q. Xie acknowledges support from University of Technology Sydney Chancellor's Postdoctoral Research Fellowship and University of Technology Sydney Faculty of Science Seed Funding (Round 2 in 2021).

Acknowledgments: The authors would like to thank Xiaoyu Zhu and Yuhe Zhao for their help in collecting information and creating the word cloud graph.

Conflicts of Interest: The authors declare no conflict of interest.

References

1. Costa, G.B.; Mendes, K.R.; Viana, L.B.; Almeida, G.V.; Mutti, P.R.; e Silva, C.M.S.; Bezerra, B.G.; Marques, T.V.; Ferreira, R.R.; Oliveira, C.P.; et al. Seasonal Ecosystem Productivity in a Seasonally Dry Tropical Forest (Caatinga) Using Flux Tower Measurements and Remote Sensing Data. *Remote Sens.* **2022**, *14*, 3955. [CrossRef]
2. Costa, G.B.; Santos e Silva, C.M.; Mendes, K.R.; dos Santos, J.G.M.; Neves, T.T.A.T.; Silva, A.S.; Rodrigues, T.R.; Silva, J.B.; Dalmagro, H.J.; Mutti, P.R.; et al. WUE and CO_2 Estimations by Eddy Covariance and Remote Sensing in Different Tropical Biomes. *Remote Sens.* **2022**, *14*, 3241. [CrossRef]
3. Medeiros, R.; Andrade, J.; Ramos, D.; Moura, M.; Pérez-Marin, A.; dos Santos, C.; da Silva, B.; Cunha, J. Remote Sensing Phenology of the Brazilian Caatinga and Its Environmental Drivers. *Remote Sens.* **2022**, *14*, 2637. [CrossRef]
4. Kim, A.R.; Lim, C.H.; Lim, B.S.; Seol, J.; Lee, C.S. Phenological Changes of Mongolian Oak Depending on the Micro-Climate Changes Due to Urbanization. *Remote Sens.* **2021**, *13*, 1890. [CrossRef]
5. Ji, Y.; Jin, J.; Zhan, W.; Guo, F.; Yan, T. Quantification of Urban Heat Island-Induced Contribution to Advance in Spring Phenology: A Case Study in Hangzhou, China. *Remote Sens.* **2021**, *13*, 3684. [CrossRef]
6. Zhang, S.; Bai, Y.; Zhang, J. Remote Sensing-Based Quantification of the Summer Maize Yield Gap Induced by Suboptimum Sowing Dates over North China Plain. *Remote Sens.* **2021**, *13*, 3582. [CrossRef]
7. Ma, D.; Wu, X.; Ma, X.; Wang, J.; Lin, X.; Mu, C. Spatial, Phenological, and Inter-Annual Variations of Gross Primary Productivity in the Arctic from 2001 to 2019. *Remote Sens.* **2021**, *13*, 2875. [CrossRef]
8. Wang, J.; Sun, H.; Xiong, J.; He, D.; Cheng, W.; Ye, C.; Yong, Z.; Huang, X. Dynamics and Drivers of Vegetation Phenology in Three-River Headwaters Region Based on the Google Earth Engine. *Remote Sens.* **2021**, *13*, 2528. [CrossRef]
9. Guo, J.; Liu, X.; Ge, W.; Ni, X.; Ma, W.; Lu, Q.; Xing, X. Specific Drivers and Responses to Land Surface Phenology of Different Vegetation Types in the Qinling Mountains, Central China. *Remote Sens.* **2021**, *13*, 4538. [CrossRef]

10. Yang, Y.; Qi, N.; Zhao, J.; Meng, N.; Lu, Z.; Wang, X.; Kang, L.; Wang, B.; Li, R.; Ma, J.; et al. Detecting the Turning Points of Grassland Autumn Phenology on the Qinghai-Tibetan Plateau: Spatial Heterogeneity and Controls. *Remote Sens.* **2021**, *13*, 4797. [CrossRef]
11. Liu, Y.; Zhou, W.; Gao, S.; Ma, X.; Yan, K. Phenological Responses to Snow Seasonality in the Qilian Mountains Is a Function of Both Elevation and Vegetation Types. *Remote Sens.* **2022**, *14*, 3629. [CrossRef]
12. Cui, X.; Xu, G.; He, X.; Luo, D. Influences of Seasonal Soil Moisture and Temperature on Vegetation Phenology in the Qilian Mountains. *Remote Sens.* **2022**, *14*, 3645. [CrossRef]
13. Chen, F.; Liu, Z.; Zhong, H.; Wang, S. Exploring the Applicability and Scaling Effects of Satellite-Observed Spring and Autumn Phenology in Complex Terrain Regions Using Four Different Spatial Resolution Products. *Remote Sens.* **2021**, *13*, 4582. [CrossRef]
14. Wang, C.; Wu, Y.; Hu, Q.; Hu, J.; Chen, Y.; Lin, S.; Xie, Q. Comparison of Vegetation Phenology Derived from Solar-Induced Chlorophyll Fluorescence and Enhanced Vegetation Index, and Their Relationship with Climatic Limitations. *Remote Sens.* **2022**, *14*, 3018. [CrossRef]
15. Guo, H.; Wang, X.; Guo, Z.; Chen, S. Assessing Snow Phenology and Its Environmental Driving Factors in Northeast China. *Remote Sens.* **2022**, *14*, 262. [CrossRef]
16. Zhao, Y.; Wang, M.; Zhao, T.; Luo, Y.; Li, Y.; Yan, K.; Lu, L.; Tran, N.N.; Wu, X.; Ma, X. Evaluating the potential of H8/AHI geostationary observations for monitoring vegetation phenology over different ecosystem types in northern China. *Int. J. Appl. Earth Obs. Geoinf.* **2022**, *112*, 102933. [CrossRef]
17. Yan, D.; Zhang, X.; Nagai, S.; Yu, Y.; Akitsu, T.; Nasahara, K.N.; Ide, R.; Maeda, T. Evaluating land surface phenology from the Advanced Himawari Imager using observations from MODIS and the Phenological Eyes Network. *Int. J. Appl. Earth Obs. Geoinf.* **2019**, *79*, 71–83. [CrossRef]
18. Miura, T.; Nagai, S.; Takeuchi, M.; Ichii, K.; Yoshioka, H. Improved Characterisation of Vegetation and Land Surface Seasonal Dynamics in Central Japan with Himawari-8 Hypertemporal Data. *Sci. Rep.* **2019**, *9*, 15692. [CrossRef]
19. Cheng, Y.; Vrieling, A.; Fava, F.; Meroni, M.; Marshall, M.; Gachoki, S. Phenology of short vegetation cycles in a Kenyan rangeland from PlanetScope and Sentinel-2. *Remote Sens. Environ.* **2020**, *248*, 112004. [CrossRef]
20. Wu, S.; Wang, J.; Yan, Z.; Song, G.; Chen, Y.; Ma, Q.; Deng, M.; Wu, Y.; Zhao, Y.; Guo, Z.; et al. Monitoring tree-crown scale autumn leaf phenology in a temperate forest with an integration of PlanetScope and drone remote sensing observations. *ISPRS J. Photogramm. Remote Sens.* **2021**, *171*, 36–48. [CrossRef]
21. Wu, X.; Wen, J.; Xiao, Q.; You, D.; Gong, B.; Wang, J.; Ma, D.; Lin, X. Spatial Heterogeneity of Albedo at Subpixel Satellite Scales and its Effect in Validation: Airborne Remote Sensing Results From HiWATER. *IEEE Trans. Geosci. Remote Sens.* **2022**, *60*, 1–14. [CrossRef]
22. Wang, J.; Wu, X.; Wen, J.; Xiao, Q.; Gong, B.; Ma, D.; Cui, Y.; Lin, X.; Bao, Y. Upscaling in Situ Site-Based Albedo Using Machine Learning Models: Main Controlling Factors on Results. *IEEE Trans. Geosci. Remote Sens.* **2022**, *60*, 1–16. [CrossRef]
23. Liu, L.; Cao, R.; Shen, M.; Chen, J.; Wang, J.; Zhang, X. How Does Scale Effect Influence Spring Vegetation Phenology Estimated from Satellite-Derived Vegetation Indexes? *Remote Sens.* **2019**, *11*, 2137. [CrossRef]
24. Peng, D.; Zhang, X.; Zhang, B.; Liu, L.; Liu, X.; Huete, A.R.; Huang, W.; Wang, S.; Luo, S.; Zhang, X.; et al. Scaling effects on spring phenology detections from MODIS data at multiple spatial resolutions over the contiguous United States. *ISPRS J. Photogramm. Remote Sens.* **2017**, *132*, 185–198. [CrossRef]
25. Tian, F.; Cai, Z.; Jin, H.; Hufkens, K.; Scheifinger, H.; Tagesson, T.; Smets, B.; Van Hoolst, R.; Bonte, K.; Ivits, E.; et al. Calibrating vegetation phenology from Sentinel-2 using eddy covariance, PhenoCam, and PEP725 networks across Europe. *Remote Sens. Environ.* **2021**, *260*, 112456. [CrossRef]
26. Wen, J.; Wu, X.; Wang, J.; Tang, R.; Ma, D.; Zeng, Q.; Gong, B.; Xiao, Q. Characterizing the Effect of Spatial Heterogeneity and the Deployment of Sampled Plots on the Uncertainty of Ground "Truth" on a Coarse Grid Scale: Case Study for Near-Infrared (NIR) Surface Reflectance. *J. Geophys. Res. Atmos.* **2022**, *127*, e2022JD036779. [CrossRef]
27. Klosterman, S.; Melaas, E.; Wang, J.A.; Martinez, A.; Frederick, S.; O'Keefe, J.; Orwig, D.A.; Wang, Z.; Sun, Q.; Schaaf, C.; et al. Fine-scale perspectives on landscape phenology from unmanned aerial vehicle (UAV) photography. *Agric. For. Meteorol.* **2018**, *248*, 397–407. [CrossRef]
28. Berra, E.F.; Gaulton, R.; Barr, S. Assessing spring phenology of a temperate woodland: A multiscale comparison of ground, unmanned aerial vehicle and Landsat satellite observations. *Remote Sens. Environ.* **2019**, *223*, 229–242. [CrossRef]
29. Chen, X.; Wang, D.; Chen, J.; Wang, C.; Shen, M. The mixed pixel effect in land surface phenology: A simulation study. *Remote Sens. Environ.* **2018**, *211*, 338–344. [CrossRef]
30. Zhang, X.; Jayavelu, S.; Liu, L.; Friedl, M.A.; Henebry, G.M.; Liu, Y.; Schaaf, C.B.; Richardson, A.D.; Gray, J. Evaluation of land surface phenology from VIIRS data using time series of PhenoCam imagery. *Agric. For. Meteorol.* **2018**, *256–257*, 137–149. [CrossRef]

remote sensing

MDPI

Article

Phenological Changes of Mongolian Oak Depending on the Micro-Climate Changes Due to Urbanization

A Reum Kim [1], Chi Hong Lim [2], Bong Soon Lim [1], Jaewon Seol [1] and Chang Seok Lee [3,*]

[1] Graduate School of Seoul Women's University, Seoul Women's University, 621 Hwarang-no, Nowon-gu, Seoul 01797, Korea; dkfma@swu.ac.kr (A.R.K.); bs6238@swu.ac.kr (B.S.L.); seol_jaewon@swu.ac.kr (J.S.)

[2] Division of Ecological Survey Research, National Institute of Ecology, 1210 Geumgang-no, Maseo-myeon, Seocheon 33657, Korea; sync03@nie.re.kr

[3] Division of Chemistry and Bio-Environmental Sciences, Seoul Women's University, 621 Hwarang-no, Nowon-gu, Seoul 01797, Korea

* Correspondence: leecs@swu.ac.kr; Tel.: +82-2-970-5822

Citation: Kim, AR.; Lim, C.H.; Lim, B.S.; Seol, J.; Lee, C.S. Phenological Changes of Mongolian Oak Depending on the Micro-Climate Changes Due to Urbanization. *Remote Sens.* **2021**, *13*, 1890. https://doi.org/10.3390/rs13101890

Academic Editors: Xuanlong Ma, Xiaolin Zhu, Jiaxin Jin, Yuke Zhou and Qiaoyun Xie

Received: 5 April 2021
Accepted: 10 May 2021
Published: 12 May 2021

Abstract: Urbanization and the resulting increase in development areas and populations cause micro-climate changes such as the urban heat island (UHI) effect. This micro-climate change can affect vegetation phenology. It can advance leaf unfolding and flowering and delay the timing of fallen leaves. This study was carried out to clarify the impact of urbanization on the leaf unfolding of Mongolian oak. The survey sites for this study were established in the urban center (Mts. Nam, Mido, and Umyeon in Seoul), suburbs (Mts. Cheonggye and Buram in Seoul), a rural area (Gwangneung, Mt. Sori in Gyeonggi-do), and a natural area (Mt. Jeombong in Gangwon-do). Green-up dates derived from the analyses of digital camera images and MODIS satellite images were the earliest in the urban center and delayed through the suburbs and rural area to the natural area. The difference in the observed green-up date compared to the expected one, which was determined by regarding the Mt. Jeombong site located in the natural area as the reference site, was the biggest in the urban center and decreased through the suburbs and rural area to the natural area. Green-up dates in the rural area, suburbs, and urban center were earlier by 11.0, 14.5, and 16.3 days than the expected ones. If these results are transformed into the air temperature based on previous research results, it could be deduced that the air temperature in the urban center, suburbs, and rural area rose by 3.8 to 4.6 °C, 3.3 to 4.1 °C, and 2.5 to 3.1 °C, respectively. Green-up dates derived based on the accumulated growing degree days (AGDD) showed the same trend as those derived from the image interpretation. Green-up dates derived from the change in sap flow as a physiological response of the plant showed a difference within one day from the green-up dates derived from digital camera and MODIS satellite image analyses. The change trajectory of the curvature K value derived from the sap flow also showed a very similar trend to that of the curvature K value derived from the vegetation phenology. From these results, we confirm the availability of AGDD and sap flow as tools predicting changes in ecosystems due to climate change including phenology. Meanwhile, the green-up dates in survey sites were advanced in proportion to the land use intensity of each survey site. Green-up dates derived based on AGDD were also negatively correlated with the land use intensity of the survey site. This result implies that differences in green-up dates among the survey sites and between the expected and observed green-up dates in the urban center, suburbs, and rural area were due to the increased temperature due to land use in the survey sites. Based on these results, we propose conservation and restoration of nature as measures to reduce the impact of climate change.

Keywords: climate change; digital camera; MODIS; Mongolian oak; phenology; sap flow; urbanization

1. Introduction

Urbanization is one of the major social and scientific changes spreading around the world [1,2]. It significantly alters land surface conditions and has profound impacts on terrestrial ecosystem processes and services [3–7]. Changes in land use release greenhouse gases into the atmosphere by changing the patterns of carbon storage and accelerate climate change by breaking the balance of the carbon budget [2]. An increase in atmospheric CO_2 due to intensive use of land and fossil fuel destroys the balance of the global carbon cycle maintained in an equilibrium fashion [2,8–11]. In addition, increased development areas and populations cause changes in weather factors and affect terrestrial ecosystems [3–6,12] Inadvertent weather factor changes in urban areas form an important effect on regional climate change [13–16].

Urbanization is an important anthropogenic influence on climate and has significantly affected terrestrial ecosystems [12,17]. It can modify local climate on daily, seasonal, and annual scales [18] and increase extreme climate events [19–22]. Changes in climate resulting from urbanization, therefore, can be considered a type of climate change on a local scale [23]. Such a change in local climate in an urban area is referred to as the "urban heat island (UHI) effect". The UHI effect is characterized by elevated temperatures in urban areas, compared to the surrounding rural areas [7,24,25]. It can affect regional climate change, increase environmental pollution, elevate energy and water consumption, affect the development of meteorological events such as increased precipitation, and have a significant impact on human health [16,26,27].

In phenological research, urban areas are important areas for study because they enable an assessment of the potential future effects of climate change on plant development [17,28]. Increased temperature by UHI can affect vegetation phenology such as the start of the season (SOS) and the end of the growing season (EOS) [25,29–31]. It is very important to understand the impact of UHI on phenology because the intensity of the UHI effect is similar to the expected temperature change in the near future [7].

Current climate change has a strong impact on vegetation phenology [6,32–37], and it causes changes in the timing of plant developmental phases, affecting the carbon budget of the terrestrial ecosystem [38–41]. Phenological characteristics are closely related to variation in weather factors [6,42–45], many of which can affect vegetation phenology such as green-up, budburst, and leaf senescence [17].

Among the numerous techniques to observe phenology, using digital camera and MODIS satellite images requires less time and cost to collect data [46,47]. Time-lapse photography provides very exact temporal sampling at daily intervals for assessing phenology. Satellite remote sensing provides decades of records of vegetation phenology across larger spatial scales than other technologies [48]. Remote sensing also has the advantage of providing large temporal records of vegetation phenology over larger spatial scales than other techniques [17,48]. In particular, the method using both digital camera and MODIS satellite images has sufficient spatial resolution to obtain detailed information about vegetation and land cover types and can collect data with more flexible spatial resolution, thereby resolving the problems pointed out in the existing data collection [17,26,46,49].

Recently, beyond the level of checking phenological phases by observing the external forms of plants, a study method to confirm the plant phenological phases through physiological responses such as sap flow time series estimates was also proposed [17,50]. Water availability is regulated by the timing and periodicity of leaf production and is very important for plant growth [51,52]. According to many studies investigating the relationship between phenology and sap flow, sap flow is closely related to the change in leaf area [50,53,54], since plants draw water from the xylem by tension from the leaves during the transpiration period [55], and the amount of transpiration increases depending on the formation of leaves. Therefore, sap flow usually has a linear relationship with leaf area development [17,50,54]. Phenological events emerging through appearance may be difficult to observe accurately and precisely due to various influences [56,57], and the method of color wavelength analysis of the leaves applies well for deciduous plants, but there will be limitations to evergreen plants. In this respect, a method of monitoring physiological changes according to seasonal changes may be more versatile [58]. In addition, since the phenophase using instrumental techniques can be better specified than pure observations and qualified guesses, the onset of spring phenological stages such as leaf area development can be easily identified from sap flow measurements [50]. Ecophysiological measurements such as sap flow measurement can provide additional value in the objectification of phenological studies [50]. Therefore, sap flow has been widely utilized as a tool to diagnose the development of the phenological phases of plants [30,51,54,59].

The objectives of this study are to (1) find the phenological trajectory of vegetation through analysis of MODIS satellite and digital camera images, (2) investigate how microclimate change caused by urbanization affects vegetation phenology, (3) diagnose the phenological trajectory of vegetation through the physiological response of plants, and (4) prepare an ecosystem management strategy to adapt to climate change.

2. Materials and Methods

2.1. Study Area

To find out the response of phenological events according to the climatic condition, we selected three areas different in land use intensity, such as urban, rural, and natural areas on the same latitude. The phenological signal is evident in deciduous broad-leaved forests because the changes in the canopy are large depending on the stage of the phenology [60]. In this study, therefore, we selected the target species as *Quercus mongolica* Fisch. ex Ledeb., a dominant species in the deciduous broad-leaved forest of Korea. Since *the Q. mongolica* that belongs to the *Quercus* genus grows at the highest elevation in South Korea, it is considered to be sensitive to temperature rises due to climate change. Seven sites including Mts. Nam, Mido, and Umyeon in the urban center, Mts. Cheonggye and Buram in the suburbs, Gwangneung (Mt. Sori) in a rural area, and Mt. Jeombong in a natural area were selected for analysis (Figure 1, Table 1). The urbanization ratio was calculated from the ratio of development area to the total area within a 5 km radius from the study area by a geographic interpolation system (GIS) using the national land use map (National Geographic Information Institute, 2016). The urbanization rates of Mts. Nam, Mido, Umyeon, Cheonggye, Buram, and Sori (Gwandneung) were 76.07, 70.35, 52.44, 35.80, 49.60, and 6.38, respectively, and Mt. Jeombong was not urbanized at all as it is located in a natural area (Table 2).

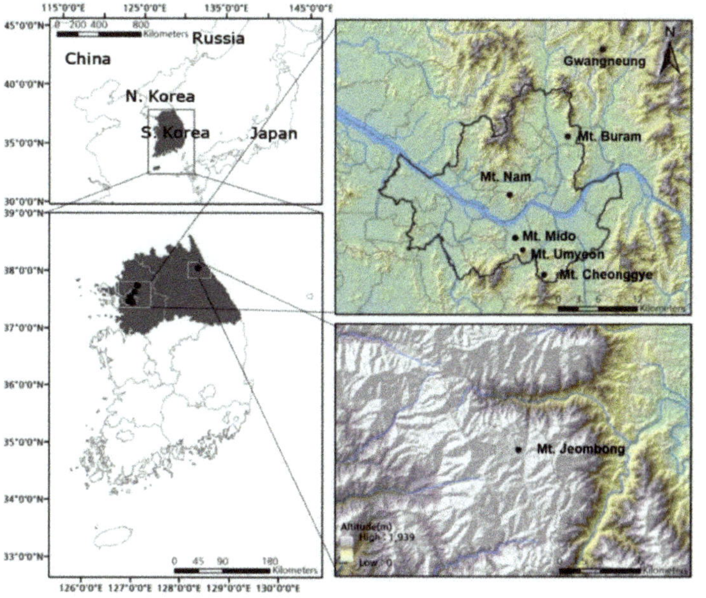

Figure 1. A map showing the location of the study sites.

Table 1. Description of study sites. *DoY*: day of year.

	Site Name	Latitude (Decimal Degree)	Longitude (Decimal Degree)	Elevation (m)	Tower Type	Data Collection Period (DOY)	
						Digital Camera	MODIS
Urban center	Mt. Nam	37.55	126.99	215	Ecological tower	56~240	2016/2/19 ~ 2016/7/18
	Mt. Mido	37.49	127.00	40		56~240	
	Mt. Umyeon	37.47	127.00	185		76~240	
Suburb	Mt. Cheonggye	37.44	127.05	276		72~240	
	Mt. Buram	37.63	127.09	115		63~240	
Rural area	Gwangneung (Mt. Sori)	37.75	127.15	345	Fire surveillance tower	96~240	
Natural area	Mt. Jeombong	38.04	128.47	830	Ecological tower	91~240	

Table 2. Urbanization rate in study sites.

	Site Name	Urbanized Area (km^2)	Urbanization Rate (%)
Urban center	Mt. Nam	59.71	76.07
	Mt. Mido	55.22	70.35
	Mt. Umyeon	41.16	52.44
Suburb	Mt. Cheonggye	28.10	35.80
	Mt. Buram	38.94	49.60
Rural area	Gwangneung (Mt. Sori)	5.01	6.38

2.2. Digital Camera and Satellite Image Acquisition

We installed digital cameras (Model Ltl-6210M, Little Acorn Outdoors, Denmark, WI, USA) near the top of each tower or tree, looking north and angled slightly downward, providing a view across the canopy. To acquire daily photos, each camera was set to record JPEG images three times per day (09:00, 12:30, and 14:30). In order to maintain consistency, only 12:30 images were used for the analysis (Table 3).

Table 3. Descriptions of the intervalometers and cameras employed. *FoV*: field of view.

Intervalometer	Daily Capture Times Interval between Captures	09:00, 12:30, 16:00 3 1/2 h
Camera	Model	Acorn Ltl-6210M
	Sensor	5 megapixel color CMOS
	Pixel size	2560 × 1920
	Channels	RGB (red, green, blue)
	Lens	F = 3.1; FoV = 52; Auto IR-Cut
	Memory card	32 GB SD
	File type	High-quality JPEG (2MP)
	Power	12 × AA; Solar panel
	Flash	Disabled

As the notion the shorter the collection cycle, the higher the clarity applies [61], we used MODIS (MODerate-Resolution Imaging Spectroradiometer) 500 m resolution land surface imagery (MOD09GA), which is supplied at daily intervals as multi-spectral satellite images. The MODIS satellite is a payload scientific instrument placed in the earth's orbit by NASA in December 1999 on the Terra (EOS AM) satellite. The MODIS satellite imagery measures the surface temperature of the land and ocean, and the distribution map of the earth's vegetation is re-synthesized with control variables such as clouds and distributed free of charge by NASA. The MODIS satellite is suitable for monitoring phenological changes because the sensor incorporates enhanced cloud detection, atmospheric correction, georeferencing, and the ability to monitor vegetation [39,62].

2.3. Digital Camera Image Analysis

An annual vegetation phenological cycle inferred from remote sensing is characterized by four stages that define the key phenological phases at annual time scales: (1) green-up, (2) maturity, (3) senescence, and (4) dormancy [17,60,63]. Phenological signals remain low values during the dormancy phase and then increase rapidly as the green-up phase begins. When the leaves reach maturity, signals no longer increase but maintain a high value. After that, the plants enter into the senescence phase, and the signals decrease radically. As the dormancy phase begins, the phenological signals return to the lowest value of the initial phase. As such, an inflection point at which the curvature rapidly changes in the phenological signal curve may be interpreted as the start date of each stage [17,63]. The formula for obtaining the curvature K value of the inflection point is as follows:

$$f(t) = a + \frac{c}{1 + exp(a + bt)} + d \quad (1)$$

$$K = \frac{f''(t)}{\left(1 + (f'(t))^2\right)^{\frac{3}{2}}} \quad (2)$$

where t is time, c is the amplitude of an increase or decrease in the green value, d is the baseline value of the dormant season, and a and b control the lower and upper limits of the function [17,64–66].

To extract phenological signals, we collected images from the digital cameras periodically and classified them into red, green, and blue bands. Using digital numbers for each band, we calculated the average excess green index (ExG) for each ROI based on the equation [64,65]

$$ExG = 2 \times \rho_{GREEN} - (\rho_{RED} + \rho_{BLUE}) \quad (3)$$

where ρ_{RED}, ρ_{GREEN}, and ρ_{BLUE} are values in red, green, and blue channels acquired from digital camera images, respectively. The region of interest (ROI) is used when digital camera images are analyzed to clarify phenological changes [67]. As the digital camera images include a mixture of the sky, landscape, and other factors, the ROI is limited to the crown layer to extract the phenological signal from the images. Furthermore, because, in these study sties, the *Q. mongolica* stands are mixed with *Quercus variabilis* Blume and *Quercus serrata* Murray stands, and other species, the ROI was limited to pure stands of *Q. mongolica*, and we tried to avoid mountains and sky [60,64] (Figure 2). In this study, we set up a number of ROIs for the images of the *Q. mongolica* community in each site and extracted the ExG index.

Figure 2. A field of view from the digital camera at the Mt. Jeombong site. Regions of interest (ROIs) 1–3 are indicated in red.

2.4. Analysis of Satellite Images

The MOD09GA datasets are comprised of seven bands, including visible light bands and near-infrared bands. The EVI in MODIS was calculated using red (Band 1: 620–670 μm), green (Band 4: 545–565 μm), blue (Band 3: 459–479 μm), and near-infrared (Band 2: 841–876 μm) based on the equations. The vegetation index uses the EVI index, which is an improvement over other indexes. The EVI index is an improved vegetation index to reduce the effects of spatial differences by using blue bands in areas with large spatial differences and is suitable for observing seasonal changes in vegetation by reflecting the characteristics of the canopy [62]. The EVI calculations used in the analysis are as follows:

$$EVI = 2.5 \times (\rho nir - \rho red)/\{\rho nir + (6 \times \rho red - 7.5 \times \rho blue) + 1\} \quad (4)$$

where ρnir, ρred, and $\rho blue$ are values in near-infrared, red, and blue bands. MODIS satellite images were reprojected to TM (transverse Mercator) coordinates because they use a sinusoidal projection. Based on the extracted data, the EVI index for each study site was derived. Then, the EVI was obtained using the smooth curve fitting method to remove variation and to gather trends because interpretation error can occur due to data errors and variation depending on weather conditions [17]. In this study, the EVI was smoothed to the 80th percentile using an exponentially weighted moving average (EWMA). The EWMA was defined as

$$S_t = \alpha \times Y_t + (1 - \alpha) \times S_{t-1} \ (t > 1, S_1 = Y_1) \quad (5)$$

where t is the day of year (DoY); S_t is the EWMA value at the DoY; Y_t is the EVI value at the DoY; and α is the smoothing coefficient.

2.5. Sap Flow Measurement

To analyze the relationship between the phenological signal and the physiological responses of plants, we collected data from a sap flow measurement instrument (model SFM1 Sap Flow Meter, ICT international, Armidale, Australia) installed in the study sites. Measured individuals were randomly selected from individuals included in the ROI. Sap flow velocity ($cm^3 \cdot hr^{-1}$) was calculated by heat pulse, and temperature was measured from the thermistor inserted 7.5 mm and 22.5 mm inside the removed bark [68]. The seasonal trajectory of sap flow was interpreted using curvature K (formula 2) based on the daily sap velocity. The transition date of the sap flow was compared with the phenological transition date obtained from the digital camera and MODIS installed at the same site.

2.6. Data Correction

The green-up date of *Q. mongolica* derived from digital camera and MODIS satellite images showed a difference depending on the study site. According to [69], increases in latitude of 1° N, longitude of 1° E, and altitude of 100 m result in delays of 2.6 ± 0.2 days, 0.6 ± 0.1 days, and 2.1 ± 0.2 days, respectively, in the leaf unfolding date. Based on this information, we corrected the difference in leaf unfolding dates due to differences in latitude, longitude, and altitude among the study areas.

On the other hand, we designated the Mt. Jeombong site as the reference site to clarify the effect of urbanization. Mt. Jeombong maintains a healthy and integrated stand of *Q. mongolica* as it is designated as a reserve by the Korea Forest Service and thereby escapes artificial interferences. Based on the Mt. Jeombong data, the expected dates of green-up were obtained by reflecting the geographic and topographic location of each study site. The effect of urbanization on plant phenology in each study area was determined by comparing the differences between the expected and observed dates of green-up in each site.

2.7. Weather Factor Collection and Analysis

To analyze the correlation between phenological events and weather factors and identify the weather conditions at the time of major phenological phases, the atmospheric temperature measurement instrument HOBO (HOBO Pro v2 Temp/RH Temp, Onset Computer,

Bourne, MA, USA) was installed. The temperature was measured every 30 minutes every day, and the data measured at 12:30 were used due to compatibility with the digital camera.

Based on the collected weather data, AGDD (accumulated growing degree days) were calculated to analyze the temperature threshold of the plant's green-up period. The formula for calculating AGDD is as follows:

$$\sum_{1}^{n} \frac{T_{max} + T_{min}}{2} - T_{\Delta}\ (T \geq 5\ ^{\circ}\mathrm{C}) \tag{6}$$

where T_{max} and T_{min} are the maximum and minimum air temperatures, respectively, and T_{Δ} is the temperature below which plant growth is zero. In this study, we assumed a minimum temperature threshold of 5 °C for enzymatic activity in *Q. mongolica* based on studies of [70,71].

3. Results

3.1. The Green-up Date of Mongolian Oak

ExG values obtained from digital cameras clearly indicate the phenological changes of Mongolian oak (Figure 3). The green-up date of Mongolian oak, derived from the curvature K value of the digital camera ExG, was the earliest in the urban area and gradually delayed moving through the suburban and rural areas toward the natural area with 94, 95, 95, 95, 97, 102, and 117 days in Mts. Nam, Mido, Umyeon, Buram, Cheonggye, Sori (Gwangneung), and Jeombong, respectively, based on the day of year (DoY) (Table 4).

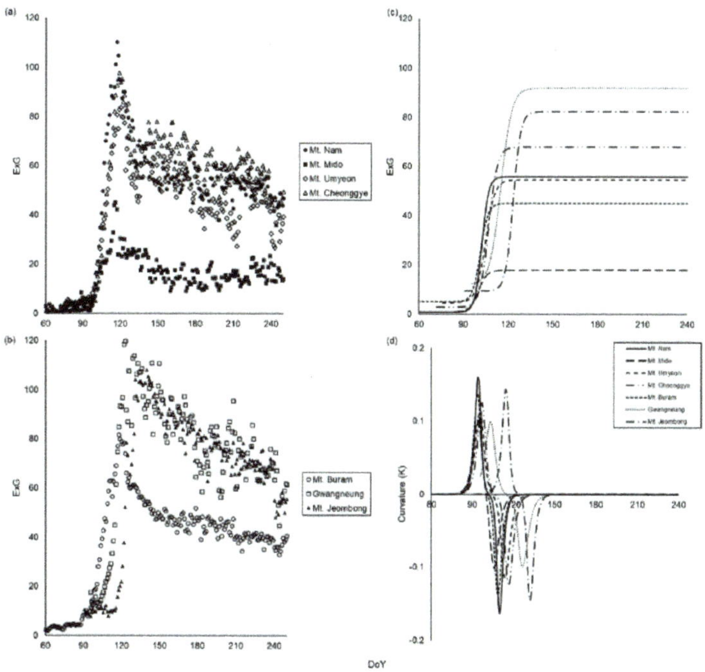

Figure 3. (**a**,**b**) show the seasonal courses of ExG in the Mongolian oak stands of each site. (**c**) shows the logistic models of green-up based on ExG, and (**d**) shows the rate of change of curvature K in study sites. The time at which the rate of change in curvature exhibits local maxima indicates the green-up date.

Table 4. Green-up dates estimated by two different criteria, ExG and EVI, and the difference between observed and expected green-up dates from ExG and EVI. *Obs*: observed, *Exp*: expected, *Diff*: difference.

	Site	ExG			EVI		
		Obs (DoY)	Exp (DoY)	Diff (Days)	Obs (DoY)	Exp (DoY)	Diff (Days)
Urban center	Mt. Nam	94	110	−16	94	110	−16
	Mt. Mido	95	113	−18	96	114	−18
	Mt. Umyeon	95	110	−15	97	112	−15
Suburb	Mt. Cheonggye	97	111	−14	96	110	−14
	Mt. Buram	95	110	−15	96	111	−15
Rural area	Gwangneung (Mt. Sori)	103	114	−11	104	115	−11
Natural area	Mt. Jeombong	118	118	0	114	114	0

EVI values obtained from MODIS images also clearly indicate the phenological changes of Mongolian oak (Figure 4). The green-up date of Mongolian oak, derived from the curvature K value of the MODIS image EVI, showed a similar trend to the result from the digital camera, with 94, 96, 97, 95, 96, 104, and 114 days in Mts. Nam, Mido, Umyeon, Buram, Cheonggye, Sori (Gwangneung), and Jeombong, respectively, based on the DoY (Table 4).

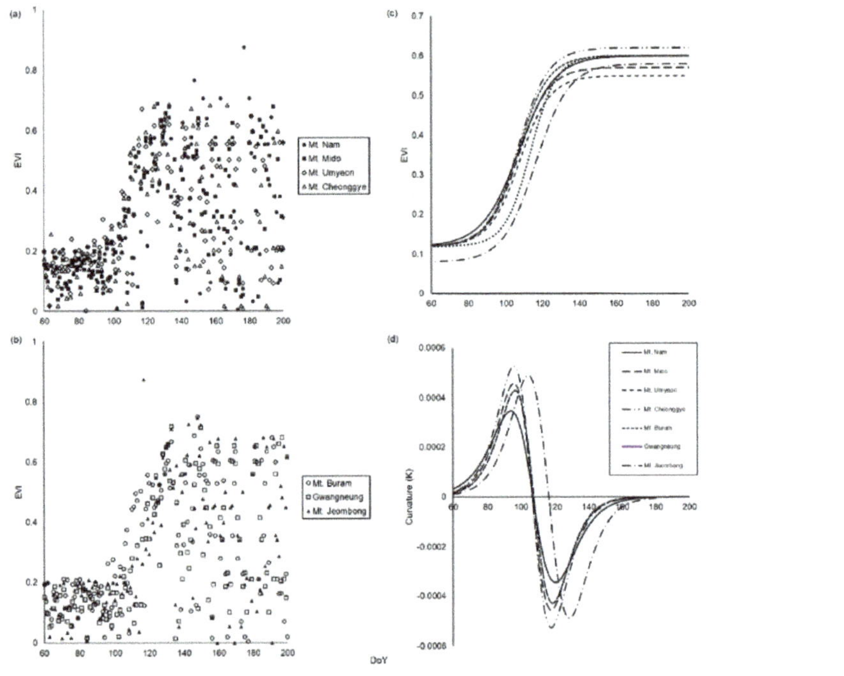

Figure 4. (**a**,**b**) show the seasonal courses of EVI in the Mongolian oak stands of each site. (**c**) shows the logistic models of green-up based on EVI, and (**d**) shows the rate of change of curvature K in study sites. The time at which the rate of change in curvature exhibits local maxima indicates the green-up date.

In addition, to clarify the differences among sites due to artificial interference, the expected dates of green-up were obtained through latitude, altitude, and elevation correction based on the natural area, Mt. Jeombong. This was compared with the actual observed dates of the study site (Table 4). The green-up date observed from the ExG of Mt. Nam was DoY 94, which is 16 days earlier than the expected DoY 110. The green-up date observed from the ExG of Mt. Mido was DoY 95, which is 18 days earlier than the expected DoY 113. The green-up date observed from the ExG of Mt. Umyeon was DoY 95, which is 15 days earlier than the expected DoY 110. The green-up date observed from the ExG of Mt. Cheonggye was DoY 97, which is 14 days earlier than the expected DoY 111. The green-up date observed from the ExG of Mt. Buram was DoY 95, which is 15 days earlier than the expected DoY 110. The green-up date observed from the ExG of Gwangneung (Mt. Sori) was DoY 103, which is 11 days earlier than the expected DoY 114 (Table 4).

The green-up date observed from the EVI of Mt. Nam was DoY 94, which is 16 days earlier than the expected DoY 110. The green-up date observed from the EVI of Mt. Mido was DoY 96, which is 18 days earlier than the expected DoY 114. The green-up date observed from the EVI of Mt. Umyeon was DoY 97, which is 15 days earlier than the expected DoY 112. The green-up date observed from the EVI of Mt. Cheonggye was DoY 96, which is 14 days earlier than the expected DoY 110. The green-up date observed from the EVI of Mt. Buram was DoY 96, which is 15 days earlier than the expected DoY 111. The green-up date observed from the EVI of Gwangneung (Mt. Sori) was DoY 104, which is 11 days earlier than the expected DoY 115 (Table 4).

As a result of analysis, the correlation between green-up dates derived from ExG and EVI values and land use intensity showed a significant negative correlation (Figure 5).

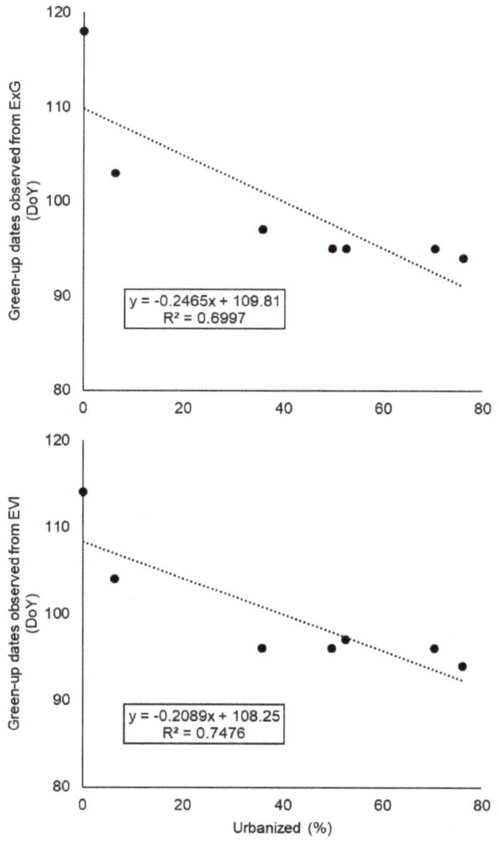

Figure 5. Relationship between urbanization ratio and green-up dates derived from ExG (**upper**) and EVI (**lower**) ($p < 0.05$).

3.2. Accumulated Growing Degree Days (AGDD)

Green-up dates estimated by AGDD in each study site are shown in Table 5. Green-up dates expressed as DoY were 94, 95, 95, 97, 96, 103, and 114 in Mts. Nam, Mido, Woomyeon, Cheonggye, Buram, Sori (Gwangneung), and Jeombong, respectively (Table 5). AGDD for green-up dates derived from the ExG of the digital camera were 160.3 °C, 158.2 °C, 156.3 °C, 162.2 °C, 162.3 °C, 156.6 °C, and 160.5 °C, and AGDD for green-up dates derived from the EVI of MODIS images were 160.3 °C, 166.9 °C, 162.6 °C, 153.2 °C, 168.3 °C, 163.3 °C, and 154.4 °C in the aforementioned site order (Table 5). As a result of analyzing the correlation between the land use intensity of the study sites and the date when the AGDD value reached 159 °C, they showed a significant negative correlation (Figure 6).

Table 5. Green-up dates estimated by AGDD and AGDD values for green-up dates derived from ExG and EVI.

	Site Name	**Green-Up Dates (DoY)**	**AGDD (°C)**	
			ExG	**EVI**
Urban center	Mt. Nam	94	160.3	160.3
	Mt. Mido	95	158.2	166.9
	Mt. Umyeon	95	156.3	162.6
Suburb	Mt. Cheonggye	97	162.2	153.2
	Mt. Buram	96	162.3	168.3
Rural area	Gwangneung (Mt. Sori)	103	156.6	163.3
Natural area	Mt. Jeombong	114	160.5	154.4
	Average		159.5	161.3

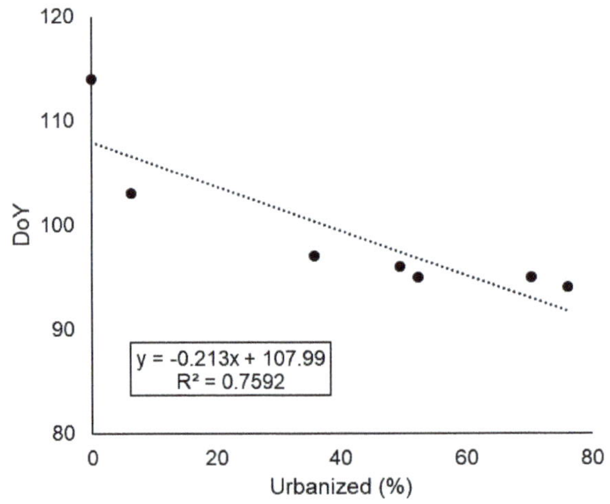

Figure 6. Relationship between dates when AGDD reached 159 °C and urbanization ratio ($p < 0.05$).

3.3. Seasonal Trajectory of the Sap Flow

The seasonal change in sap flow is expressed in Figure 7. Green-up dates expressed in DoY were 94, 96, 97, and 104 in Mts. Nam, Woomyeon, Cheonggye, and Sori (Gwangneung), respectively (Figure 8). The difference between the green-up dates derived from sap flow and the ExG of the digital camera was within one day, and the date was the same in Mts. Nam and Cheonggye. The difference between the green-up dates derived from the sap flow and the EVI of MODIS images was within one day, and the date was the same in Mt. Nam and Gwangneung (Mt. Sori) (Table 6). The trajectory change in the curvature K value derived from the sap flow of the plant showed a very similar trend to that of the curvature K value derived from the digital camera and MODIS satellite images (Figure 8).

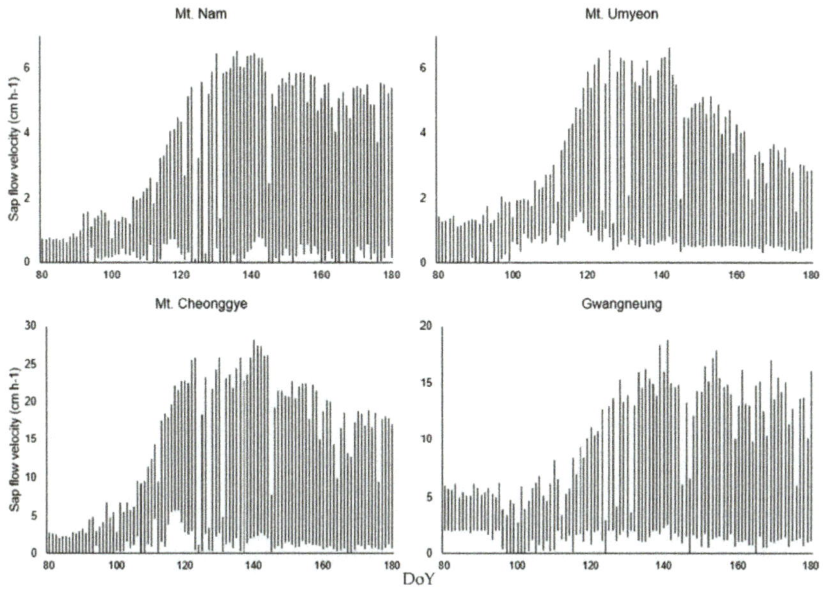

Figure 7. Changes in sap flow velocity during the study period in each site.

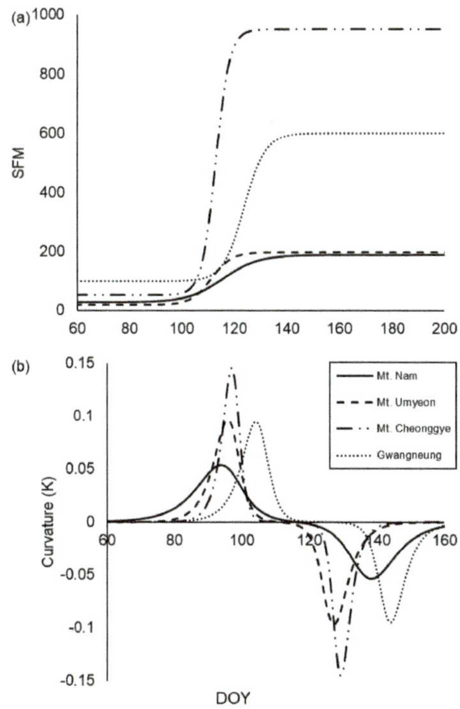

Figure 8. SFM logistic models of green-up (**a**) and curvature K (**b**) in study.

Table 6. Comparisons of green-up dates derived from vegetation phenology (ExG and EVI) and sap flow (SFM).

Site	ExG (DoY)	EVI (DoY)	SFM (DoY)	Difference (Days)	
				ExG-SFM	EVI-SFM
Mt. Nam	94	94	94	0	0
Mt. Mido	95	96	-	-	-
Mt. Umyeon	95	97	96	−1	1
Mt. Cheonggye	97	96	97	0	1
Mt. Buram	95	96	-	-	-
Gwangneung (Mt. Sori)	103	104	104	1	0
Mt. Jeombong	118	114	-	-	-

4. Discussion

4.1. Changes in the Green-up Dates Depending on Land Use Intensity

Land use by human activities releases greenhouse gases, and the increase in development areas and populations causes urban heat islands and changes the microclimate [6,37,72,73]. Massive urban sprawling can bring about more deforestation, habitat destruction, and greenhouse gas (GHG) or carbon emissions, and these factors can lead to local climate change [74]. In fact, in Korea, the urbanization rate and population of urban areas increased rapidly from 1971 to 2000, during which the daily minimum, maximum, and mean temperature increased [75]. Studies carried out in China showed that the UHI effect contributes to climate warming by about 30% [76,77].

Climate change affects the developmental phase of plants, and, thus, it can bring about significant changes in phenology [23]. Every 1 °C increase in the land surface temperature (LST) in spring and fall advanced the SOS by 9 to 11 days, and EOS was delayed by 6 to 10 days in China [78]. In eastern North America, the SOS generally advanced by three days for every 1 °C increase in the LST. These phenomena were the largest in the urban center and decreased exponentially as they headed toward the rural area [78]. The authors of [79] reported that the spring phenology of vegetation occurs earlier along the urban–rural gradient, and it occurs much earlier when close to the urban center because of the UHI effect. In the urban area of eastern North America, the SOS was advanced, on average, by seven days compared with the surrounding rural area, and EOS was delayed about eight days [12]. According to [80], on average, Boston's land surface temperatures were about 7 °C warmer, and its growing season was 18 to 22 days longer relative to the adjacent rural areas.

In this study, the observed green-up dates in the urban center, suburbs, and rural areas were earlier than the expected dates (Table 4), which is attributed to the temperature increase due to urbanization in those areas. Mts. Nam, Mido, and Umyeon, which showed the largest difference from the expected dates, are located in the urban center where the urbanization ratio is high and thus maintain a higher temperature than the surrounding areas due to the urban heat island [17]. Although Mt. Cheonggye and Mt. Buram are located in the suburbs, it is known that they are affected artificially because they are adjacent to the urban center [81]. Therefore, those sites also showed a big difference from the expected dates. Gwangneung (Mt. Sori), which is located in a rural area, has less land use intensity and population density than the urban center and suburbs, and thus the difference between the observed and expected dates of green-up is relatively small, but compared with Mt. Jeombong, a natural forest, there is an artificial influence, indicating a difference in the green-up date (Table 4). By comparing these results by landscape type according to land use intensity, this showed the biggest difference in the urban center as the difference between the observed date and the expected date of each study site was 11 days in the rural area, about 14.5 days in the suburbs, and about 16.3 days in the urban center (Table 4). According to a previous study [23], if the mean air temperature rises by 1 °C, the green-up date of *Q. mongolica* is advanced by 3.58 (based on MODIS image interpretation) and 4.33 (based on AGDD) days. If the results obtained from this study are translated into the air temperature based on previous research results [23], it could be deduced that the air temperature in the urban center, suburbs, and rural area rose by 3.8 to 4.6 °C, 3 to 4.1 °C, and 2.5 to 3.1 °C, respectively. In fact, the air temperature in the urban center of Seoul was reported to be about 5 °C different from the outskirt of the city [82].

According to [83], early flowering plays an important role in determining plant reproduction and pollen limitations by increasing the probability of experiencing frost damage. In addition, a delay or shortening of the flowering period can have a significant influence on the pollination process by affecting the available time of pollen and the sharing of pollen [84]. Differences in the timing of phenological events between urban and rural areas can lead to reproductive isolation, especially with plants that have a short flowering period [29]. Different responses of plant phenology between urban and rural can be blocked or restrict gene flow among meta-populations and meta-communities in rural–urban transects, and in addition, these different responses are likely to accelerate species polarization [85].

The UHI effect caused by urbanization can be confirmed through AGDD. AGDD values are highly correlated with the date of green-up and flowering and can be used as indicators of vegetation phenology [86]. The AGDD threshold for the green-up of *Q. mongolica* is about 159 °C [86], and this study shows that the higher the land use strength at the study site, the faster the AGDD threshold is reached (Figure 6). These results indicate that *Q. mongolica* reaches green-up faster because the AGDD value reaches the threshold earlier due to the increase in temperature from the UHI effect. Furthermore, the results prove that the green-up of plants is accelerating due to climate change caused by urbanization.

Urbanization, along with its consequence, climate change, is occurring at an unprecedented rate [87,88]. This rapid, uncontrollable acceleration of urbanization has led to worsening environmental degradation, resulting in issues such as pollution and unpredictable climate patterns, among many other indirect consequences [19–22,88–91]. The environmental change occurring in the urban world does not only affect the cities themselves—the climate impact of urbanization is spreading out on a global scale [1,2,87]. The loss of vegetation due to urbanization leads to several consequences. Not only does the area lose its richness in biodiversity but also its circulation of water, nitrogen, and other elements would be affected [2,5,8–11,92]. At the same time, as the areal size of greenery space decreases, CO_2 emissions rise, which leads to further warming of the area and, consequently, the world [2,93].

In phenological research, urban areas are an important study field because they enable an assessment of the future potential impacts of climate change on plant development [17,28]. The investigation of urban phenology is important because cities with their amplified temperatures can serve as a proxy for future conditions, and thus future phenology can be estimated from current information [29]. In this respect, this study, which indicates that vegetation phenology was advanced due to the urbanization effect, provides information on how vegetation phenology changes when the temperature increases in the future.

4.2. *Diagnosis of Phenological Changes by Analyzing Sap Flow*

Traditionally, the study of vegetation phenology focuses on monitoring and analyses of the timing of phenological events [17,60]. Phenological observations are mainly performed through visual observation. Consequently, most phenological studies were conducted with easily observable things such as green-up, leaf bud break, first flowering, and leaf fall [6,17,43,44]. Recently, in addition to the method of checking phenological phases by observing the visual observation, a study method to confirm phenological phases through physiological responses such as photosynthesis has also been proposed [17,50,94,95].

At scales from organs to ecosystems, many processes, particularly those related to the cycling of carbon (productivity and growth), water (evapotranspiration and runoff), and nutrients (decomposition and mineralization), are directly mediated by phenology, and the seasonality of these processes is implicitly phenological [30]. Sap flow, which is well known as a harbinger of spring, is a physiological process driven by phenological change. Sap flow becomes active and increases contemporarily with leaf development, and thereby sap flow and the leaf area index denote similar early spring patterns [54]. Simultaneously with leaf development, transpiration has to progress [96] to participate in forming the leaves and the follow-up of tree radial growth. Therefore, phenology is tightly connected to the ecophysiological processes of deciduous tree species [51]. Stem volume changes and sap flow provided valuable additional information specifying the tree development during both spring and autumn phenological stages. During the leaf expansion phase, the diameter of trees decreased in the deciduous trees. There is a close relationship between the use of stem water storage and leaf phenology. Sap flow was detected in the branches and the main stem of trees without leaf transpiration. These sap flow patterns observed in branches and stems, along with changes in VWC (volumetric water content) in sapwood and in the stem diameter, may be associated with the movement of water and carbohydrates necessary for the process of developing new leaves [59].

In this study, both the green-up date and the change trajectory of the curvature K value derived from the sap flow were similar to the green-up date and the change trajectory of the curvature K value derived from the digital camera and MODIS satellite images (Table 6). These results show that vegetation phenology observed through the appearance of plants is reflected in the sap flow as a physiological reaction within the plant body. In fact, according to [54], sap flux density and leaf unfolding showed a linear relationship, and in the late stage of leaf development, a decrease in sap flow was observed due to the reduced transpirational demand.

The results of this study, which show that physiological responses in plants are similar to the vegetation phenology, can be evaluated as the results of a step forward in phenological studies, which have mainly been observed through the appearance of plants. In particular, considering that phenological events emerging in appearance may be difficult to observe accurately and precisely due to various influences [56,57], observation of phenological events through physiological responses could be used as a tool to verify the response of vegetation according to various environmental changes including climate. Furthermore, it is expected that the sap flow of plants could be used more diversely as a tool to reinforce monitoring of vegetation phenology by collecting sap flow data of plants in various spatiotemporal scales and comparing and analyzing them with seasonal data of phenology collected using remote sensing techniques.

4.3. Ecosystem Management Strategy to Adapt Climate Change

Climate change has already become a reality, and even if we try to balance greenhouse gas emissions and absorption, it seems that we will soon be in danger of being hit by greenhouse gases already emitted [97]. An IPCC-led international agreement system is pushing to contain the amount of greenhouse gases currently emitted as much as possible, but it is expected that the absolute volume will increase in the coming years as the emissions of developing countries such as China, India, Brazil, and Russia increase explosively [98]. Ecosystems have experienced environmental changes such as climate change in the past and have adapted to these changes [99], but the rapid climate change that is happening will be far beyond the speed at which species and ecosystems can adapt and will have a variety of effects, including the extinction of many species [98,100]. In this regard, in parallel with efforts to reduce greenhouse gas emissions, we need to find countermeasures to adapt to future climate change [100].

We have interpreted the cause of climate change with an emphasis on the increased use of fossil fuels up to date. However, as the results of this study show, the response of ecosystems is closely related to the land use intensity of the site. The observed evidence shows that the effects of urban heat islands were greater than those from climate change that greenhouse gases cause in some locations [101,102]. The concentration of CO_2 is also steadily increasing at a global level, but it is showing a distinct seasonal variation that is high in winter and low in summer [8,103], which is the result of temperate forests acting as a source of absorption [104]. All environmental problems, including climate change, have sources of both emission and absorption. Therefore, we can mitigate climate change by reducing CO_2 emissions, but we can also mitigate climate change by increasing absorption sources. Sound nature helps adapt to climate change by absorbing and storing carbon. Since about 20% of greenhouse gas emissions are estimated to be due to deforestation, forest conservation and restoration can store a considerable amount of carbon [105–107]. Achieving sustainable land use by preserving and restoring nature can be a climate change adaptation measure that can mitigate climate change. As the IUCN suggests, preserving and restoring nature to achieve sustainable land use can be an adaptation measure to mitigate climate change. Vegetation achieved through ecological restoration can function as a true adaptation measure by displaying ecosystem service functions such as climate control and carbon dioxide absorption. In this respect, systematic and wise land use planning is required to achieve efficient adaptation to climate change. In fact, the balance of the carbon cycle and the air temperature increase coefficient were shown to depend on the land use pattern of local areas, and the carbon budget by region also showed such trend [108].

Even at the site scale, we can use vegetation to conserve energy and create thermally pleasant environments by encouraging evapotranspirational cooling, and shading from the hot summer sun [109–111]. As we understand the ecological functions that create surface climates and the specific landscape features that alter these functions, we can make the climate favorable for us by taking advantage of natural landscape processes [109,111–118]. This is a vital theme within the land use planning field, which advocates understanding local environmental features as part of the site planning process [117,119] and creating designs that are in harmony with the environment, especially in terms of energy and water conservation [120,121]. Therefore, we recommend conservation and restoration of natural ecosystems as a strategy that enables humankind to adapt to climate change impacts [113,122,123]

5. Conclusions

Climate change is rapidly progressing. The most important anthropogenic influences on climate are the emission of greenhouse gases and changes in land use such as urbanization, but the importance of the latter is increasingly being highlighted.

Urbanization is one of the major social changes that have spread around the world. Urbanization is happening rapidly at an unprecedented rate, and the increases in development areas and population growth due to it are causing changes in weather factors and affecting the ecosystem. As phenology is a significant diagnostic tool for the biological impacts from climate change, it could be an indicator for clarifying the effect of urbanization. In this study, the relationship between the phenology response of Mongolian oak and land use intensity was investigated by determining the green-up date of plants through digital camera image and MODIS satellite image analyses. We confirmed that the green-up date of Mongolian oak was advanced due to the temperature rise resulting from urbanization. The change was in proportion to the degree of urbanization and thereby was the largest in the urban center and tended to decrease moving through the rural area to the natural area. By comparing by landscape type according to land use intensity, this showed the biggest difference in the urban center as the difference between the observed date and the expected date of each study site was 11.0 days in the rural area, about 14.5 days in the suburbs, and about 16.3 days in the urban center. If we translate the results into the air temperature based on previous research results, it could be deduced that the air temperature in the urban center, suburbs, and rural area rose by 3.8 to 4.6 °C, 3.3 to 4.1 °C, and 2.5 to 3.1 °C, respectively.

This trend was also identified by AGDD, which determine the physiological activity of the plant depending on the seasonal changes, and the sap flow, one of the physiological responses of the plant. The higher the intensity of land use, the faster the green-up threshold is reached. From this result, we were able to confirm the availability of AGDD and sap flow in predicting changes in ecosystems due to climate change including phenology.

On the other hand, the change in sap flow was almost consistent with that of the green-up date in the change trajectory as the difference was within one day. As most studies on plant phenology have focused on external changes of plants, observing the seasonal change in plants through this physiological response is meaningful in terms of expanding the scope of research in the field.

Furthermore, the significant difference in the plant phenology response in proportion to land use intensity on the same latitude in the same climate zone can be important evidence for proving the impact of urbanization as a factor in causing climate change. This result is expected to contribute significantly to developing future climate change adaptation strategies.

Author Contributions: Conceptualization, A.R.K., C.H.L. and C.S.L.; methodology, A.R.K., C.H.L. and C.S.L.; software, A.R.K. and C.H.L.; validation, B.S.L., J.S. and C.S.L.; formal analysis, A.R.K. and C.H.L.; investigation, A.R.K. and C.H.L.; resources, C.H.L. and C.S.L.; data curation, B.S.L. and J.S.; writing—original draft preparation, A.R.K.; writing—review and editing, C.H.L. and C.S.L.; visualization, A.R.K.; supervision, C.S.L.; project administration, C.S.L. All authors have read and agreed to the published version of the manuscript.

Funding: This research received no external funding.

Institutional Review Board Statement: Not applicable.

Informed Consent Statement: Not applicable.

Data Availability Statement: No new data were created or analyzed in this study. Data sharing is not applicable to this article.

Conflicts of Interest: The authors declare no conflict of interest.

References

1. Lyu, R.; Clarke, K.C.; Zhang, J.; Jia, X.; Feng, J.; Li, J. The impact of urbanization and climate change on ecosystem services: A case study of the city belt along the Yellow River in Ningxia, China. *Comput. Environ. Urban Syst.* **2019**, *77*, 101351. [CrossRef]
2. Fu, Q.; Xu, L.; Zheng, H.; Chen, J. Spatiotemporal dynamics of carbon storage in response to urbanization: A case study in the Su-Xi-Chang region, China. *Processes* **2019**, *7*, 836. [CrossRef]
3. Foley, J.A.; DeFries, R.; Asner, G.P.; Barford, C.; Bonan, G.; Carpenter, S.R.; Chapin, F.S.; Coe, M.T.; Daily, G.C.; Gibbs, H.K.; et al. Global consequences of land use. *Science* **2005**, *309*, 570–574. [CrossRef] [PubMed]
4. Grimm, N.B.; Faeth, S.H.; Golubiewski, N.E.; Redman, C.L.; Wu, J.; Bai, X.; Briggs, J.M. Global change and the ecology of cities. *Science* **2008**, *319*, 756–760. [CrossRef] [PubMed]
5. Seto, K.C.; Güneralp, B.; Hutyra, L.R. Global forecasts of urban expansion to 2030 and direct impacts on biodiversity and carbon pools. *Proc. Natl. Acad. Sci. USA* **2012**, *109*, 16083–16088. [CrossRef]
6. Qiu, T.; Song, C.; Li, J. Impacts of urbanization on vegetation phenology over the past three decades in Shanghai, China. *Remote Sens.* **2017**, *9*, 970. [CrossRef]
7. Meng, L.; Mao, J.; Zhou, Y.; Richardson, A.D.; Lee, X.; Thornton, P.E.; Ricciuto, D.M.; Li, X.; Dai, Y.; Shi, X.; et al. Urban warming advances spring phenology but reduces the response of phenology to temperature in the conterminous United States. *Proc. Natl. Acad. Sci. USA* **2020**, *117*, 4228–4233. [CrossRef]
8. Amthor, J.S. Terrestrial higher-plant response to increasing atmospheric [CO2] in relation to the global carbon cycle. *Glob. Chang. Biol.* **1995**, *1*, 243–274. [CrossRef]
9. Houghton, R.A. Land-use change and the carbon cycle. *Glob. Chang. Biol.* **1995**, *1*, 275–287. [CrossRef]
10. Socolow, R.; Hotinski, R.; Greenblatt, J.B.; Pacala, S. Solving the climate problem: Technologies available to curb CO_2 emissions. *Environ. Sci. Policy Sustain. Dev.* **2004**, *46*, 8–19. [CrossRef]
11. Kashiwagi, H. Atmospheric carbon dioxide and climate change since the Late Jurassic (150 Ma) derived from a global carbon cycle model. *Palaeogeogr. Palaeoclimatol. Palaeoecol.* **2016**, *454*, 82–90. [CrossRef]
12. Zhang, X.; Friedl, M.A.; Schaaf, C.B.; Strahler, A.H.; Schneider, A. The footprint of urban climates on vegetation phenology. *Geophys. Res. Lett.* **2004**, *31*, L12209. [CrossRef]
13. Changnon, S.A. Inadvertent Weather Modification in Urban Areas: Lessons for Global Climate Change. *Bull. Am. Meteorol. Soc.* **1992**, *73*, 619–627. [CrossRef]
14. Tayanc, M.; Toros, H. Urbanization effects on regional climate change in the case of four large cities of Turkey. *Clim. Chang.* **1997**, *35*, 501–524. [CrossRef]

15. Liu, J.; Niyogi, D. Meta-analysis of urbanization impact on rainfall modification. *Sci. Rep.* **2019**, *9*, 1–14.
16. Chang, Y.; Xiao, J.; Li, X.; Frolking, S.; Zhou, D.; Schneider, A.; Weng, Q.; Yu, P.; Wang, X.; Li, X.; et al. Exploring diurnal cycles of surface urban heat island intensity in Boston with land surface temperature data derived from GOES-R geostationary satellites. *Sci. Total Environ.* **2021**, *763*, 144224. [CrossRef] [PubMed]
17. Lim, C.H.; An, J.H.; Jung, S.H.; Nam, G.B.; Cho, Y.C.; Kim, N.S.; Lee, C.S. Ecological consideration for several methodologies to diagnose vegetation phenology. *Ecol. Res.* **2018**, *33*, 363–377. [CrossRef]
18. Krehbiel, C.; Henebry, G.M. A comparison of multiple datasets for monitoring thermal time in urban areas over the US Upper Midwest. *Remote. Sens.* **2016**, *8*, 297. [CrossRef]
19. Ha, K.J.; Ha, E.H.; YOO, C.S.; Jeon, E.H. Temperature Trends and Extreme Climate since 1909 at Big Four Cities of Korea. *J. Korean Meteorol. Soc.* **2004**, *41*, 1–16.
20. Alexander, L.; Zhang, V.X.; Peterson, T.C.; Caesar, J.; Gleason, B.; Klein Tank, A.; Haylock, M.; Collins, D.; Vazquez-Aguirre, J.L. Global observed changes in daily climate extremes of temperature and precipitation. *J. Geophys. Res. Atmos.* **2006**, *111*, D05109. [CrossRef]
21. Yang, X.; Ruby Leung, L.; Zhao, N.; Zhao, C.; Qian, Y.; Hu, K.; Liu, X.; Chen, B. Contribution of urbanization to the increase of extreme heat events in an urban agglomeration in east China. *Geophys. Res. Lett.* **2017**, *44*, 6940–6950. [CrossRef]
22. Zhao, N.; Jiao, Y.; Ma, T.; Zhao, M.; Fan, Z.; Yin, X.; Liu, Y.; Yue, T. Estimating the effect of urbanization on extreme climate events in the Beijing-Tianjin-Hebei region, China. *Sci. Total Environ.* **2019**, *688*, 1005–1015. [CrossRef] [PubMed]
23. Lim, C.H.; Jung, S.H.; Kim, A.R.; Kim, N.S.; Lee, C.S. Monitoring for Changes in Spring Phenology at Both Temporal and Spatial Scales Based on MODIS LST Data in South Korea. *Remote Sens.* **2020**, *12*, 3282. [CrossRef]
24. Oke, T.R. The urban energy balance. *Prog. Phys. Geogr.* **1988**, *12*, 471–508. [CrossRef]
25. Zipper, S.C.; Schatz, J.; Singh, A.; Kucharik, C.J.; Townsend, P.A.; Loheide, S.P., II. Urban heat island impacts on plant phenology: Intra-urban variability and response to land cover. *Environ. Res. Lett.* **2016**, *11*, 054023. [CrossRef]
26. Orimoloye, I.R.; Mazinyo, S.P.; Nel, W.; Kalumba, A.M. Spatiotemporal monitoring of land surface temperature and estimated radiation using remote sensing: Human health implications for East London, South Africa. *Environ. Earth Sci.* **2018**, *77*, 1–10. [CrossRef]
27. Willie, Y.A.; Pillay, R.; Zhou, L.; Orimoloye, I.R. Monitoring spatial pattern of land surface thermal characteristics and urban growth: A case study of King Williams using remote sensing and GIS. *Earth Sci. Inform.* **2019**, *12*, 447–464. [CrossRef]
28. Jochner, S.; Caffarra, A.; Menzel, A. Can spatial data substitute temporal data in phenological modelling? A survey using birch flowering. *Tree Physiol.* **2013**, *33*, 1256–1268. [CrossRef] [PubMed]
29. Jochner, S.; Menzel, A. Urban phenological studies–past, present, future. *Environ. Pollut.* **2015**, *203*, 250–261. [CrossRef] [PubMed]
30. Richardson, A.D.; Keenan, T.F.; Migliavacca, M.; Ryu, Y.; Sonnentag, O.; Toomey, M. Climate change, phenology, and phenological control of vegetation feedbacks to the climate system. *Agric. For. Meteorol.* **2013**, *169*, 156–173. [CrossRef]
31. Keenan, T.F.; Gray, J.; Friedl, M.A.; Toomey, M.; Bohrer, G.; Hollinger, D.Y.; Munger, J.W.; O'Keefe, J.; Schmid, H.P.; Wing, I.S.; et al. Net carbon uptake has increased through warming-induced changes in temperate forest phenology. *Nat. Clim. Chang.* **2014**, *4*, 598–604. [CrossRef]
32. Bradley, N.L.; Leopold, A.C.; Ross, J.; Huffaker, W. Phenological changes reflect climate change in Wisconsin. *Proc. Natl. Acad. Sci. USA* **1999**, *96*, 9701–9704. [CrossRef] [PubMed]
33. Peñuelas, J.; Filella, I. Responses to a warming world. *Science* **2001**, *294*, 793–795. [CrossRef]
34. Kim, N.S. Characteristics of Spatio-temporal Distribution of Phenology Fluctuation By Using Modis Images. *Symp. Korean Geogr. Soc.* **2011**, *5*, 73–77.
35. Walther, G.R.; Post, E.; Convey, P.; Menzel, A.; Parmesan, C.; Beebee, T.J.C.; Fromentin, J.M.; Hoegh-Guldberg, O.; Bairlein, F. Ecological responses to recent climate change. *Nature* **2002**, *416*, 389–395. [CrossRef] [PubMed]
36. Ovaskainen, O.; Skorokhodova, S.; Yakovleva, M.; Sukhov, A.; Kutenkov, A.; Kutenkova, N.; Shcherbakov, A.; Meyke, E.; del Mar Delgado, M. Community-level phenological response to climate change. *Proc. Natl. Acad. Sci. USA* **2013**, *110*, 13434–13439. [CrossRef]
37. Julitta, T.; Cremonese, E.; Migliavacca, M.; Colombo, R.; Galvagno, M.; Siniscalco, C.; Rossini, M.; Fava, F.; Cogliatia, S.; Cellab, U.M.; et al. Using digital camera images to analyse snowmelt and phenology of a subalpine grassland. *Agric. For. Meteorol.* **2014**, *198*, 116–125. [CrossRef]
38. Keeling, C.D.; Chin, J.F.S.; Whorf, T.P. Increased activity of northern vegetation inferred from atmospheric CO_2 measurements. *Nature* **1996**, *382*, 146–149. [CrossRef]
39. Kang, S.; Running, S.W.; Lim, J.H.; Zhao, M.; Park, C.R.; Loehman, R. A regional phenology model for detecting onset of greenness in temperate mixed forests, Korea: An application of MODIS leaf area index. *Remote Sens. Environ.* **2003**, *86*, 232–242. [CrossRef]
40. Badeck, F.W.; Bondeau, A.; Böttcher, K.; Doktor, D.; Lucht, W.; Schaber, J.; Sitch, S. Responses of spring phenology to climate change. *New Phytol.* **2004**, *162*, 295–309. [CrossRef]
41. Gray, S.B.; Brady, S.M. Plant developmental responses to climate change. *Dev. Biol.* **2016**, *419*, 64–77. [CrossRef]
42. Reed, B.C.; Brown, J.F.; VanderZee, D.; Loveland, T.R.; Merchant, J.W.; Ohlen, D.O. Measuring phenological variability from satellite imagery. *J. Veg. Sci.* **1994**, *5*, 703–714. [CrossRef]
43. Schwartz, M.D.; Reiter, B.E. Changes in North American spring. *Int. J. Climatol.* **2000**, *20*, 929–932. [CrossRef]

44. Peñuelas, J.; Filella, I.; Comas, P. Changed plant and animal life cycles from 1952 to 2000 in the Mediterranean region. *Glob. Chang. Biol.* **2002**, *8*, 531–544. [CrossRef]
45. Cao, R.; Shen, M.; Zhou, J.; Chen, J. Modeling vegetation green-up dates across the Tibetan Plateau by including both seasonal and daily temperature and precipitation. *Agric. For. Meteorol.* **2018**, *249*, 176–186. [CrossRef]
46. Ide, R.; Oguma, H. Use of digital cameras for phenological observations. *Ecol. Inform.* **2010**, *5*, 339–347. [CrossRef]
47. Xiang, Q.; Zhou, Y.; Liu, J. Monitoring mangrove phenology using camera images. In *IOP Conference Series: Earth and Environmental Science, Proceedings of the 2019 International Conference on Resources and Environmental Research, Qingdao, China, 25–27 October 2019*; IOP Publishing: Qingdao, China, 2019; Volume 432, p. 012001.
48. Fitchett, J.M.; Grab, S.W.; Thompson, D.I. Plant phenology and climate change: Progress in methodological approaches and application. *Prog. Phys. Geog.* **2015**, *39*, 460–482. [CrossRef]
49. Zhang, Y.; Chen, D.; Wang, S.; Tian, L. A promising trend for field information collection: An air-ground multi-sensor monitoring system. *Inf. Process. Agric.* **2018**, *5*, 224–233. [CrossRef]
50. Urban, J.; Bednářová, E.; Plichta, R.; Kučera, J. Linking phenological data to ecophysiology of European beech. *Acta Hortic. (ISHS)* **2013**, *991*, 293–299. [CrossRef]
51. Paloschi, R.A.; Ramos, D.M.; Ventura, D.J.; Souza, R.; Souza, E.; Morellato, L.P.C.; Nóbrega, R.L.B.; Coutinho, Í.A.C.; Verhoef, A.; Körting, T.S.; et al. Environmental Drivers of Water Use for Caatinga Woody Plant Species: Combining Remote Sensing Phenology and Sap Flow Measurements. *Remote Sens.* **2021**, *13*, 75. [CrossRef]
52. Alberton, B.; da Silva Torres, R.; Sanna Freire Silva, T.; Rocha, H.R.D.; Moura, M.S.B.; Morellato, L.P.C. Leafing patterns and drivers across seasonally dry tropical communities. *Remote Sens.* **2019**, *11*, 2267. [CrossRef]
53. Rojas-Jiménez, K.; Holbrook, N.M.; Gutiérrez-Soto, M.V. Dry-season leaf flushing of *Enterolobium cyclocarpum* (ear-pod tree): Above–and belowground phenology and water relations. *Tree Physiol.* **2007**, *27*, 1561–1568. [CrossRef] [PubMed]
54. Lavrič, M.; Eler, K.; Ferlan, M.; Vodnik, D.; Gričar, J. Chronological sequence of leaf phenology, xylem and phloem formation and sap flow of *Quercus pubescens* from abandoned karst grasslands. *Front. Plant Sci.* **2017**, *8*, 314. [CrossRef] [PubMed]
55. Dixon, H.H.; Joly, J. On the ascent of sap. *Philos. Trans. R. Soc. Lond. B* **1895**, *186*, 563–576.
56. Sparks, T.H.; Huber, K.; Croxton, P.J. Plant development scores from fixed-date photographs: The influence of weather variables and recorder experience. *Int. J. Biometeorol.* **2006**, *50*, 275–279. [CrossRef] [PubMed]
57. Richardson, A.D.; Hufkens, K.; Milliman, T.; Frolking, S. Intercomparison of phenological transition dates derived from the PhenoCam Dataset V1.0 and MODIS satellite remote sensing. *Sci. Rep.* **2018**, *8*, 1–12.
58. Link, P.; Simonin, K.; Maness, H.; Oshun, J.; Dawson, T.; Fung, I. Species differences in the seasonality of evergreen tree transpiration in a Mediterranean climate: Analysis of multiyear, half-hourly sap flow observations. *Water Resour. Res.* **2014**, *50*, 1869–1894. [CrossRef]
59. Di Francescantonio, D.; Villagra, M.; Goldstein, G.; Campanello, P.I. Leaf phenology and water-use patterns of canopy trees in Northern Argentinean subtropical forests. *Tree Physiol.* **2018**, *38*, 1841–1854. [CrossRef]
60. Richardson, A.D.; Jenkins, J.P.; Braswell, B.H.; Hollinger, D.Y.; Ollinger, S.V.; Smith, M.L. Use of digital webcam images to track spring green-up in a deciduous broadleaf forest. *Oecologia* **2007**, *152*, 323–334. [CrossRef]
61. Kim, N.S.; Lee, C.S. Prediction for climate change: Climate change and prediction for change of vegetation distribution due to it in the Korean peninsula. *Ser. Long Term Ecol. Res.* **2014**, *7*, 102–112.
62. Kim, N.S.; Lee, H.C.; Cha, J.Y. A Study on Changes of Phenology and Characteristics of Spatial Distribution Using MODIS Images. *J. Korea Soc. Environ. Restor. Technol.* **2013**, *16*, 59–69. [CrossRef]
63. Zhang, X.; Friedl, M.A.; Schaaf, C.B.; Strahler, A.H.; Hodges, J.C.; Gao, F.; Reed, B.C.; Huete, A. Monitoring vegetation phenology using MODIS. *Remote Sens. Environ.* **2003**, *84*, 471–475. [CrossRef]
64. Richardson, A.D.; Braswell, B.H.; Hollinger, D.Y.; Jenkins, J.P.; Ollinger, S.V. Near-surface remote sensing of spatial and temporal variation in canopy phenology. *Ecol. Appl.* **2009**, *19*, 1417–1428. [CrossRef] [PubMed]
65. Hufkens, K.; Friedl, M.; Sonnentag, O.; Braswell, B.H.; Milliman, T.; Richardson, A.D. Linking near-surface and satellite remote sensing measurements of deciduous broadleaf forest phenology. *Remote Sens. Environ.* **2012**, *117*, 307–321. [CrossRef]
66. Klosterman, S.T.; Hufkens, K.; Gray, J.M.; Melaas, E.; Sonnentag, O.; Lavine, I.; Mitchell, L.; Norman, R.; Friedl, M.A.; Richardson, A.D. Evaluating remote sensing of deciduous forest phenology at multiple spatial scales using PhenoCam imagery. *Biogeosciences* **2014**, *11*, 4305–4320. [CrossRef]
67. Walker, J.; De Beurs, K.; Wynne, R.; Gao, F. Evaluation of Landsat and MODIS data fusion products for analysis of dryland forest phenology. *Remote Sens. Environ.* **2012**, *117*, 381–393. [CrossRef]
68. Burgess, S.S.; Adams, M.A.; Turner, N.C.; Beverly, C.R.; Ong, C.K.; Khan, A.A.; Bleby, T.M. An improved heat pulse method to measure low and reverse rates of sap flow in woody plants. *Tree Physiol.* **2001**, *21*, 589–598. [CrossRef]
69. Richardson, A.D.; Hufkens, K.; Li, X.; Ault, T.R. Testing Hopkins' bioclimatic law with PhenoCam data. *Appl. Plant Sci.* **2019**, *7*, e01228. [CrossRef]
70. Kira, T. *New Climatic Zonation in Eastern Asia as a Basis of Agricultural Geography*; Kyoto Imperial University: Kyoto, Japan, 1945; p. 45. (In Japanese)
71. Schenker, G.; Lenz, A.; Körner, C.; Hoch, G. Physiological minimum temperatures for root growth in seven common European broad-leaved tree species. *Tree Physiol.* **2014**, *34*, 302–313. [CrossRef]
72. Dale, V.H. The relationship between land-use change and climate change. *Ecol. Appl.* **1997**, *7*, 753–769. [CrossRef]

73. Pielke, R.A. Land use and climate change. *Science* **2005**, *310*, 1625–1626. [CrossRef] [PubMed]
74. Olsson, L.; Barbosa, H.; Bhadwal, S.; Cowie, A.; Delusca, K.; Flores-Renteria, D.; Hermans, K.; Jobbagy, E.; Kurz, W.; Li, D.; et al. Land Degradation. In *Climate Change and Land: An IPCC Special Report on Climate Change, Desertification, Land Degradation, Sustainable Land Management, Food Security, and Greenhouse Gas Fluxes in Terrestrial Ecosystems*; IPCC: Geneva, Switzerland, 2019; pp. 345–436.
75. Chung, U.; Choi, J.; Yun, J.I. Urbanization effect on the observed change in mean monthly temperatures between 1951–1980 and 1971–2000 in Korea. *Clim. Chang.* **2004**, *66*, 127–136. [CrossRef]
76. Ren, G.; Zhou, Y.; Chu, Z.; Zhou, J.; Zhang, A.; Guo, J.; Liu, X. Urbanization effects on observed surface air temperature trends in North China. *J. Clim.* **2008**, *21*, 1333–1348. [CrossRef]
77. Jones, P.D.; Lister, D.H.; Li, Q. Urbanization effects in large-scale temperature records, with an emphasis on China. *J. Geophys. Res. Atmos.* **2008**, *113*, D16122. [CrossRef]
78. Zhou, D.; Zhao, S.; Zhang, L.; Liu, S. Remotely sensed assessment of urbanization effects on vegetation phenology in China's 32 major cities. *Remote Sens. Environ.* **2016**, *176*, 272–281. [CrossRef]
79. Tian, J.; Zhu, X.; Shen, Z.; Wu, J.; Xu, S.; Liang, Z.; Wang, J. Investigating the urban-induced microclimate effects on winter wheat spring phenology using Sentinel-2 time series. *Agric. For. Meteorol.* **2020**, *294*, 108153. [CrossRef]
80. Melaas, E.K.; Wang, J.A.; Miller, D.L.; Friedl, M.A. Interactions between urban vegetation and surface urban heat islands: A case study in the Boston metropolitan region. *Environ. Res. Lett.* **2016**, *11*, 054020. [CrossRef]
81. Lee, C.S.; Cho, Y.C.; Lee, A.N. Restoration planning for the Seoul metropolitan area, Korea. In *Ecology, Planning, and Management of Urban Forests*; Springer: New York, NY, USA, 2008; pp. 393–419.
82. Lee, C.S.; Jung, S.; Lim, B.S.; Kim, A.R.; Lim, C.H.; Lee, H. Forest Decline Under Progress in the Urban Forest of Seoul, Central Korea. In *Forest Degradation Around the World*; IntechOpen: London, UK, 2019.
83. Gezon, Z.J.; Inouye, D.W.; Irwin, R.E. Phenological change in a spring ephemeral: Implications for pollination and plant reproduction. *Glob. Chang. Biol.* **2016**, *22*, 1779–1793. [CrossRef]
84. Primack, R.B.; Higuchi, H.; Miller-Rushing, A.J. The impact of climate change on cherry trees and other species in Japan. *Biol. Conserv.* **2009**, *142*, 1943–1949. [CrossRef]
85. Neil, K.; Wu, J. Effects of urbanization on plant flowering phenology: A review. *Urban Ecosyst.* **2006**, *9*, 243–257. [CrossRef]
86. Lim, C.H.; Jung, S.H.; Kim, N.S.; Lee, C.S. Deduction of a meteorological phenology indicator from reconstructed MODIS LST imagery. *J. For. Res.* **2019**, *2019*, 1–12. [CrossRef] [PubMed]
87. Bazrkar, M.H.; Zamani, N.; Eslamian, S.; Eslamian, A.; Dehghan, Z. Urbanization and Climate Change. In *Handbook of Climate Change Adaptation*; Leal Filho, W., Ed.; Springer: Berlin/Heidelberg, Germany, 2015.
88. Wang, W.; Wu, T.; Li, Y.; Xie, S.; Han, B.; Zheng, H.; Ouyang, Z. Urbanization Impacts on Natural Habitat and Ecosystem Services in the Guangdong-Hong Kong-Macao Megacity. *Sustainability* **2020**, *12*, 6675. [CrossRef]
89. Huong, H.T.L.; Pathirana, A. Urbanization and climate change impacts on future urban flooding in Can Tho city, Vietnam. *Hydrol. Earth Syst. Sci.* **2013**, *17*, 379–394. [CrossRef]
90. D'amato, G.; Pawankar, R.; Vitale, C.; Lanza, M.; Molino, A.; Stanziola, A.; Sanduzzi, A.; Vatrella, A.; D'amato, M. Climate change and air pollution: Effects on respiratory allergy. *Allergy Asthma Immunol. Res.* **2016**, *8*, 391–395. [CrossRef]
91. Miller, J.D.; Hutchins, M. The impacts of urbanisation and climate change on urban flooding and urban water quality: A review of the evidence concerning the United Kingdom. *J. Hydrol. Reg. Stud.* **2017**, *12*, 345–362. [CrossRef]
92. Zhang, W.; Xu, H. Effects of land urbanization and land finance on carbon emissions: A panel data analysis for Chinese provinces. *Land Use Policy* **2017**, *63*, 493–500. [CrossRef]
93. UNEP. *Guidelines for Social Life Cycle Assessment of Products*; United Nations Environment Program: Paris, France, 2009.
94. Gu, L.; Post, W.M.; Baldocchi, D.; Andy Black, T.; Verma, S.B.; Vesala, T.; Wofsy, S.C. Phenology of Vegetation Photosynthesis. In *Phenology: An Integrative Environmental Science*; Schwartz, M.D., Ed.; Springer: Dordrecht, The Netherlands, 2003; pp. 467–485.
95. Shen, M.; Tang, Y.; Desai, A.R.; Gough, C.; Chen, J. Can EVI-derived land-surface phenology be used as a surrogate for phenology of canopy photosynthesis? *Int. J. Remote Sens.* **2014**, *35*, 1162–1174. [CrossRef]
96. Střelcová, K.; Priwitzer, T.; Minďáš, J. *Fenologické Fázy a Transpirácia Buka Lesného v Horskom Zmiešanom Lese*; Fenologická Odezva Proměnlivosti Podnebí: Brno, Czech Republic, 2006.
97. Hardy, J.T. *Climate Change: Causes, Effects, and Solutions*; John Wiley & Sons: Hoboken, NJ, USA, 2003.
98. Ritchie, H.; Roser, M. Fossil fuels. In *Our World in Data*; Available online: https://ourworldindata.org/fossil-fuels?utm_content=link5 (accessed on 30 March 2021).
99. Primack, R.B. *A Primer of Conservation Biology*, 4th ed.; Sinauer Associates, Inc.: Sunderland, MA, USA, 2008.
100. IPCC. *Climate Change 2014: Synthesis Report. Contribution of Working Groups I, II and III to the Fifth Assessment Report of the Intergovernmental Panel on Climate Change*; Core Writing Team, Pachauri, R.K., Meyer, L.A., Eds.; IPCC: Geneva, Switzerland, 2014.
101. Stone, B. Urban and rural temperature trends in proximity to large US cities: 1951–2000. *Int. J. Climatol.* **2007**, *27*, 1801–1807. [CrossRef]
102. Fujibe, F. Urban warming in Japanese cities and its relation to climate change monitoring. *Int. J. Climatol.* **2011**, *31*, 162–173. [CrossRef]
103. Park, S. Cloud and cloud shadow effects on the MODIS vegetation index composites of the Korean Peninsula. *Int. J. Remote Sens.* **2013**, *34*, 1234–1247. [CrossRef]

104. Barbour, M.G. *Terrestrial Plant Ecology;* The Benjamin/Cummings: San Francisco, CA, USA, 1999; p. 649.
105. The World Bank. *Biodiversity, Climate Change and Adaptation: Nature-Based Solutions from the World Bank Portfolio;* The World Bank: Washington, DC, USA, 2008; pp. 1–112.
106. Olander, L.P.; Boyd, W.; Lawlor, K.; Madeira, E.M.; Niles, J.O. *International Forest Carbon and the Climate Change Challenge: Issues and Options;* Nicholas Institute for Environmental Policy Solution: Durham, UK, 2009.
107. UNEP. *The Role of Ecosystems in Developing a Sustainable Green Economy;* Ecosystem Management Policy Series, Policy Brief 2–2010; United Nations Environment Program: Nairobi, Kenya, 2010.
108. Lee, C.S. Climate change adaptation realizing through the ecological restoration. In *Climate Change and Ecology: Series of Long Term Ecological Research 7;* Lee, C.S., Ed.; NIE: Seocheon, Korea, 2014; pp. 165–180.
109. Chang, C.R.; Li, M.H.; Chang, S.D. A preliminary study on the local cool-island intensity of Taipei city parks. *Landsc. Urban Plan.* **2007**, *80*, 386–395. [CrossRef]
110. Kleerekoper, L.; Van Esch, M.; Salcedo, T.B. How to make a city climate-proof, addressing the urban heat island effect. *Resour. Conserv. Recycl.* **2012**, *64*, 30–38. [CrossRef]
111. Doick, K.J.; Hutchings, T. *Air Temperature Regulation by Urban Trees and Green Infrastructure;* Forestry Commission: Edinburgh, UK, 2013; Volume 12, pp. 1–10.
112. Rahn, J.J. *Making the Weather Work for You: A Practical Guide for Gardener and Farmer;* Garden Way Publishing: Charlotte, VT, USA, 1979.
113. MEA (Millennium Ecosystem Assessment). Millennium Ecosystem Assessment Synthesis Report. 2005. Available online: http://pdf.wri.org/mea_synthesis.pdf (accessed on 28 February 2018).
114. Ryu, Y.; Kim, S.; Lee, D. The influence of wind flows on thermal comfort in the Daechung of a traditional Korean house. *Build. Environ.* **2009**, *44*, 18–26. [CrossRef]
115. Bajracharya, B.; Childs, I.; Hastings, P. Climate change adaptation through land use planning and disaster management: Local government perspectives from Queensland. In Proceedings of the 17th Pacific Rim Real Estate Society Conference, Gold Coast, Australia, 16–19 January 2011; PRRES: Gold Coast, Australia, 2011.
116. Obiakor, M.O.; Ezeonyejiaku, C.D.; Mogbo, T.C. Effects of vegetated and synthetic (impervious) surfaces on the microclimate of urban area. *J. Appl. Sci. Environ. Manag.* **2012**, *16*, 85–94.
117. Ren, Z.; He, X.; Zheng, H.; Zhang, D.; Yu, X.; Shen, G.; Guo, R. Estimation of the relationship between Urban Park characteristics and Park cool island intensity by remote sensing data and Field measurement. *Forests* **2013**, *4*, 868–886. [CrossRef]
118. Müller, N.; Kuttler, W.; Barlag, A.B. Counteracting urban climate change: Adaptation measures and their effect on thermal comfort. *Theor. Appl. Climatol.* **2014**, *115*, 243–257. [CrossRef]
119. Collins, N.; Smith, G.; Allen, J.B. *A Guide for Incorporating Adaptation to Climate Change into Land-Use Planning*, version 1; MCIP CEF Consultants: Halifax, NS, Canada, 2005.
120. Arnfield, A.J. Two decades of urban climate research, A review of turbulence, exchanges of energy and water, and the urban heat island. *Int. J. Climatol.* **2003**, *23*, 1–26. [CrossRef]
121. Hamin, E.M.; Gurran, N. Urban form and climate change: Balancing adaptation and mitigation in the U.S. and Australia. *Habitat Int.* **2009**, *33*, 238–245. [CrossRef]
122. EU Ad Hoc Expert Working Group on Biodiversity and Climate Change. Towards a Strategy on Climate Change. In *Ecosystem Services and Biodiversity;* European Commission: Brussels, Belgium, 2009; Available online: http://ec.europa.eu/environment/nature/pdf/discussion_paper_climate_change.pdf (accessed on 28 February 2021).
123. European Commission. *Science for Environment Policy Thematic Issue: Ecosystem-Based Adaptation;* European Commission: Brussels, Belgium, 2013; Available online: http://ec.europa.eu/environment/integration/research/newsalert/pdf/37si_en.Pdf (accessed on 28 February 2021).

remote sensing

MDPI

Article

Dynamics and Drivers of Vegetation Phenology in Three-River Headwaters Region Based on the Google Earth Engine

Jiyan Wang [1], Huaizhang Sun [1], Junnan Xiong [1,2,*], Dong He [3,4], Weiming Cheng [2], Chongchong Ye [1], Zhiwei Yong [1] and Xianglin Huang [5]

[1] School of Civil Engineering and Geomatics, Southwest Petroleum University, Chengdu 610500, China; wangjiyan@swpu.edu.cn (J.W.); 201822000598@stu.swpu.edu.cn (H.S.); 201822000599@stu.swpu.edu.cn (C.Y.); 201822000071@stu.swpu.edu.cn (Z.Y.)
[2] State Key Laboratory of Resources and Environmental Information System, Institute of Geographic Sciences and Natural Resources Research, Chinese Academy of Sciences, Beijing 100101, China; chengwm@lreis.ac.cn
[3] International Institute for Earth System Science, Nanjing University, Nanjing 210023, China; DG1927011@smail.nju.edu.cn
[4] Jiangsu Provincial Key Laboratory of Geographic Information Science and Technology, Nanjing University, Nanjing 210023, China
[5] School of Earth Science and Engineering, Hohai University, Nanjing 211106, China; xlhuang@hhu.edu.cn
* Correspondence: Xiongjn@swpu.edu.cn; Tel.: +86-28-83037604

Abstract: Phenology shifts over time are known as the canary in the mine when studying the response of terrestrial ecosystems to climate change. Plant phenology is a key factor controlling the productivity of terrestrial vegetation under climate change. Over the past several decades, the vegetation in the three-river headwaters region (TRHR) has been reported to have changed greatly owing to the warming climate and human activities. However, uncertainties related to the potential mechanism and influence of climatic and soil factors on the plant phenology of the TRHR are poorly understood. In this study, we used harmonic analysis of time series and the relative and absolute change rate on Google Earth Engine to calculate the start (SOS), end (EOS), and length (LOS) of the growing season based on MOD09A1 datasets; the results were verified by the observational data from phenological stations. Then, the spatiotemporal patterns of plant phenology for different types of terrain and basins were explored. Finally, the potential mechanism involved in the influence of climatic and soil factors on the phenology of plants in the TRHR were explored based on the structural equation model and Pearson's correlation coefficients. The results show the remotely sensed monitoring data of SOS ($R^2 = 0.84$, $p < 0.01$), EOS ($R^2 = 0.72$, $p < 0.01$), and LOS ($R^2 = 0.86$, $p < 0.01$) were very similar to the observational data from phenological stations. The SOS and LOS of plants possessed significant trends toward becoming advanced (*Slope* < 0) and extended (*Slope* > 0), respectively, from 2001 to 2018. The SOS was the earliest and the LOS was the longest in the Lancang River Basin, while the EOS was the latest in the Yangtze River Basin owing to the impact of climate change and soil factors. Meanwhile, the spatial patterns of SOS, EOS, and LOS have strong spatial heterogeneity at different elevations, slopes, and aspects. In addition, the results show that the drivers of plant phenology have basin-wide and stage differences. Specifically, the influence of soil factors on plant phenology in the Yangtze River Basin was greater than that of climatic factors, but climatic factors were key functional indicators of LOS in the Yellow and Lancang river basins, which directly or indirectly affect plant LOS through soil factors. This study will be helpful for understanding the relationship between the plant phenology of the alpine wetland ecosystem and climate change and improving the level of environmental management.

Keywords: plant phenology; spatiotemporal patterns; structural equation model; Google Earth Engine; Three-River Headwaters region

Citation: Wang, J.; Sun, H.; Xiong, J.; He, D.; Cheng, W.; Ye, C.; Yong, Z.; Huang, X. Dynamics and Drivers of Vegetation Phenology in Three-River Headwaters Region Based on the Google Earth Engine. *Remote Sens.* **2021**, *13*, 2528. https://doi.org/10.3390/rs13132528

Academic Editors: Xiaolin Zhu, Xuanlong Ma, Jiaxin Jin, Yuke Zhou and Qiaoyun Xie

Received: 6 May 2021
Accepted: 25 June 2021
Published: 28 June 2021

1. Introduction

The global climate has been warming gradually over the past several decades, which has important impacts on vegetation phenology in ecological systems [1–3]. Vegetation phenology acts as a sensitive and precise indicator that responds to climate warming and has become an important topic in the fields of climate and ecology [4,5]. Studies have shown that changes in spring and autumn plant phenology caused by climate change can differentially alter the length of the growing season and affect water, carbon, and energy fluxes between the atmosphere and the terrestrial biosphere [6]. Increased carbon uptake stimulated by an extended growing season has the potential to mitigate climate change [7]. Therefore, elucidating the trends in plant phenology can improve our understanding of the influence of climate change on ecosystem productivity, carbon cycling, and energy flow.

Although many studies have investigated plant phenology, little attention has been paid to alpine wetland ecosystems [8,9]. As the largest alpine wetland ecosystem in the world, the Three-River Headwaters region (TRHR) is considered the premonitory region of global climate change. It is worth noting that increasing human activities and global climate warming have led to severe ecological degradation in the TRHR, such as vegetation degradation, soil erosion, desertification, lake and wetland decline, and glacial retreat [10,11]. Because of the unique geographical location and climate of TRHR, a large number of researchers have studied this area. For example, Han et al. studied the relationship between plant greening and climate factors based on plant phenological site data, and the results showed that the trend for the time of plant greening was ahead–postpone–ahead–postpone [12]. Li explored the phenology response of plant to hydrothermal conditions from 1999 to 2010 based on SPOT NDVI, and the results indicated that the increase of cumulative precipitation and temperature of response time make SOS delayed [13]. Chen et al. used SPOT NDVI to explore the spatiotemporal patterns of plant phenology during 2000–2013, and the results showed that the SOS advanced, EOS delayed, and LOS extended [14]. Hence, it is a good idea to select the TRHR as a study area to explore the changes in plant phenology under climate change, which will improve our understanding of changes in plant phenology in alpine wetland ecosystems. Increased warming trends and frequent extreme events caused by climate change have produced significant impacts on many ecosystems, such as changes in vegetation phenology, grassland degradation, wetland shrinkage, and encroachment upon farmlands [15]. Currently, many research studies have focused on the response of vegetation phenology to specific climate factors, including temperature, precipitation, and shortwave radiation. The results indicated that interaction between temperature, shortwave radiation, and water has caused various impacts on vegetation activities in different regions [16–18]. For example, the SOS arrived 2.5 days earlier, and the EOS was delayed by 1 day for every 1 °C increase in the temperature across 19 European countries [19]. The onset time of 70.1% of vegetation in the growth season was delayed by 2.7 days because of winter precipitation in boreal forests [20]. Shortwave radiation plays a potential role in regulating vegetation growth in humid tropical or subtropical regions [21]. However, many factors can affect vegetation growth. Some changes in vegetation growth are caused by changes in climatic factors, but the soil factor (i.e., total soil C, N, and K) also affects vegetation dynamics because of the effects of soil conditions on the production of new cells that control plant photosynthesis [22]. For example, increasing the N input to land terrestrial ecosystems can promote vegetative growth and accelerate respiration in plants and soil microorganisms [23]. In fact, plants are very sensitive to resource conditions and tend to adjust their growth rates according to changing environments at different time scales [24]. A change in soil nutrient availability and mobility can change the photosynthetic rate of vegetation, which ultimately determines the difference in vegetative growth [25]. Therefore, it is important to understand the underlying mechanisms of how soil resources affect vegetative growth, especially under global climate change. Furthermore, traditional multivariate analysis ignored the total effects associated with the interaction between variables and only focused on the direct effects of predictors on the response variables [24]. Simultaneously, the interaction between variables often has a greater impact on response variables. Hence, it is necessary to analyze the direct or indirect effects of a particular variable on another variable to study the factors that influence plant phenology.

In this study, we extracted plant phenological information based on MOD09A1 datasets with Google Earth Engine; the accuracy of the extracted plant phenology results was verified by using the station data of plant phenology; our main aims are: (1) investigating spatiotemporal characteristics of plant phenology (2) and analyzing the potential influence mechanism of climate and soil factors on the plant phenology of TRHR.

2. Materials and Methods

2.1. Processing

The flow chart of research ideas for this paper is as follows (Figure 1). First, we calculated the plant phenology according to the following steps: (1) the NDVI of the TRHR was calculated from MOD09A1 datasets in Google Earth Engine; (2) next, bare soil, sparse vegetation, and evergreen forest pixels were eliminated according to certain requirements; (3) then, the NDVI datasets were smoothed by harmonic analysis of time series (HANTS); (4) we used relative and absolute rates of change to calculate plant phenology (SOS and EOS) based on the NDVI datasets smoothed by HANTS; (5) the phenological data obtained by remote sensing monitoring were verified by using the observation data of phenological stations. Then, we analyzed the spatiotemporal dynamic pattern of plant phenology on different types of terrain and basins. Finally, we explored the potential influence mechanism of climate and soil factors on the phenology of the TRHR based on the structural equation model (SEM) and Pearson correlation coefficients.

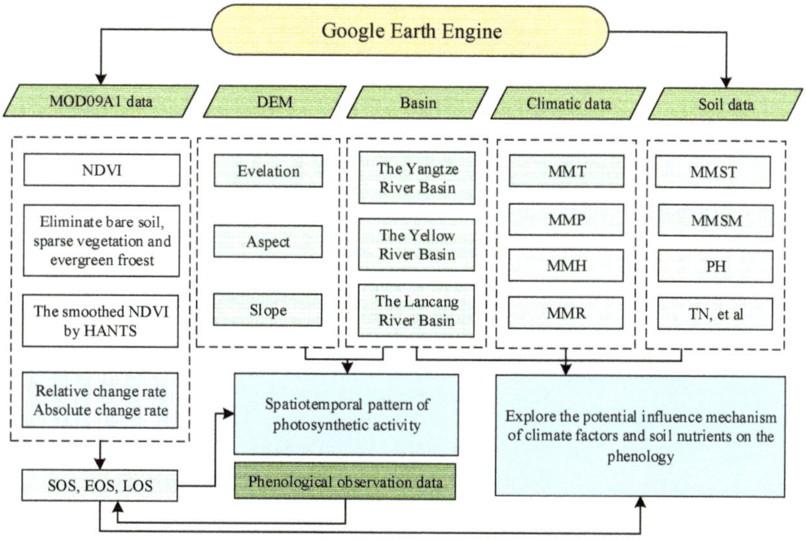

Figure 1. Flow chart of research ideas for this paper. NDVI, HANTS, SOS, EOS, LOS, DEM, MMT, MMP, MMH, MMR, MMST, MMSM, pH, and TN indicate the normalized difference vegetation index, harmonic analysis of time series, start of the growing season, end of the growing season, length of the growing season, digital elevation model, monthly mean temperature, monthly mean precipitation, monthly mean relative humidity, monthly mean shortwave radiation, monthly mean soil temperature, monthly mean soil moisture, pH (H_2O), and total N, respectively.

2.2. Study Area

The TRHR (31°39′N–37°10′N, 89°24′E–102°27′E) is located in the hinterland of the Tibetan Plateau and in southern Qinghai Province of China (Figure 2). As the source area for the Yellow, Lancang, and Yangtze rivers, the TRHR supplies approximately 40 billion m^3 of water downstream every year. It serves as an important source of freshwater resources in China and Asia and is often referred to as the "Chinese water tower" [26]. The TRHR spans 22 counties and covers an area of about 3.95×10^5 km^2, and the elevation increases from 1987.5 m in the southeast to 6714.5 m in the northwest, with an average elevation over 4000 m. In 2010, the main land-use types in the TRHR were grassland (68.4%), desert (16.0%), wetland (9.4%), shrub (4.6%), and forest (0.3%), where alpine steppe and alpine meadow were the main types of grassland [27]. The TRHR has major extensive wetlands in China, with abundant lake, river, glacier, and mountain snow resources, and supports the largest alpine wetland ecosystem in the world. Moreover, the TRHR is an important ecological functional zone, a typical ecologically fragile area in China, and is quite sensitive to climate change.

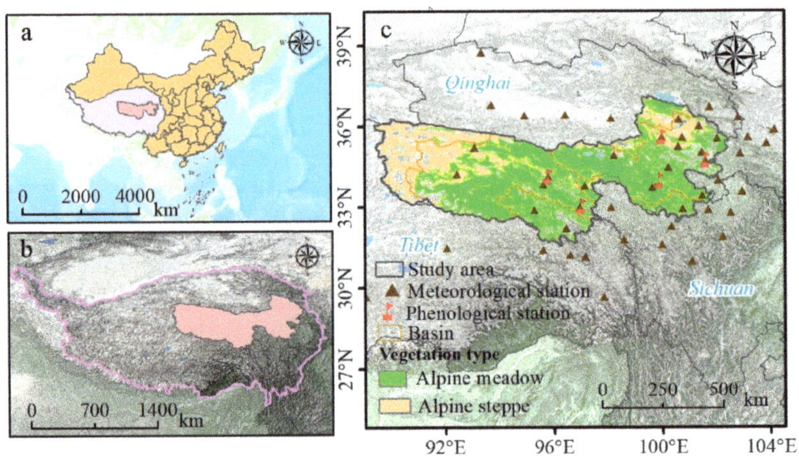

Figure 2. Maps of the study area: (**a**) Tibetan Plateau in China; (**b**) Three-River Headwaters region on the Tibetan Plateau; (**c**) locations of meteorological, phenological stations, alpine steppe, and alpine meadow in and near the Three-River Headwaters region.

The TRHR experiences a typical plateau continental climate with large daily temperature differences, small annual temperature differences, intense radiation, and a large number of sunshine hours [28]. The TRHR has cool and dry winters with wet and warm summers, mainly caused by the influence of the Asian monsoon and high elevation [29]. Meanwhile, the annual average precipitation in the THRH gradually increased from northwest (262.2 mm) to southeast (772.8 mm), primarily concentrated between June and September owing to the influence of the warm and humid air currents in the southern Bay of Bengal [28,30]. Furthermore, the annual average temperature, sunshine hours, and evaporation of TRHR ranging −5.6 to 7.8 °C, 2300 to 2900 h, and 730–1700 mm, respectively [14,26].

2.3. Data Sources

2.3.1. MOD09A1 Data

The main vegetation types in the TRHR are alpine steppe and alpine meadow. These do not have high amounts of vegetation coverage, with the highest values of NDVI being less than 0.6. Therefore, there are no areas where vegetation is so saturated that the NDVI cannot be accurately expressed. Hence, NDVI was selected in this paper for use in analyzing plant phenological characteristics. Based on a previous study, this paper selected MOD09A1 data products because they have a high temporal resolution, which can provide us with better detailed information related to vegetation growth. The NDVI time-series data came from the Google Earth Engine (https://developers.google.com/earth-engine/datasets/catalog/MODIS_006_MOD09A1 (accessed on 20 July 2020)), with a temporal resolution of 8 days and a spatial resolution of 500 × 500 m. In order to improve the calculated phenological results accurately, we used the following rules for data processing. (1) NDVI was calculated based on the MOD09A1 band using Equations (1) and (2). To eliminate the influence of bare soil, sparse vegetation, and evergreen forest, the pixels of NDVI data in this study had to meet the following requirements: (a) the average value of NDVI should be more than 0.2 in April–October; (b) the maximum value of annual NDVI shall exceed 0.30; (c) the annual maximum value shall occur from July to September; and (d) the average value of NDVI in winter shall be less than 0.4. (3) The NDVI data with a temporal resolution of 1 day was obtained by using HANTS to fit the data, which were processed in steps 1 and 2 (Figure 3a).

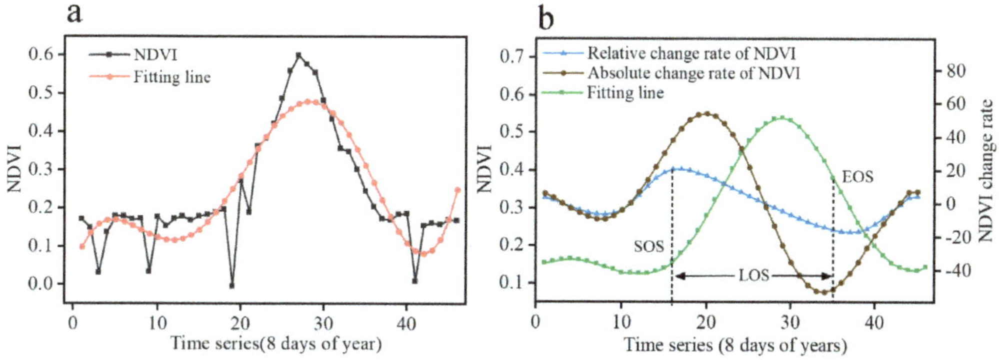

Figure 3. (**a**) Fitting effect of the harmonic analysis of time series; (**b**) definition of the normalized difference vegetation index (NDVI), threshold of the start of the growing season (SOS), and end of the growing season (EOS). The green line is the NDVI time-series change curve of 8 days after smooth treatment. The vertical axis of the left side is the NDVI value, and the vertical axis of the right side is the change rate of NDVI. For comparison, the change rate of NDVI has been zoomed in integer times, the ratio of absolute change value is 1000, and the ratio of relative change rate is 100.

2.3.2. Phenological Observation Data

Vegetation phenological observation data (2001–2013) of the THRH used in this study were extracted from the ten-day datasets on crop growth and farmland soil moisture in China, which were obtained from the Chinese Meteorological Administration (http://data.cma.cn/ (accessed on 12 October 2020)). We selected the phenological stations according to the principle that the vegetation type around each station is grassland. Finally, we selected five phenological observation stations (Figure 2c).

2.3.3. Climate Datasets

The meteorological data were selected from the monthly cumulative precipitation, monthly mean relative humidity, and monthly mean air temperature from April to October during 2001 to 2018 for 51 nationally standard meteorological stations in and near the TRHR (Figure 2c), which were provided by the Chinese Meteorological Administration (http://data.cma.cn/ (accessed on 5 July 2020)). Some observational data were missing and had non-uniformity characteristics owing to the influence of changes in meteorological stations and in instruments used to observe. Thus, the regression equation of time series and the homogeneity test of variance were used to fill in the missing values and test for data homogeneity at first in this paper. The commonly used spatial interpolation methods include inverse distance weighted, co-kriging, and thin plate splines (TPS). After comparative experiments, the monthly accumulated precipitation and monthly mean relative humidity were interpolated by the co-kriging method in ArcGIS10.5 software (ESRI, Redlands, CA, USA), with 500 × 500 m resolution. Furthermore, the TPS method of

Anusplin software (Centre for Resource and Environmental Studies, Australian National University, Canberra, Australia) was adopted to interpolate the monthly mean temperature at a resolution of 500 × 500 m.

In this study, time-series shortwave radiation data were acquired from the European Centre for Medium-Range Weather Forecasts website (https://cds.climate.copernicus.eu/cdsa-pp#!/dataset/reanalysis-era5-land-monthly-means?tab=overview (accessed on 5 July 2020)), with a temporal resolution of one month and a spatial resolution of 0.1° × 0.1°.

2.3.4. Soil Characteristics Database

In order to explore the influence of soil physical and chemical attributes on plant phenology, we used a database of soil characteristics that was produced by the Land–Atmosphere Interaction Research Group at Sun Yat-sen University (http://globalchange.bnu.edu.cn/home (accessed on 25 October 2020)). The database included information on total N (g/100 g), total P (g/100 g), total K (g/100 k), soil organic matter (g/100 g), alkali-hydrolysable N (mg/kg), available P (mg/kg), available K (mg/kg), cation exchange capacity (me/100 g), porosity (cm^3/100 cm^3), bulk density (g/cm^3), and pH (H_2O). Furthermore, soil moisture and soil temperature data were obtained from Google Earth Engine (https://developers.google.com/earth-engine/datasets/catalog/NASA_FLDAS_NOAH01_C_GL_M_V001#bands (accessed on 25 October 2020)) with a spatial resolution of 0.1° × 0.1° and temporal resolution of one month.

2.3.5. Digital Elevation Model

Digital Elevation Model (DEM) data were collected from the Advanced Spaceborne Thermal Emission and Reflection Radiometer Global Digital Elevation Model (ASTER GDEM) Version 3, which was provided by the US National Aeronautics and Space Administration's Earth Data website (https://earthdata.nasa.gov/ (accessed on 30 October 2020)), with a spatial resolution of 30 m. For this study, the DEM data were processed with ArcGIS 10.5 to obtain the slope, elevation, and aspect.

2.4. Methods

2.4.1. Extraction of Plant Phenological Information

(1) *NDVI*

In this study, on the basis of *NDVI* that is estimated by the MOD09A1 band information, we calculated the SOS and EOS using the method of the relative and absolute rates of *NDVI* change, respectively. The *NDVI* is defined by [31]:

$$NDVI = (\rho_{NIR} - \rho_{Red})/(\rho_{NIR} + \rho_{Red}), \quad (1)$$

where ρ_{NIR} and ρ_{Red} are the spectral reflectance values calculated in the near-infrared and red bands, respectively.

(2) Determination of the SOS and EOS

We used the maximum relative and minimum absolute rates of change in *NDVI* to calculate the SOS and EOS based on previous studies [14]. The equations of these rates of change can be expressed by:

$$NDVI_{rate\_rel} = \frac{NDVI_{t+1} - NDVI_t}{NDVI_t},\ t \in [1, 2, \ldots, 365], \quad (2)$$

$$NDVI_{rate\_abs} = NDVI_{t+1} - NDVI_t,\ t \in [1, 2, \ldots, 365], \quad (3)$$

where $NDVI_{rate\_rel}$ and $NDVI_{rate\_abs}$ are the relative and absolute rates of change, respectively.

The specific calculation process is as follows. First, we calculated the time (*T*) when the maximum value appears based on the *NDVI* time-series data. The *NDVI* curve was divided into a rising (0, T) and a descending (T, 365) stage. Second, based on Equations (2) and (3), the maximum relative and minimum absolute rates of change were calculated by using the *NDVI* time-series data. Then, the thresholds of SOS and EOS were determined based on the maximum relative and minimum absolute rates of change, respectively. Third, if the *NDVI* value was greater than the SOS threshold at time 0 to T, the corresponding date of the year was regarded as the SOS. Similarly, if the *NDVI* value of some pixels was less than the EOS threshold at time T to 365, the corresponding day of the year plus one was regarded as the EOS (Figure 3b).

2.4.2. The Spatiotemporal Pattern of Plant Phenology

(1) Linear Regression Analysis

We adopted a linear regression analysis to analyze the monotonic trend of the vegetation phenology and indicators [32,33]. The trend slope in a multi-year regression equation represents the amount of inter-annual change and can be found using the least squares method as follows:

$$Slope = \frac{n \cdot \sum_{t=1}^{n} t \cdot X_t - \sum_{t=1}^{n} t \sum_{t=1}^{n} X_t}{n \cdot \sum_{t=1}^{n} t^2 - \left(\sum_{t=1}^{n} t\right)^2}, \quad (4)$$

where *Slope* refers to the inter-annual trend, n is the number of years of the study, and the X_t is the value of this variable in the t-th year. When the slope is positive or negative, this indicates an increasing or decreasing trend, respectively.

(2) Standard Deviation Analysis

Standard deviation is a measure of the degree of data dispersion that can reflect the stability or fluctuation of variables [34]. For this study, the stability or fluctuation of plant phenology was calculated by standard deviation based on the pixel scale. The calculation formula is as follows:

$$S_i = \sqrt{\frac{1}{n} \sum_{i=1}^{n} (X_i - \overline{X})^2}, \quad (5)$$

where S_i indicates the standard deviation of an X dataset. When the S_i value is larger, the distribution of the data is more discrete and has a larger range of fluctuation. In contrast, when the S_i value is smaller, the distribution of the data is more concentrated and the range of fluctuation is smaller.

2.4.3. Driving Force Analysis

(1) Pearson Correlation Coefficient

For this paper, we used correlation analysis to determine the relationship between the plant phenology (SOS, EOS, and LOS) and other factors. A higher value indicates a stronger correlation; otherwise, it means a weaker correlation [28,35]. The relevant formula is as follows:

$$R_{xy} = \frac{\sum_{i=1}^{n} [[x_i - \overline{x}] \cdot [y_i - \overline{y}]]}{\sqrt{\sum_{i=1}^{n} [[x_i - \overline{x}]^2 \cdot [y_i - \overline{y}]^2]}}, \quad (6)$$

where R_{xy} is the correlation coefficient between x and y, n is the number of years during the study, x_i and y_i are the two sets of variables, and $\overline{x}$ and $\overline{y}$ are the mean values of variables.

(2) Structural Equation Model

SEM is a method used to analyze the relationship between variables based on a covariance matrix of variables, which includes maximum likelihood, synthesis of factor, and path analyses [24]. It pre-sets the dependence relationship between the factors in the system based on the researcher's prior knowledge, which can judge the strength of the relationship between the factors and can fit and judge the overall model. In addition, SEM has several advantages. For example, the direct or indirect effects of a particular variable on another variable can be partitioned by SEM, and SEM estimates and reports the total path coefficient to present the strengths of these multiple effects [36]. Since the change in SOS and EOS eventually lead to the change in LOS, this paper only used SEM to explore the potential influence mechanism of climate and soil factors on LOS in the TRHR.

3. Results

3.1. The Verification of Vegetation Phenological Results

For this study, a regional plant phenological dataset was developed based on data acquired from 2001 to 2018. Figure 4 shows that the remote sensing monitoring data of SOS (R^2 = 0.84, $p < 0.01$), EOS (R^2 = 0.72, $p < 0.01$), and LOS (R^2 = 0.86, $p < 0.01$) have strong similarity with the phenological observation data. Specifically, the times of SOS monitored by remote sensing and observed by phenological stations are distributed near a straight line ($Y = X$). However, the times of EOS and LOS observed by remote sensing and phenological stations are generally distributed above the straight line ($Y = X$). This showed that the time product of SOS is highly consistent with the values observed at phenological stations, but the time product of EOS is delayed when compared with that of phenological stations; in addition, the LOS product of remote sensing monitoring is longer than observed at the phenological station (Figure 4).

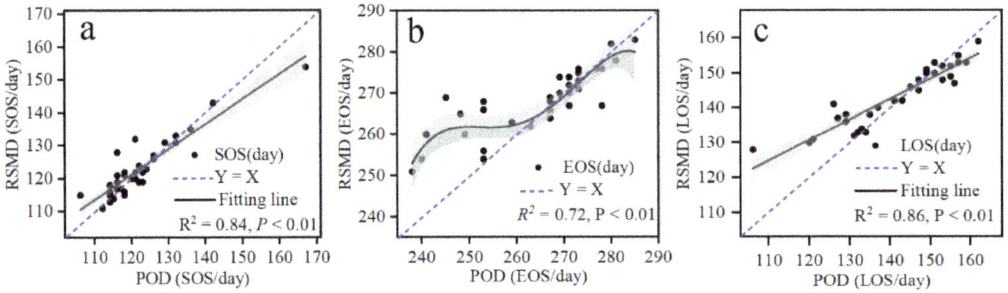

Figure 4. The comparison between remote sensing monitoring data (RSMD) and phenological observation data (POD): (**a**) start of the growing season (SOS); (**b**) end of the growing season (EOS); (**c**) length of the growing season (LOS).

3.2. Spatiotemporal Pattern of Plant Phenology

During the study period, the spatiotemporal trends and standard deviations of SOS, EOS, and LOS had a heterogeneous geographical distribution from 2001 to 2018. The spatial distribution of the multiyear mean SOS primarily occurred between day 100 and 150, and the multiyear average SOS arrived before day 100 in the low-elevation river valley areas of the Yellow and Lancang river basins and appeared after day 150 in some high-elevation or high-latitude areas of the Yangtze River Basin (Figure 5a,d). Similarly, the high value (>16 day/year) of standard deviation for SOS principally occurred in the Lancang River Basin and the southwestern part of the Yangtze River Basin, with the lowest value (<8 day/year) in the center of the Yangtze River Basin and the southeastern part of the Yellow River Basin (Figure 5g). We also found that the Yellow River Basin had the earliest SOS, and the time is in advance (Figure 5j). Furthermore, the spatial distribution of the multiyear average EOS was mainly observed from day 265 to 283, the multiyear average EOS arrived before day 265 in the northeast of the Yellow River Basin, and appeared after day 280 in the center of the Yangtze River Basin (Figure 5b,e). The high value of the standard deviation of EOS was mainly in the Yangtze and Lancang river basins (Figure 5h). In addition, we also compared the temporal trend of EOS in different basins; the earliest EOS was in the Lancang River Basin and the latest in the Yangtze River Basin (Figure 5k). Last, the spatial distribution of the multiyear average LOS was mainly between day 120 and 160, while the multiyear average LOS was longer than day 150 in some areas of the Yellow and Lancang river basins (Figure 5c,f). The high value of the standard deviation of LOS was mainly distributed in the Lancang and Yangtze river basins (Figure 5i). Furthermore, we also found that the Lancang River Basin had the longest LOS, which is becoming longer over time (Figure 5l).

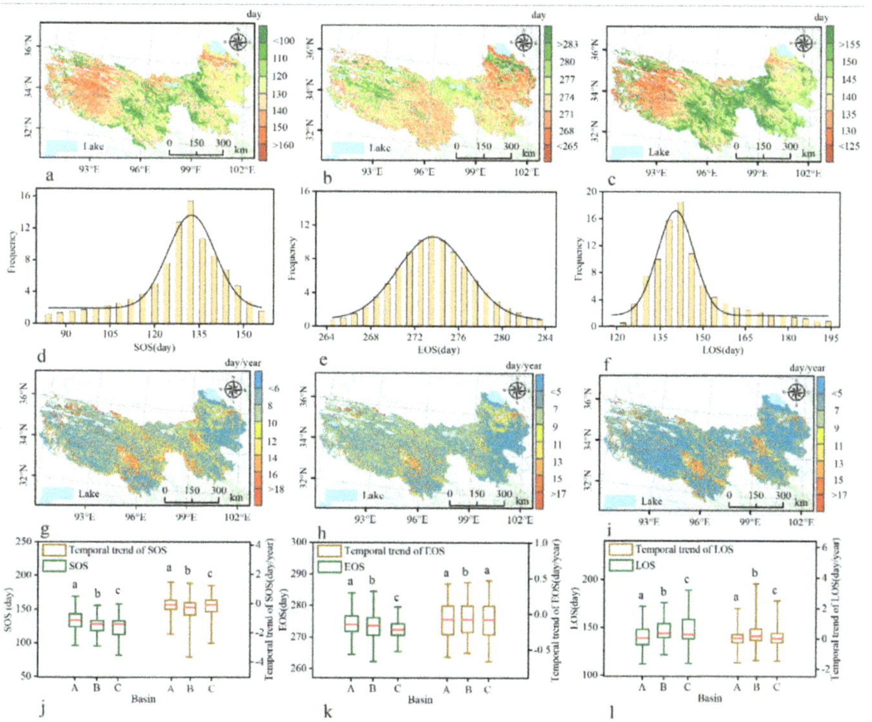

Figure 5. Spatiotemporal patterns of vegetation phenology: **(a–c)** the spatial pattern of a multi-year average of the start (SOS), end (EOS), and length (LOS) of the growing season on the Three-River Headwaters from 2001 to 2018; **(d–f)** time-frequency distribution of SOS, EOS, and LOS, respectively; **(g–i)** standard deviation for the SOS, EOS, and LOS, respectively; **(j–l)** temporal variation characteristics of vegetation phenology of (A) Yangtze, (B) Yellow, and (C) Lancang river basins in SOS, EOS, and LOS, respectively. The different letters above the box plots indicate significant differences among different basins at $p < 0.05$. The green boxplots indicate the overall distribution characteristics of SOS, EOS, and LOS values in different basins. The yellow boxplots indicate the overall distribution characteristics of the trend of SOS, EOS, and LOS values in different basins.

For this study, SOS, EOS, and LOS have different distributions at different elevations, slopes, and aspects in the THRH (Figure 6). Specifically, the SOS generally showed an upwards (0.001 day/m, $R^2 = 0.17$, $p > 0.01$) trend with an increase in elevation (Figure 6a). This phenomenon indicates that with an increase in elevation, the SOS is delayed. In contrast, EOS and LOS decreased (0.002 day/m, $R^2 = 0.34$, $p > 0.01$ and 0.003 day/m, $R^2 = 0.84$, $p < 0.01$, respectively) as elevation increased, which represents that the time of EOS and the LOS advance and shorten with an increase in elevation, respectively (Figure 6b–c). Furthermore, SOS and EOS decreased significantly (0.32 day/°, $R^2 = 0.93$, $p < 0.01$ and 1 day/°, $R^2 = 0.85$, $p < 0.01$, respectively) with an increase in slope (Figure 6d–e). This indicates that the time of SOS and EOS advance with an increase in slope. However, the relationship between LOS and slope was the opposite of that between SOS or EOS and slope. The LOS was prolonged (0.2 day/°, $R^2 = 0.94$, $p < 0.01$) with an increase in slope (Figure 6f). Last, we find the north-facing slopes had the lowest value of SOS but had the highest value of EOS and LOS. The results showed that the times of SOS, EOS, and LOS were the earliest, latest, and longest, respectively, on the north slope.

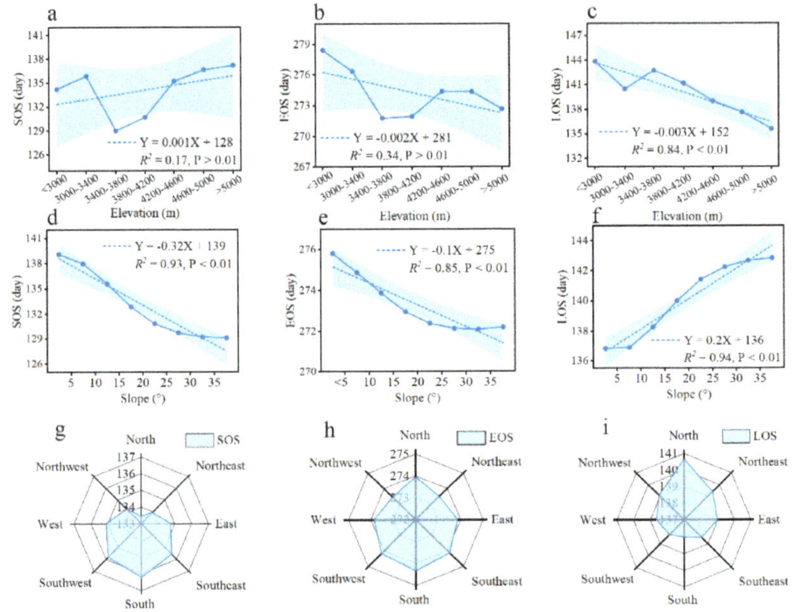

Figure 6. The relationship between different terrain factors and the start (SOS), end (EOS), and length (LOS) of the growing season: distribution and change characteristics at different elevations (**a–c**), slopes (**d–f**), and aspects (**g–i**).

3.3. Linking Climatic and Soil Factors to Plant Phenology

The correlation coefficients between plant phenology connected to the principal climate characteristics along with soil physical and chemical factors were significant at $p < 0.01$ (Table 1). In the Yangtze River Basin, our results show that the SOS was positively correlated with monthly mean shortwave radiation (MMR; 0.73**), pH (0.50**), and total phosphorus (TK) (0.44**) but negatively correlated with monthly mean precipitation (MMP; −0.68**), available nitrogen (AN; −0.39**), monthly mean relative humidity (MMH; −0.38**), and monthly mean soil moisture (MMSM; −0.37**). Furthermore, the correlation coefficients between EOS and MMP, pH, TK, and MMR were −0.45**, 0.40**, 0.41**, and 0.44**. Finally, we found that LOS was significantly negatively correlated with pH (−0.46**) and TK (−0.37**), but LOS was significantly positively correlated with MMR (0.53**), MMH (0.52**), and AN (0.38**) during the growing season.

In the Yellow River Basin, significant positive relationships were observed between SOS and monthly mean temperature (MMT; 0.50**), monthly mean soil temperature (MMST; 0.48**), MMR (0.50**), and MMSM (0.31**). However, the EOS was significantly negatively correlated with MMSM (−0.39**) and AN (−0.32**) and significantly positively correlated with pH (0.37**). In addition, we found that LOS was significantly negatively correlated with MMR (−0.55**), MMT (−0.46**), and MMST (−0.43**).

In the Lancang River Basin, the results indicated that there were significant positive correlations between the SOS and AK (0.50**), MMT (0.65**), MMST (0.55**), MMR (0.53**), and MMSM (0.43**). In addition, we found that the EOS was significantly negatively correlated with MMST (−0.41**), MMT (−0.43**), MMR (−0.36**), and MMP (−0.33**). Meanwhile, we also found that the correlation coefficients between LOS and MMH, MMT, MMST, MMR, and AK were 0.41**, −0.69**, −0.65**, −0.58**, and −0.50** (Table 1).

Table 1. The relationships between plant phenology and the principal climate factors along with soil physical and chemical factors at different stages (SOS, EOS, and LOS) in the Yangtze River Basin, Yellow River Basin, and Lancang River Basin.

	MMST	MMH	MMT	MMP	MMSM	AN	AP	TN	SOM	CEC	TP	AK	POR	MMR	PH	TK	BD
SOS [a]	−0.06 **	−0.38 **	−0.13 **	−0.68 **	−0.37 **	−0.39 **	−0.31 **	−0.24 **	−0.22 **	−0.23 **	0.04 **	0.09 **	−0.03 *	0.73 **	0.50 **	0.44 **	0.06 **
SOS [b]	0.48 **	−0.16 **	0.50 **	−0.04 **	0.31 **	−0.04 **	0.10 **	0.09 **	0.09 **	0.12 **	0.14 **	0.15 **	−0.19 **	0.50 **	0.09 **	0.10 **	−0.05 **
SOS [c]	0.55 **	−0.36 **	0.65 **	0	0.43 **	0.31 **	0.35 **	0.29 **	0.23 **	0.26 **	0.05	0.50 **	−0.26 **	0.53 **	−0.09 **	0.12 **	−0.08 **
EOS [a]	0.22 **	−0.16 **	0.10 **	−0.45 **	−0.27 **	−0.27 **	−0.17 **	−0.17 **	−0.16 **	−0.19 **	0.08 **	0.05 **	0.02	0.44 **	0.40 **	0.41 **	0.02
EOS [b]	0.24 **	−0.23 **	0.20 **	−0.23 **	−0.39 **	−0.32 **	0	−0.12 **	−0.07 **	−0.07 **	0.05 **	0.04 **	−0.21 **	0.03	0.37 **	0.01	−0.06 **
EOS [c]	−0.41 **	0.08 **	−0.43 **	−0.33 **	−0.22 **	−0.12 **	0.13 **	0.01	0.04	0.04	0.09 **	−0.12 **	0.24 **	−0.36 **	0.07 **	−0.26 **	−0.18 **
LOS [a]	0.17 **	0.52 **	0.21 **	0.21 **	−0.02	0.38 **	0.33 **	0.24 **	0.21 **	0.21 **	−0.01	−0.09 **	0.06 **	0.53 **	−0.46 **	−0.37 **	−0.07 **
LOS [b]	−0.43 **	0.10 **	−0.46 **	0.02	−0.02	−0.05 **	−0.10 **	−0.13 **	−0.11 **	−0.14 **	−0.13 **	−0.15 **	0.14 **	−0.55 **	0.01	−0.10 **	0.04 **
LOS [c]	−0.65 **	0.41 **	−0.69 **	−0.08 **	−0.16 **	−0.32 **	−0.33 **	−0.28 **	−0.22 **	−0.25 **	−0.03	−0.50 **	0.28 **	−0.58 **	0.10 **	−0.15 **	0.06 *

MMR, MMP, MMT, MMH, MMSM, and MMST indicate monthly mean shortwave radiation, precipitation, temperature, relative humidity, soil moisture, and soil temperature, respectively. AK, AN, AP, BD, CEC, pH, POR, SOM, TK, TN, and TP represent available K, alkali-hydrolysable N, available P, bulk density, cation exchange capacity, pH (H_2O), porosity, soil organic matter, total K, total N, and total P, respectively. ** and * indicate significance coefficients of less than $p < 0.01$ and $p < 0.05$, respectively. [a,b] and [c] represent the Yangtze River Basin, Yellow River Basin, and Lancang River Basin, respectively.

The mechanisms involved in patterns in the length of the plant growing season in different basins were explored using SEM. In general, the effect of soil factors on LOS is greater than that of climate factors in the Yangtze River Basin. Specifically, AP, pH, and TN had a significant effect on the LOS ($p < 0.01$), with scores of 0.30, −0.65, and −0.77, respectively. However, the impact scores of MMR and MMH on LOS were only 0.35 and 0.33 (Figure 7a). Path analyses identified that climate factors, as a key functional indicator of the LOS in the Yellow River Basin, had either direct or indirect effects via edaphic factors. Specifically, the MMR (scored at −0.55), MMT (scored at −0.30), and MMST (scored at 0.54) had significant effects on the LOS (Figure 7b). Furthermore, in the Lancang River Basin, the effects of each variable on LOS were different (ranging from −0.52 to 0.25), which suggests that the LOS might be co-determined by both the soil and climatic factors (Figure 7c). This assumption was confirmed in that soil factors were significantly affected by climatic factors. Specifically, the AK (scored at 0.41), AP (scored at 0.27), and AN (scored at 0.32) were significantly ($p < 0.01$) influenced by MMT. Furthermore, AK and AP had a significant interaction (scored at 0.58).

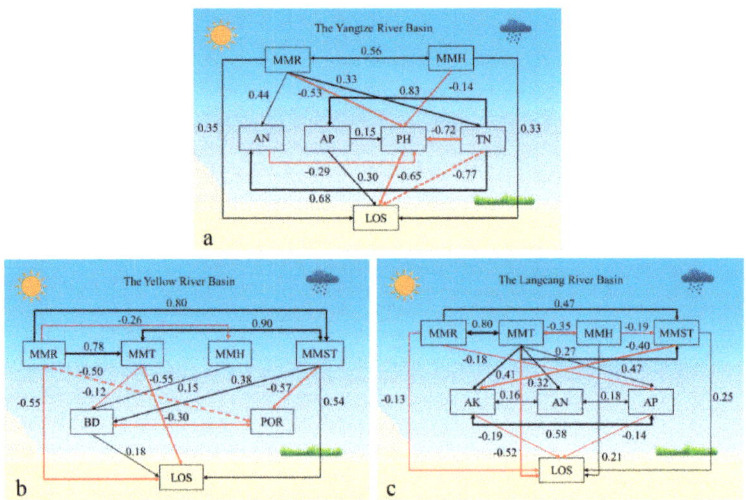

Figure 7. Mechanisms involved in the patterns of the length of the plant growing season in different basins. Structural equation modeling (SEM) was used to analyze the total effects of variables. The black and red solid lines represent positive and negative standardized SEM coefficients, respectively, while the line thickness indicates the magnitude of these coefficients for the Yangtze (**a**), Yellow (**b**), and Lancang (**c**) river basins, respectively. MMR, MMT, MMH, and MMST represent monthly mean shortwave radiation, temperature, relative humidity, and soil temperature, respectively. AN, AP, pH, TN, BD, POR, and AK represent alkali-hydrolysable N, available P, pH (H_2O), total N, bulk density, porosity, and available K, respectively.

4. Discussion

4.1. Spatial–Temporal Patterns of Plant Phenology

The time of the SOS experienced a significant downtrend (*slope* < 0), but the LOS increased over time (*slope* > 0) during 2001–2018 in the Yellow River Basin (Figure 5j,l). The research showed that with the increase of annual mean precipitation, temperature, relative humidity, shortwave radiation, soil temperature, and soil moisture in the Yellow River Basin during the SOS and LOS period, the time of SOS and LOS became earlier and longer (Figures S3 and S5). The favorable water and heat environment provided important resources for vegetative growth [37,38]. Water supply determines whether the photosynthesis occurs normally with an adequate CO_2 concentration and sufficient light [24,39]. Meanwhile, water is also an indispensable intermediary used to ensure nutrient substance transport [24,40]. Therefore, the increased humidity, precipitation, and soil moisture played a crucial role in the advance of the timing of the SOS and the prolonged nature of the LOS. In addition to water, temperature is also an indispensable factor in vegetative growth. An increase in temperature could facilitate vegetative growth if plants do not encounter water limitation [4,40]. Furthermore, climate warming can stimulate the enzymatic activities involved in photosynthesis [24,41], accelerate the mineralization and decomposition of organic matter [42], and extend the length of the vegetative growing season [7,43]. In general, the improvement of water and heat conditions visibly promoted the growth of vegetation in the TRHR.

The multi-year (2001–2018) average time of the SOS, EOS, and LOS in the TRHR presented a discrepant geographical pattern. In general, the time of the plant SOS was earliest in the Lancang and western Yellow river basins, while the time of the plant EOS was latest in the middle of the Yangtze River Basin. The duration of the vegetation growing season was longest in the Lancang and western Yellow river basins (Figure 5a–c). This phenomenon is closely related to the distribution of climate in the TRHR. The TRHR's climate is dominated by the East Asian monsoon, because the Himalayan Mountain Range obstructs the Indian monsoon [44,45] and causes a gradual reduction in precipitation, relative humidity, and soil moisture from southeast to northwest (Figures S2–S5). Likewise, vegetative growth is easily affected by climate change in the TRHR as also supported by previous studies [46,47]. Furthermore, the time of the plant SOS was delayed with an increase in elevation, but the times of the EOS or LOS were advanced or shortened, respectively, with an increase in elevation (Figure 6a–c). Possible reasons include the following: the ecosystems of high elevation areas are fragile, and the vegetative growth is vulnerable to extreme weather, such as extreme low temperature and frost. Another possibility is that the perennial snowfall occurring in high elevation regions causes low temperatures, which weaken the activity of soil microorganisms [14]. In contrast, with an increase in slope, the time of the SOS and EOS are in advance, while the LOS is prolonged (Figure 6d–f). The main reason for this result is that the areas with high slopes were mainly concentrated in the Lancang and the south part of the Yellow river basins, which have lower elevations (Figure S1). This provides reliable water and temperatures to guarantee the normal operation of vegetative photosynthesis. Finally, the vegetation of shady slopes started growing earlier in the growing season, ended later, and so had the longest growing season (Figure 6g–i), mainly due to the strong illumination and high temperature that accelerated soil organic matter mineralization and caused sunny slopes to have less soil moisture [48]. However, the shady slopes have soft solar radiation, moist soil, less moisture evaporation, and higher soil fertility [49,50].

4.2. The Response of the Plant Phenology to Climate Change

Our results illustrate that the variations in soil resources (e.g., pH and soil total N) that support vegetative growth, together with the climatic conditions that were suitable for vegetative growth, co-explained the phenological differences in plants from different basins. Specifically, the Yangtze River Basin is affected by the East Asian monsoon and elevation (Figure S1a), insufficient water supply, and relatively low temperatures, and low levels of soil nutrients constrained the growth of vegetation at the start of the growing season (Figures S2–S3). The air and soil temperatures are relatively low with less precipitation and soil moisture from April to May, which is not enough to support the transport of nutrients in plants, soil nutrient absorption by roots, and photosynthesis [24,51]. With the increased temperature, winter snow, permafrost, and glaciers have begun to melt slowly, and mineralization and decomposition of organic matter are accelerated [52,53] so that warmer temperatures provide plants with earlier opportunities to germinate [32]. However, the Yellow and Lancang river basins have higher temperatures and more shortwave radiation and precipitation than other areas owing to the lower elevation (Figures S1–S5). The increase in precipitation significantly increased soil carbon and N content, making it easier for plants to absorb nutrients owing to an increase in leaf stomatal conductance and photosynthesis [24,54]. This may explain why temperature is more important for seed germination than water at the start of the growing season in the Yangtze River Basin, while water and heat are equally important for seed germination in the Yellow and Lancang river basins. In addition, the Yangtze River Basin lies in a high elevation area, which has thin air, strong solar radiation, and a long sunshine season (Figure S3). At high elevations, the decomposition of soil litter slows, which promotes the accumulation of organic matter due to the low temperatures caused by snow cover, which, in turn, slows the activity of soil microorganisms [14]. Thus, the growing season ends relatively late in the Yangtze River Basin. An interesting question arises: why were the soil factors having a greater impact on LOS in the Yangtze River Basin when compared with the Yellow and Lancang river basins (Table 1 and Figure 7)? Here, we propose one explanation. The precipitation, air temperature, relative humidity, soil moisture, and soil temperature showed a decreasing trend from southeast (the Yellow River Basin) to northwest (the Yangtze River Basin) in the TRHR because of the influence of the monsoon and elevation (Figures S2–S5). This situation led to low air temperature and less precipitation in the Yangtze River Basin, which does not provide enough energy for the growth of plants. At this time, the melting of permafrost and glaciers and the mineralization of soil organic matter provide energy for plant growth. However, the Yellow and the Lancang river basins have high air temperature, soil temperature, precipitation, and soil moisture, which can provide sufficient energy for plant growth.

4.3. Limitations of the Current Study

Despite the achievements in this study, large uncertainties still exist. In addition to NDVI, multiple vegetation indices can be used to reflect vegetation dynamics, such as EVI and LAI [1,34]. Note that the calculated plant phenology results may be vary based on the differences in resolution and quality of datasets using different vegetation indices. Furthermore, the present smoothing methods of remote sensing time series data have great differences in the model structure, which may result in great differences among the extracted plant phenology results [55,56]. Meanwhile, although the smoothing method used for the remote sensing time series is the same, different smoothing parameters also cause different results. Although the guidelines for some smoothing methods suggested using default parameter values when they were proposed, the best parameter values may vary because of the different growth trajectories of vegetation at specific sites, which lead to a difference in plant phenology in various regions and with different vegetation types [57,58]. Moreover, many methods can be used to extract plant phenological information, and they all have a certain level of applicability. Therefore, different methods may lead to different conclusions regarding the same question [56]. As mentioned above, it is necessary to further check whether the plant phenological results calculated from different datasets, smoothing methods of remote sensing time series, smoothing parameters, and phenology extraction methods provide the same or similar results and to improve the credibility of the results. Furthermore, there are many factors that affect plant phenology. Some changes in phenology are caused by climatic and soil factors; other decisive factors have shown effects on plant phenology, such as flash floods and extreme drought. Hence, more attention should be paid to the relationship between plant phenology and natural disasters in future studies.

5. Conclusions

In the present study, we calculated plant phenology information in the TRHR based on the MOD09A1 dataset using the method of HANTS and the relative and absolute rates of change on Google Earth Engine. Meanwhile, the extracted plant phenology results were verified using plant phenology station data. Then, we explored the spatiotemporal patterns of plant phenology based on linear regression and standard deviation analyses. Finally, the potential influence mechanism of climatic and soil factors on phenology was analyzed using Pearson correlation coefficients and an SEM model. The verification of plant phenological results shows that our results were well-correlated with observational data acquired by phenological stations; the determination coefficients of SOS, EOS, and LOS stages were 0.84, 0.72, and 0.86, respectively. The temporal variation of the SOS and LOS indicated that the SOS advanced while the LOS extended. As for spatial patterns, the SOS was the earliest and the LOS was the longest in the Lancang River Basin, while the EOS was the latest in the Yangtze River Basin. Furthermore, the spatial distributions of SOS, EOS, and LOS have strong spatial heterogeneity at different elevations, slopes, and aspects. The potential influence mechanism of climatic and soil factors on the phenology indicated that plant phenology in the Yangtze River Basin is mainly affected by soil factors, while that in the Yellow and Lancang river basins is mainly impacted by climatic factors. The results of this study revealed the spatiotemporal patterns of plant phenology of the TRHR and emphasize the important role of soil factors, precipitation, and temperature in controlling plant phenological dynamics. These findings might help to reveal the mechanisms of potential impacts on plant phenology in alpine wetland ecosystems and provide a theoretical basis for ecosystem management.

Supplementary Materials: The following are available online at https://www.mdpi.com/2072-4292/13/13/2528/s1, Figure S1: Topographic features of Three-River Headwaters region: (a) elevation; (b) slope; (c) aspect; and (d) topographic relief, Figure S2. The monthly means for (a) soil temperature, (b) soil moisture, (c) relative humidity, (d) temperature, (e) precipitation, (f) shortwave radiation in the Yangtze (A), Yellow (B), and Lancang (C) river basins in different periods. Horizontal lines in box plots denote the 95th, 75th, 50th, 25th, and 5th percentiles from top to bottom; the rectangles represent the average values, Figure S3. Spatial pattern of (a, d, g, j, m, and p), standard deviation (b, e, h, k, n, and q) and temporal trend (c, f, i, l, o, and r) for the monthly mean temperature, precipitation, relative humidity, shortwave radiation, soil temperature, and soil moisture at the start of the growing season, Figure S4. Spatial pattern of (a, d, g, j, m, and p), standard deviation (b, e, h, k, n, and q), and temporal trend (c, f, i, l, o, and r) for the monthly mean temperature, precipitation, relative humidity, shortwave radiation, soil temperature, and soil moisture at the end of the growing season, Figure S5. Spatial pattern (a, d, g, j, m, and p), standard deviation (b, e, h, k, n, and q), and temporal trend (c, f, i, l, o, and r) for the monthly mean temperature, precipitation, relative humidity, shortwave radiation, soil temperature, and soil moisture in length of the growing season.

Author Contributions: Conceptualization, J.W. and H.S.; methodology, H.S.; software, H.S.; validation, J.W. and H.S.; formal analysis, H.S.; investigation, H.S.; resources, J.X.; data curation, H.S. and D.H.; writing—original draft preparation, H.S.; writing—review and editing, J.W., J.X., D.H., W.C., C.Y., Z.Y. and X.H.; visualization, H.S.; supervision, J.W.; project administration, J.W.; funding acquisition, J.W. and J.X. All authors have read and agreed to the published version of the manuscript.

Funding: The study has been funded by the National Natural Science Foundation of China (41701428), Key R & D project of Sichuan Science and Technology Department (Grant No. 2021YFQ0042), the Strategic Priority Research Program of Chinese Academy of Sciences (XDA20030302), the Science and Technology Project of Xizang Autonomous Region (XZ201901-GA-07), National Key R&D Program of China (2020YFD1100701).

Institutional Review Board Statement: Not applicable.

Informed Consent Statement: Not applicable.

Data Availability Statement: The data used to support the findings of this study are included within the article.

Conflicts of Interest: The authors declare no conflict of interest.

References

1. O'Neill, B.C.; Oppenheimer, M.; Warren, R.; Hallegatte, S.; Kopp, R.E.; Pörtner, H.O.; Scholes, R.; Birkmann, J.; Foden, W.; Licker, R.; et al. IPCC reasons for concern regarding climate change risks. *Nat. Clim. Chang.* **2017**, *7*, 28–37. [CrossRef]
2. Zhang, G.; Zhang, Y.; Dong, J.; Xiao, X. Green-up dates in the Tibetan Plateau have continuously advanced from 1982 to 2011. *Proc. Natl. Acad. Sci. USA* **2013**, *110*, 4309–4314. [CrossRef]
3. Brandt, L.A.; Butler, P.R.; Handler, S.D.; Janowiak, M.K.; Shannon, P.D.; Swanston, C.W. Integrating science and management to assess forest ecosystem vulnerability to climate change. *J. For.* **2017**, *115*, 212–221. [CrossRef]

4. Piao, S.; Friedlingstein, P.; Ciais, P.; Viovy, N.; Demarty, J. Growing season extension and its impact on terrestrial carbon cycle in the Northern Hemisphere over the past 2 decades. *Glob. Biogeochem. Cycles* **2007**, *21*. [CrossRef]
5. Gusewell, S.; Furrer, R.; Gehrig, R.; Pietragalla, B. Changes in temperature sensitivity of spring phenology with recent climate warming in Switzerland are related to shifts of the preseason. *Glob. Chang. Biol.* **2017**, *23*, 5189–5202. [CrossRef]
6. Richardson, A.D.; Keenan, T.F.; Migliavacca, M.; Ryu, Y.; Sonnentag, O.; Toomey, M. Climate change, phenology, and phenological control of vegetation feedbacks to the climate system. *Agric. For. Meteorol.* **2013**, *169*, 156–173. [CrossRef]
7. Dragoni, D.; Schmid, H.P.; Wayson, C.A.; Potter, H.; Grimmond, C.S.B.; Randolph, J.C. Evidence of increased net ecosystem productivity associated with a longer vegetated season in a deciduous forest in south-central Indiana, USA. *Glob. Chang. Biol.* **2011**, *17*, 886–897. [CrossRef]
8. Geng, X.; Fu, Y.H.; Hao, F.; Zhou, X.; Zhang, X.; Yin, G.; Vitasse, Y.; Piao, S.; Niu, K.; de Boeck, H.J.; et al. Climate warming increases spring phenological differences among temperate trees. *Glob. Chang. Biol.* **2020**, *26*, 5979–5987. [CrossRef]
9. Gao, M.; Wang, X.; Meng, F.; Liu, Q.; Li, X.; Zhang, Y.; Piao, S. Three-dimensional change in temperature sensitivity of northern vegetation phenology. *Glob. Chang. Biol.* **2020**, *26*, 5189–5201. [CrossRef]
10. Shen, X.; An, R.; Feng, L.; Ye, N.; Zhu, L.; Li, M. Vegetation changes in the Three-River Headwaters Region of the Tibetan Plateau of China. *Ecol. Indic.* **2018**, *93*, 804–812. [CrossRef]
11. Zhang, W.; Jin, H.; Shao, H.; Li, A.; Li, S.; Fan, W. Temporal and spatial variations in the leaf area index and its response to topography in the Three-River Source Region, China from 2000 to 2017. *ISPRS Int. J. Geo Inf.* **2021**, *10*, 33. [CrossRef]
12. Binghong, H.; Bingrong, Z.; Henghe, Z.; Mingming, S.; Ying, S.; Decao, N.; Hua, F. Relationships between grassland vegetation turngreen and climate factors in the Three-river Resource region. *Acta Ecol. Sin.* **2019**, *39*, 5635–5641.
13. Qiang, L. Phenology response of vegetation to hydrothermal condition in Three-river Source Region for the last 12 years. *Arid Zone Res.* **2016**, *33*, 150–158.
14. Chen, T.; Yi, G.; Zhang, T.; Wang, Q.; Bie, X. A method for determining vegetation growth process using remote sensing data: A case study in the Three-River Headwaters Region, China. *J. Mt. Sci.* **2019**, *16*, 2001–2014. [CrossRef]
15. Han, H.; Bai, J.; Ma, G.; Yan, J. Vegetation phenological changes in multiple landforms and responses to climate change. *ISPRS Int. J. Geo-Inf.* **2020**, *9*, 111. [CrossRef]
16. Niu, B.; Zhang, X.; Piao, S.; Janssens, I.A.; Fu, G.; He, Y.; Zhang, Y.; Shi, P.; Dai, E.; Yu, C.; et al. Warming homogenizes apparent temperature sensitivity of ecosystem respiration. *Sci. Adv.* **2021**, *7*, eabc7358. [CrossRef]
17. Chen, A.; Huang, L.; Liu, Q.; Piao, S. Optimal temperature of vegetation productivity and its linkage with climate and elevation on the Tibetan Plateau. *Glob. Chang. Biol.* **2021**, *27*, 1942–1951. [CrossRef]
18. Cong, N.; Shen, M.; Yang, W.; Yang, Z.; Zhang, G.; Piao, S. Varying responses of vegetation activity to climate changes on the Tibetan Plateau grassland. *Int. J. Biometeorol.* **2017**, *61*, 1433–1444. [CrossRef]
19. Menzel, A.; Sparks, T.H.; Estrella, N.; Koch, E.; Aasa, A.; Ahas, R.; Alm-Kübler, K.; Bissolli, P.; Braslavská, O.G.; Briede, A.; et al. European phenological response to climate change matches the warming pattern. *Glob. Chang. Biol.* **2006**, *12*, 1969–1976. [CrossRef]
20. Yun, J.; Jeong, S.-J.; Ho, C.-H.; Park, C.-E.; Park, H.; Kim, J. Influence of winter precipitation on spring phenology in boreal forests. *Glob. Chang. Biol.* **2018**, *24*, 5176–5187. [CrossRef]
21. Nemani, R.R.; Keeling, C.D.; Hashimoto, H.; Jolly, W.M.; Piper, S.C.; Tucker, C.J.; Myneni, R.B.; Running, S.W. Climate-driven increases in global terrestrial net primary production from 1982 to 1999. *Science* **2003**, *300*, 1560–1563. [CrossRef]
22. Chapin, F.S., III; Matson, P.A.; Vitousek, P. *Principles of Terrestrial Ecosystem Ecology*; Springer Science & Business Media: Berlin, Germany, 2002; pp. 369–397.
23. Yue, K.; Peng, Y.; Peng, C.; Yang, W.; Peng, X.; Wu, F. Stimulation of terrestrial ecosystem carbon storage by nitrogen addition: A meta-analysis. *Sci. Rep.* **2016**, *6*. [CrossRef]
24. Sun, J.; Zhou, T.C.; Liu, M.; Chen, Y.C.; Liu, G.H.; Xu, M.; Shi, P.L.; Peng, F.; Tsunekawa, A.; Liu, Y.; et al. Water and heat availability are drivers of the aboveground plant carbon accumulation rate in alpine grasslands on the Tibetan Plateau. *Glob. Ecol. Biogeogr.* **2019**, *29*, 50–64. [CrossRef]
25. Gower, S.T.; Kucharik, C.J.; Norman, J.M. Direct and indirect estimation of leaf area index, f APAR, and net primary production of terrestrial ecosystems. *Remote Sens. Environ.* **1999**, *70*, 29–51. [CrossRef]
26. Jiang, C.; Zhang, L. Ecosystem change assessment in the Three-river Headwater Region, China: Patterns, causes, and implications. *Ecol. Eng.* **2016**, *93*, 24–36. [CrossRef]
27. Xianfeng, L.; Jinshui, Z.; Xiufang, Z.; Yaozhong, P.; Yanxu, L.; Donghai, Z.; Zhihui, L. Spatiotemporal changes in vegetation coverage and its driving factors in the Three-River Headwaters Region during 2000–2011. *J. Geogr. Sci.* **2014**, *24*, 288–302.
28. Zheng, D.; Wang, Y.; Hao, S.; Xu, W.; Lv, L.; Yu, S. Spatial-temporal variation and tradeoffs/synergies analysis on multiple ecosystem services: A case study in the Three-River Headwaters region of China. *Ecol. Indic.* **2020**, *116*, 106494. [CrossRef]
29. Zheng, Y.; Han, J.; Huang, Y.; Fassnacht, S.R.; Xie, S.; Lv, E.; Chen, M. Vegetation response to climate conditions based on NDVI simulations using stepwise cluster analysis for the Three-River Headwaters region of China. *Ecol. Indic.* **2018**, *92*, 18–29. [CrossRef]
30. Zeng, N.; He, H.; Ren, X.; Zhang, L.; Zeng, Y.; Fan, J.; Li, Y.; Niu, Z.; Zhu, X.; Chang, Q. The utility of fusing multi-sensor data spatio-temporally in estimating grassland aboveground biomass in the three-river headwaters region of China. *Int. J. Remote Sens.* **2020**, *41*, 7068–7089. [CrossRef]

31. Piao, S.; Cui, M.; Chen, A.; Wang, X.; Ciais, P.; Liu, J.; Tang, Y. Altitude and temperature dependence of change in the spring vegetation green-up date from 1982 to 2006 in the Qinghai-Xizang Plateau. *Agric. For. Meteorol.* **2011**, *151*, 1599–1608. [CrossRef]
32. Li, L.; Zhang, Y.; Liu, L.; Wu, J.; Wang, Z.; Li, S.; Zhang, H.; Zu, J.; Ding, M.; Paudel, B. Spatiotemporal patterns of vegetation greenness change and associated climatic and anthropogenic drivers on the Tibetan Plateau during 2000–2015. *Remote Sens.* **2018**, *10*, 1525. [CrossRef]
33. Xu, X.; Liu, H.; Lin, Z.; Jiao, F.; Gong, H. Relationship of abrupt vegetation change to climate change and ecological engineering with multi-timescale analysis in the Karst Region, Southwest China. *Remote Sens.* **2019**, *11*, 1564. [CrossRef]
34. He, D.; Huang, X.; Tian, Q.; Zhang, Z. Changes in vegetation growth dynamics and relations with climate in inner Mongolia under more strict multiple pre-processing (2000–2018). *Sustainability* **2020**, *12*, 2534. [CrossRef]
35. Sun, H.; Wang, J.; Xiong, J.; Bian, J.; Jin, H.; Cheng, W.; Li, A. Vegetation change and its response to climate change in Yunnan Province, China. *Adv. Meteorol.* **2021**, *2021*, 8857589. [CrossRef]
36. Shipley, B. Cause and correlation in biology. *Ann. Bot.* **2002**, *90*, 777–778.
37. Zhang, Y.; Gao, J.; Liu, L.; Wang, Z.; Ding, M.; Yang, X. NDVI-based vegetation changes and their responses to climate change from 1982 to 2011: A case study in the Koshi River Basin in the middle Himalayas. *Glob. Planet. Chang.* **2013**, *108*, 139–148. [CrossRef]
38. Kong, D.; Zhang, Q.; Singh, V.P.; Shi, P. Seasonal vegetation response to climate change in the Northern Hemisphere (1982–2013). *Glob. Planet. Chang.* **2017**, *148*, 1–8. [CrossRef]
39. Ye, C.; Sun, J.; Liu, M.; Xiong, J.; Zong, N.; Hu, J.; Huang, Y.; Duan, X.; Tsunekawa, A. Concurrent and lagged effects of extreme drought induce net reduction in vegetation carbon uptake on Tibetan Plateau. *Remote Sens.* **2020**, *12*, 2347. [CrossRef]
40. Yao, Y.; Wang, X.; Li, Y.; Wang, T.; Shen, M.; Du, M.; He, H.; Li, Y.; Luo, W.; Ma, M.; et al. Spatiotemporal pattern of gross primary productivity and its covariation with climate in China over the last thirty years. *Glob. Chang. Biol.* **2018**, *24*, 184–196. [CrossRef]
41. Davidson, E.A.; Janssens, I.A. Temperature sensitivity of soil carbon decomposition and feedbacks to climate change. *Nature* **2006**, *440*, 165–173. [CrossRef]
42. Kim, D.G.; Vargas, R.; Bond-Lamberty, B.; Turetsky, M.R. Effects of soil rewetting and thawing on soil gas fluxes: A review of current literature and suggestions for future research. *Biogeosciences* **2012**, *9*, 2459–2483. [CrossRef]
43. Bao, G.; Qin, Z.; Bao, Y.; Zhou, Y.; Li, W.; Sanjjav, A. NDVI-based long-term vegetation dynamics and its response to climatic change in the Mongolian Plateau. *Remote Sens.* **2014**, *6*, 8337–8358. [CrossRef]
44. Duan, A.; Wu, G.; Liu, Y.; Ma, Y.; Zhao, P. Weather and climate effects of the Tibetan Plateau. *Adv. Atmos. Sci.* **2012**, *29*, 978–992. [CrossRef]
45. Immerzeel, W.W.; Bierkens, M.F.P. Asian water towers: More on monsoons—Response. *Science* **2010**, *330*, 584–585.
46. Huang, K.; Zhang, Y.; Zhu, J.; Liu, Y.; Zu, J.; Zhang, J. The influences of climate change and human activities on vegetation dynamics in the Qinghai-Tibet Plateau. *Remote Sens.* **2016**, *8*, 876. [CrossRef]
47. Guojin, P.; Xuejia, W.; Meixue, Y. Using the NDVI to identify variations in, and responses of, vegetation to climate change on the Tibetan Plateau from 1982 to 2012. *Quat. Int.* **2016**, *444*, 87–96. [CrossRef]
48. Pu, Y.; Long, G.; Liu, S.; Lu, C.; Kang, Q. Research progress in slope-directive variation of mountain soils. *Chin. J. Soil Sci.* **2007**, *38*, 753–757. [CrossRef]
49. Gao, T.; Li, J.; Lu, J.; Zheng, W.; Chen, J.; Wang, J.; Duan, F. Soil nutrient and fertility of different slope directions in the *Abies georgei* var. smithii forest in Sejila Mountain. *Acta Ecol. Sin.* **2020**, *40*, 1331–1341.
50. Liu, M. Response of plant element content and soil factors to the slope gradient of alpine meadows in Gannan. *Acta Ecol. Sin.* **2017**, *37*, 8275–8284.
51. Makiranta, P.; Laiho, R.; Mehtatalo, L.; Strakova, P.; Sormunen, J.; Minkkinen, K.; Penttila, T.; Fritze, H.; Tuittila, E.S. Responses of phenology and biomass production of boreal fens to climate warming under different water-table level regimes. *Glob. Chang. Biol.* **2018**, *24*, 944–956. [CrossRef]
52. Reichstein, M.; Bahn, M.; Ciais, P.; Frank, D.; Mahecha, M.D.; Seneviratne, S.I.; Zscheischler, J.; Beer, C.; Buchmann, N.; Frank, D.C.; et al. Climate extremes and the carbon cycle. *Nature* **2013**, *500*, 287–295. [CrossRef]
53. Gang, C.; Zhou, W.; Chen, Y.; Wang, Z.; Sun, Z.; Li, J.; Qi, J.; Odeh, I. Quantitative assessment of the contributions of climate change and human activities on global grassland degradation. *Environ. Earth Sci.* **2014**, *72*, 4273–4282. [CrossRef]
54. Keenan, T.; Sabate, S.; Gracia, C. The importance of mesophyll conductance in regulating forest ecosystem productivity during drought periods. *Glob. Chang. Biol.* **2010**, *16*, 1019–1034. [CrossRef]
55. Zeng, L.; Wardlow, B.D.; Xiang, D.; Hu, S.; Li, D. A review of vegetation phenological metrics extraction using time-series, multispectral satellite data. *Remote Sens. Environ.* **2020**, *237*, 111511. [CrossRef]
56. Li, N.; Zhan, P.; Pan, Y.; Zhu, X.; Li, M.; Zhang, D. Comparison of remote sensing time-series smoothing methods for grassland spring phenology extraction on the Qinghai-Tibetan Plateau. *Remote Sens.* **2020**, *12*, 3383. [CrossRef]
57. Zhu, W.; Pan, Y.; He, H.; Wang, L.; Mou, M.; Liu, J. A changing-weight filter method for reconstructing a high-quality NDVI time series to preserve the integrity of vegetation phenology. *IEEE Trans. Geosci. Remote Sens.* **2012**, *50*, 1085–1094. [CrossRef]
58. Spiess, A.N.; Deutschmann, C.; Burdukiewicz, M.; Himmelreich, R.; Klat, K.; Schierack, P.; Rodiger, S. Impact of smoothing on parameter estimation in quantitative DNA amplification experiments. *Clin. Chem* **2015**, *61*, 379–388. [CrossRef]

remote sensing

MDPI

Article

Spatial, Phenological, and Inter-Annual Variations of Gross Primary Productivity in the Arctic from 2001 to 2019

Dujuan Ma [1], Xiaodan Wu [1,*], Xuanlong Ma [1], Jingping Wang [1], Xingwen Lin [2] and Cuicui Mu [1,3,4]

[1] College of Earth and Environmental Sciences, Lanzhou University, Lanzhou 730000, China; madj19@lzu.edu.cn (D.M.); xlma@lzu.edu.cn (X.M.); jpwang2020@lzu.edu.cn (J.W.); mucc@lzu.edu.cn (C.M.)
[2] College of Geography and Environmental Sciences, Zhejiang Normal University, Jinhua 321004, China; linxw@zjnu.edu.cn
[3] Southern Marine Science and Engineering Guangdong Laboratory, Zhuhai 519000, China
[4] University Corporation for Polar Research, Beijing 100875, China
* Correspondence: wuxd@lzu.edu.cn

Abstract: Quantifying the spatial, seasonal (phenological), and inter-annual variations of gross primary productivity (GPP) in the Arctic is critical for comprehending the terrestrial carbon cycle and its feedback to climate warming in this region. Here, we evaluated the accuracy of the MOD17A2H GPP product using the FLUXNET 2015 dataset in the Arctic, then explored the spatial patterns, seasonal variations, and interannual trends of GPP, and investigated the dependence of the spatiotemporal variations in GPP on land cover types, latitude, and elevation from 2001 to 2019. The results showed that MOD17A2H was consistent with in situ measurements (R = 0.8, RMSE = 1.26 g C m^{-2} d^{-1}). The functional phenology was also captured by the MOD17A2H product (R = 0.62, RMSE = 9 days) in the Arctic. The spatial variation of the seasonal magnitude of GPP and its interannual trends is partly related to land cover types, peaking in forests and lowest in grasslands. The interannual trend of GPP decreased as the latitude and elevation increased, except for the latitude between 62°~66° N and elevation below 700 m. Our study not only revealed the variation of GPP in the Arctic but also helped to understand the carbon cycle over this region.

Keywords: GPP; carbon cycle; arctic; phenology; photosynthesis

Citation: Ma, D.; Wu, X.; Ma, X.; Wang, J.; Lin, X.; Mu, C. Spatial, Phenological, and Inter-Annual Variations of Gross Primary Productivity in the Arctic from 2001 to 2019. *Remote Sens.* **2021**, *13*, 2875. https://doi.org/10.3390/rs13152875

Academic Editors: Shin Nagai and Rasmus Fensholt

Received: 26 May 2021
Accepted: 16 July 2021
Published: 22 July 2021

1. Introduction

Climate change is causing permafrost melting [1], shrub cover expansion, growing season lengthening, and consequently, carbon flux changes in the Arctic [2]. Furthermore, the carbon cycle is also influenced by changes in vegetation phenology [3]. GPP, which is considered the biggest carbon flux of terrestrial ecosystems [4], not only plays a vital role in offsetting the concentration of greenhouse gases and mitigating global warming to a certain extent [5] but also builds a bridge between terrestrial and air carbon. In the context of the Arctic, the rate of climate warming is almost twice the global average, a phenomenon known as Arctic amplification [2,6–10]. Therefore, quantifying the spatial, seasonal (phenological), and inter-annual variations of Arctic GPP is critical for comprehending the carbon cycle and its feedback to climate warming.

Quantifying global or local GPP has received a great deal of attention in recent studies. Utilizing satellite-based near-infrared reflectance (NIRv) as the proxy of GPP and the revised light-use-efficiency model (i.e., EC-LUE model), Wang et al. [11] and Zheng et al. [12] explored the global spatial patterns of GPP with a spatial resolution of 0.05 degrees. However, the annual average estimates of GPP were not consistent during the same period. Wang et al. [11] reported a range of 128.3 ± 4.0 Pg C $year^{-1}$ while Zheng et al. [12] reported a range of 106.2 ± 2.9 Pg C yr^{-1}. Some studies have detected the GPP in the Arctic, but most paid attention to specific ecosystems (e.g., streams and moss communities) [13,14] and few efforts [12,15] have been devoted to investigating the specific situation of the

Arctic GPP. Here, the MOD17A2H product was selected because it is one of the major official GPP products and has been most widely used in detecting the carbon cycle of terrestrial ecosystems [4,16]. Additionally, its finer resolution (500 m) can reveal detailed GPP variations in the Arctic.

Satellite products generally suffer from the uncertainty that results from complex data acquisition processes and limitations of retrieval algorithms. For this reason, different datasets lead to disparate results. MOD17 is based on the light-use-efficiency (LUE) concept, which is difficult to parameterize since it is influenced by land cover types, phenophases and different types of environmental stress. Furthermore, the maximal values of LUE are specified in the look-up tables for the same biota types, which might introduce uncertainties in GPP [17]. Although MOD17A2H, the v.6 product of MOD17, has overcome the limitation of the proposed year and filling method, the core issues caused by its algorithm still exist. In addition, MODIS (Moderate Resolution Imaging Spectroradiometer) products are inferred based on surface reflectance, which is only available when the relative accuracy of MODIS reflectance products can be determined [18]. Therefore, evaluating its performance is necessary before characterizing the spatiotemporal pattern of GPP. There have been many validation studies regarding the performance of MODIS GPP products at the global scale [19–22] and their accuracies over different biomes (e.g., grassland and forest) have been quantified [23]. However, the validation pixels belonging to specific biomes are often combined together and there are few specialized studies that quantify the accuracy of MOD17A2H in the Arctic. In fact, the performance of the MODIS GPP algorithm shows reasonable variations with climate regions and factors [18,19], species [24], and latitude [25]. Furthermore, the phenology (e.g., the peak timing of GPP) derived from satellite products is often mismatched in scale with in situ data [26]. Several studies have assessed the performance of phenological patterns of MOD17A2H in different regions or biomes [17,27]. However, it is unclear whether MOD17A2H is suitable for the Arctic. Therefore, there is a pressing need to investigate the accuracy of MOD17A2H in the Arctic.

The objective of this work was to utilize MOD17A2H to explore the spatial distribution and phenological characteristics of GPP in the Arctic. In particular, the goal was to (1) evaluate the performance of MOD17A2H in different conditions in the Arctic; (2) identify the spatial distribution and phenological characteristics of GPP, and detect the variation of GPP with land cover types, latitude, and elevation; and (3) detect the interannual trends of GPP in the Arctic and its relation with land cover types, latitude, and elevation. This article begins by describing the study area and the experimental data (Section 2). Section 3 explains the validation and trend detection methods. Section 4 provides the results and discussion of validation, spatial distribution, and phenological characteristics, as well as interannual trends of GPP. Finally, Section 5 presents a brief conclusion.

2. Study Area and Experimental Data

2.1. Study Area

The study region covers the area from 50° N to 90° N and is characterized by long cold winters and short summers. There is very little precipitation and the temperature is low. As a result of the harsh environments, there are few vegetation types in the Arctic. As shown in Figure 1a, water occupies more than half of the area in the Arctic, and the dominant vegetation type are shrubland (dominated by woody perennials), savannas (tree cover 10–60%), and grasslands (dominated by herbaceous annuals (<2 m)). Savannas are distributed in relatively low latitudes, while most of the shrublands are located at high latitudes. Grasslands are mainly scattered west of Greenland. The average altitude of the Arctic is below 1000 m. Areas with higher elevations are mainly distributed in the northwestern region of Canada, the northeastern region of Russia, as well as the Greenland Island and surrounding areas.

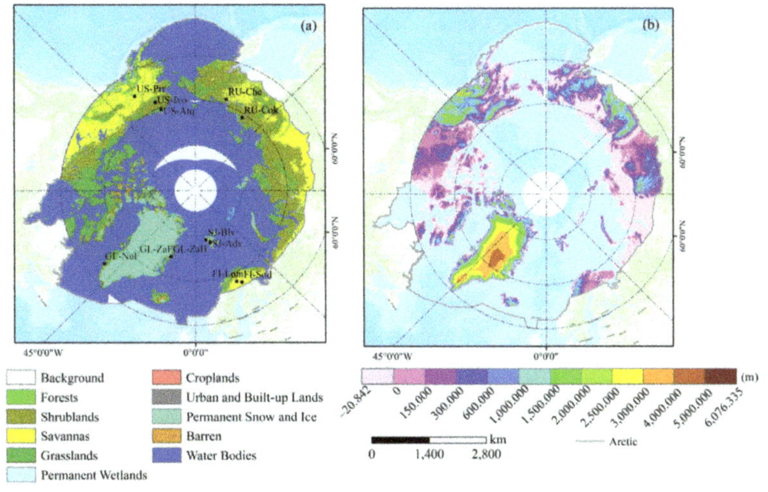

Figure 1. (**a**) Location of the FLUXNET sites used in this study. The description of these sites is provided in Table 1. The base map is the land cover types of the Arctic. (**b**) The elevation distribution of the Arctic.

Table 1. Basic information regarding the FLUXNET sites of the Arctic.

Site_ID	Site_Name	Country	Latitude (° N)	Longitude (° E)	Land Cover	N
FI-Lom	Lompolojankka	Finland	67.9972	24.2092	WET	71
GL-NuF	Nuuk Fen	Greenland	64.1308	−251.3861	WET	105
GL-ZaF	Zackenberg Fen	Greenland	74.4814	−20.5545	WET	42
RU-Che	Cherski	Russia	68.6130	161.3414	WET	33
SJ-Adv	Adventdalen	Svalbard and Jan Mayen	78.1860	15.9230	WET	23
US-Atq	Atqasuk	USA	70.4696	−157.4089	WET	88
US-Ivo	Ivotuk	USA	68.4865	−155.7503	WET	68
FI-Sod	Sodankyla	Finland	67.3624	26.6386	ENF	371
US-Prr	Poker Flat Research Range Black Spruce Forest	USA	65.1237	−147.4876	ENF	98
GL-ZaH	Zackenberg Heath	Greenland	74.4733	−20.5503	GRA	137
RU-Cok	Chokurdakh	Russia	70.8291	147.4943	OSH	107
SJ-Blv	Bayelva, Spitsbergen	Svalbard and Jan Mayen	78.9217	11.8311	SNO	13

WET: wetlands; ENF: evergreen needleleaf forests; GRA: grasslands; OSH: open shrublands; SNO: permanent snow and ice. N means the numbers of the data points of the sites after quality control.

2.2. Data

2.2.1. FLUXNET Data

FLUXNET 2015 is the latest version of the FLUXNET dataset. Compared with previous datasets, FLUXNET v.2015 improves the protocols of data quality and the pipeline of data processing [28]. The Net Ecosystem Exchange (NEE) in the FLUXNET 2015 dataset was gap-filled with the marginal distribution sampling (MDS) method [29]. It was then partitioned into Ecosystem Respiration (RECO) and GPP using the daytime fluxes method (_DT) [30] and the nighttime fluxes method (_NT) [29]. The quality flags in FLUXNET 2015 are given values ranging from 0 to 1, indicating the percentage of high quality gap-filled and measured data [12]; 1 represents the highest quality and 0 represents the poorest quality [31].

As shown in Table 1, there are 12 sites in the FLUXNET datasets located in the study area, including 7 wetland sites, 2 forest sites, a grassland site, a shrublands site, and a permanent snow and ice site. This study used GPP (GPP_NT_VUT_REF) estimated from the night-time method with a daily temporal scale. The selection of GPP followed two criteria: (1) the quality flags were larger than 0.5; and (2) the difference between the GPP derived using the night-time method (GPP_NT_VUT_REF) and the day-time method (GPP_DT_VUT_REF) was lower than 50%. After filtering the data based on the two criteria, daily GPP were temporally aggregated to generate the 8-day averaged GPP, matching the temporal resolution of MOD17A2H.

2.2.2. Satellite Data

MOD17A2H (Collection 6) is a standard satellite product with a spatial resolution of 500 m and a temporal resolution of 8 days. It is calculated based on the light use efficiency (LUE) approach by Monteith [31]:

$$\text{GPP} = \varepsilon \times \text{fPAR} \times \text{PAR} \tag{1}$$

where ε, fPAR, and PAR denote the radiation use efficiency coefficient (RUE), the fraction of incident PAR absorbed by the surface, and photosynthetically active radiation, respectively [32].

According to Running et al. [32], the GPP values of MOD17A2H refer to the sum of the GPP during an 8-day period. In this study, we averaged the total GPP to generate the 8-day averaged GPP.

2.2.3. Land Cover

The MCD12Q1 product provides global land cover type data at a spatial resolution of 500 m at an annual time step from 2001 to 2019. It is based on the supervised classification of MODIS reflectance data with six different classification schemes, including the IGBP (Annual International Geosphere-Biosphere Programmer), which was widely utilized due to its high accuracy and widespread acceptance [33]. Thus, the IGBP classification method was utilized in this study. The land cover data from 2001 to 2019 were chosen to produce a spatially continuous dataset via mosaic. The filling data of MCD12Q1 was removed to reduce their effect on the results. Evergreen needleleaf forests, evergreen broadleaf forests, deciduous needleleaf forests, deciduous broadleaf forests, and mixed forests were grouped into forests. Closed shrublands and open shrublands were grouped together as shrublands. Woody savannas and savannas were combined into savannas.

2.2.4. DEM (Digital Elevation Model)

Multi-Error-Removed Improved-Terrain (MERIT) DEM is an improvement of SRTM3 (Shuttle Radar Topography Mission v.3) DEM, with a spatial resolution of 3 arc-second (~90 m). It removes multiple error components from the SRTM3 DEM, including stripe noise caused by the sensor error, speckle noise of surface reflectance, absolute bias derived from the limited control points of the ground, and tree height bias where the canopies were incorrectly classified as the land surface [34,35]. MERIT was chosen because its accuracy is higher than that of SRTM and NASADEM (NASA Digital Elevation Model) [36] and because of the data availability in the Arctic. In order to match the resolution of MOD17A2H, the DEM dataset was resampled to 500 m using the bilinear method.

3. Methods

3.1. Accuracy Assessment

Although the validation based on in situ leaves issues of scale unresolved, which might introduce uncertainties to the verification, it is still an important method in regions lacking long-term validation data [37–39]. Here, the direct comparison method was utilized because it is simple and easy to implement. To avoid the influence of data noise, geometric mismatch, and spatial heterogeneity on the validation results, the average of the 3×3 pixels of MOD17A2H centered around tower coordinates in situ was used to match with in situ as suggested by Ueyama et al. [40].

Statistical indices, including the coefficient of correlation (R), root mean square error (RMSE), and Bias were used to indicate the accuracy of MOD17A2H [41]. R measures the consistency between MOD17A2H and in situ data, while RMSE measures the average absolute error of the MOD17A2H over a single in situ. Bias describes the average deviation between MOD17A2H and in situ measurements:

$$R = \sum_{l=1}^{h}(P_l - \overline{P})(O_l - \overline{O}) / \sqrt{\sum_{l=1}^{h}(P_l - \overline{P})^2 \sum_{l=1}^{h}(O_l - \overline{O})^2}, \quad (2)$$

$$\text{RMSE} = \sqrt{\sum_{l=1}^{h}(P_l - O_l)^2 / h}, \quad (3)$$

$$\text{Bias} = \sum_{l=1}^{h}(P_l - O_l) / h \quad (4)$$

where P_l and O_l are the MOD17A2H and in situ-based GPP on the l th time period, respectively. $\overline{P}$ and $\overline{O}$ are the averaged value of the MOD17A2H and in situ-based GPP time series, respectively. h is the total number of time periods.

The assessment was twofold. First, the performances of the MOD17A2H product were assessed separately over each site. Second, by combining the data points of all sites within each specific land cover type, the accuracies of MOD17A2H over different land cover types were assessed and compared.

3.2. Comparison of Phenological Patterns between In Situ and MOD17A2H

The phenological patterns in ecosystem GPP are important in the terrestrial carbon cycle and have significant ecological implications. To understand if the MOD17A2H satellite GPP product can capture the functional phenology, which has been defined as the interaction and close association between plant functional traits and phenology [42], of in situ GPP, we first filled the data gap in the original 8-day GPP time series using a linear interpolation method. The gap-filled GPP time series was further smoothed using the Savitzky-Golay (SavGol) filter with a window size of 9 time steps and a second-order polynomial, which not only eliminated noise but also preserved the basic phenological attributes [43]. Finally, we extracted the timing of maximum GPP (day of year, DOY) during the photosynthetically active period for each site from both the in situ and MOD17A2H data. Agreement between the peak timings extracted from the in situ GPP and those extracted from the MOD17A2H was used as an indicator of the performance of the MOD17A2H GPP product in representing the functional phenology patterns of arctic ecosystems.

3.3. The Spatial Distribution Characteristics Identification and Trend Detection

The spatial distribution of GPP was explored. First, the pixel-wise multiyear averaged monthly GPP was calculated to check the GPP spatial distribution and the phenology patterns in different months. Second, the annual-maximum and annual-averaged GPP were identified to detect the distribution of GPP in the Arctic.

A pixel-based simple linear regression, in which time is the independent variable and GPP is the dependent variable, was applied to detect the trend of GPP. In addition, the significance of the interannual trend was evaluated utilizing the Mann–Kendall (MK) test [44], and the significant trends ($p < 0.025$) of the Arctic were retained.

In this study, both MOD17A2H data and the linear regression function were provided by Google Earth Engine (GEE), which is a cloud-based computing platform for planetary-scale data analysis, mapping, and modeling, providing free access to numerous global datasets and advanced computational capabilities [45]. GEE was employed for the following reasons: it provides easy access to the MOD17A2H datasets and other related datasets such as land cover types and elevation; it enables rapid exploration of long time series datasets without downloading them; and it provides a library of functions such as linear regression function, which are applied for data analysis and result display.

4. Results and Discussion

4.1. Validation MOD17A2H Based on In Situ

The results of in situ and MOD17A2H GPP were compared in different land cover types. Figure 2 shows the time series of MOD17A2H and the in situ-based GPP over wetlands; their scatter plots are presented in Figure 3. As shown in Figure 2, missing data occurs frequently in the time series, especially in winter and early spring. This is attributed to the weak photosynthetic activity of vegetation and the lower data coverage during this period. Both MOD17A2H and in situ-based GPP show reasonable seasonal and annual variability over wetlands (Figure 2). However, the agreement between them is significantly different from site to site.

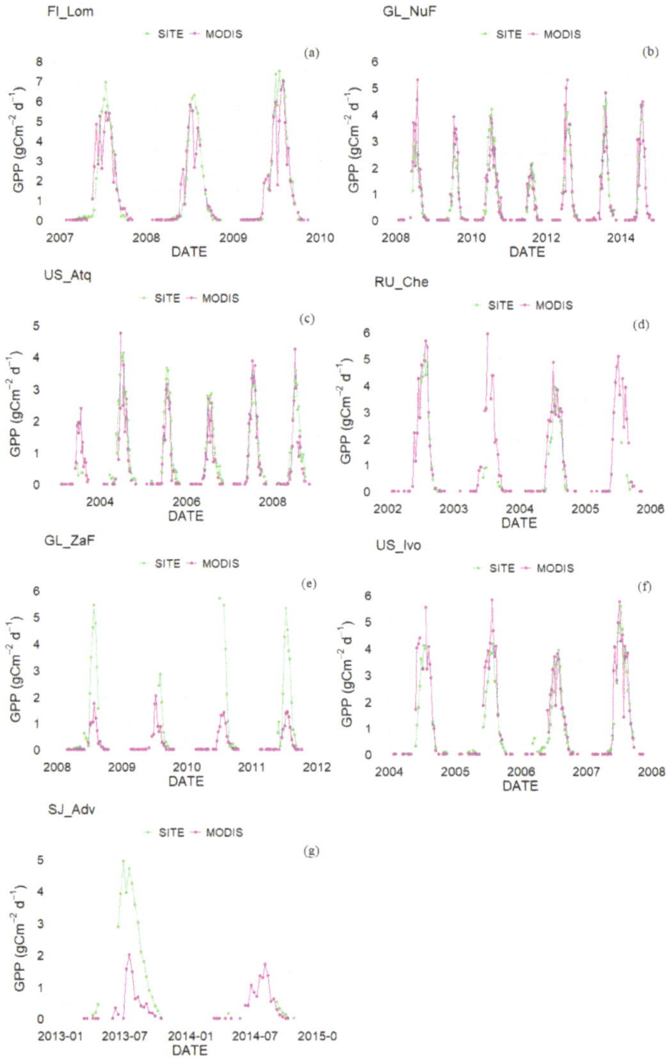

Figure 2. Time series comparison between site-based GPP and MOD17A2H over wetland sites, (**a**–**g**) refers to FI_Lom, GL_NuF, US_Atq, RU_Che, GL_ZaF, US_Ivo and SJ_Adv, respectively.

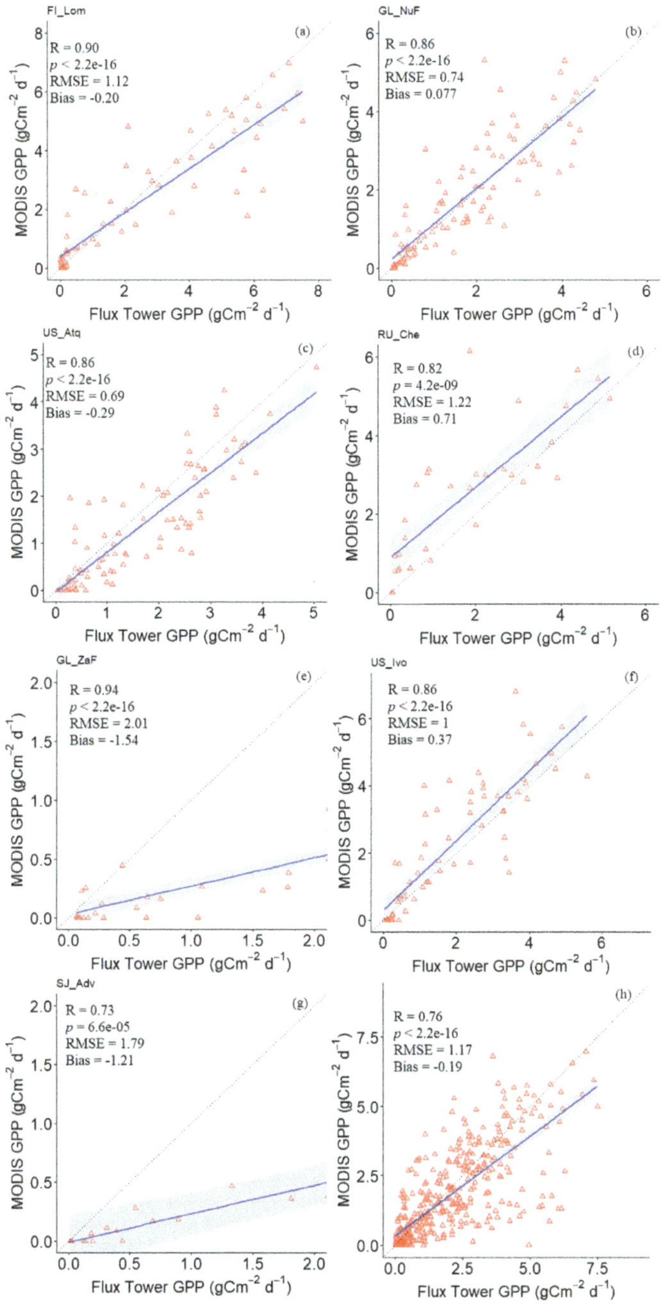

Figure 3. Scatter plots of site-based GPP against MOD17A2H over different wetland sites (**a–g**) and the overall scatter plots by combining all wetland sites (**h**). Flux tower GPP means in situ GPP. The gray belt refers to the confidence interval of 95%.

The comparison results over wetlands can be divided into two groups according to the performance of MOD17A2H: (1) FI_Lom, US_Atq, GL_ZaF, and SJ_Adv; and (2) GL_NuF,

RU_Che and US_Ivo. MOD17A2H was found to underestimate in group (1) and overestimate in group (2). Nevertheless, the degree of underestimation varies significantly from site to site and is low for FI_lom and US_Atq, both in magnitude and temporal variation trend (Figure 2a,c), with RMSEs of 1.12 and 0.69 g C m^{-2} d^{-1} and R of 0.9 and 0.86, respectively (Figure 3a,c). For GL_ZaF and SJ_Adv, the agreement is worse, as the RMSEs are 2.01 and 1.79 g C m^{-2} d^{-1}, respectively (Figure 3e,g), and the bias is very large, with values of −1.54 and −1.21, respectively. The large discrepancy between MOD17A2H and in situ GPP may be caused by their spatial scale mismatch. Furthermore, it should be noted that the data points (less than 50) over these two sites are very limited. For the second group, MOD17A2H is generally consistent with in situ measurements, with the RMSEs of 0.74, 1.22, and 1.0 g C m^{-2} d^{-1}, and R of 0.86, 0.82, and 0.86 over GL_NuF, RU_Che, and US_Ivo, respectively.

As indicated by Figure 3, the MOD17A2H performs well over wetlands when excluding the sites with limited data points (i.e., GL_ZaF and SJ_Adv). One interesting observation was that the latitudes of GL_ZaF and SJ_Adv are higher than 74° N while those of the other sites are lower than 74° N. In addition, the GL_NuF site where the best agreement between MOD17A2H and in situ occurs has the lowest latitude of all the wetland sites. Therefore, it can be inferred that the accuracy of MOD17A2H may be related to latitude, as it is higher over low latitudes but lower over high latitudes. When it comes to the overall performance of MOD17A2H over wetlands (Figure 3h), the overall RMSE and R are 1.17 g C m^{-2} d^{-1} and 0.76, respectively. MOD17A2H slightly underestimates GPP, with a bias of −0.19. Based on the results above, MOD17A2H can be considered capable of revealing the spatial distribution characteristics and the temporal trend of GPP over wetlands in the Arctic.

Figures 4 and 5 show the evaluation results of MOD17A2H over forest sites (i.e., FI_Sod and US_Prr). MOD17A2H underestimates GPP at FI_Sod but overestimates GPP at US_Prr. The extent of misestimation differs between the two sites and is weak for FI_Sod but strong for US_Prr. Over FI_Sod, MOD17A2H agrees well with in situ measurements (Figure 4a), with the RMSE of 1.33 g C m^{-2} d^{-1} and bias of −0.77 (Figure 5a). However, for US_Prr, their agreement is worse (Figure 4b), with the RMSE of 2.05 g C m^{-2} d^{-1} and bias of 1.19 (Figure 5b). It is important to note that the extent of misestimating MOD17A2H varies from year to year. This is especially true over US_Prr, where the overestimation of MOD17A2H is more significant from 2012 to 2014 (Figure 4b). Although the two sites feature forests, their locations are different (Table 1), indicating that the accuracy of MOD17A2H is also influenced by other factors. As shown in Figure 5c, although MOD17A2H slightly underestimates GPP over forests (Bias = −0.36), overall, MOD17A2H was consistent with site measurements over forests, with RMSE and R of 1.51 g C m^{-2} d^{-1} and 0.79, respectively.

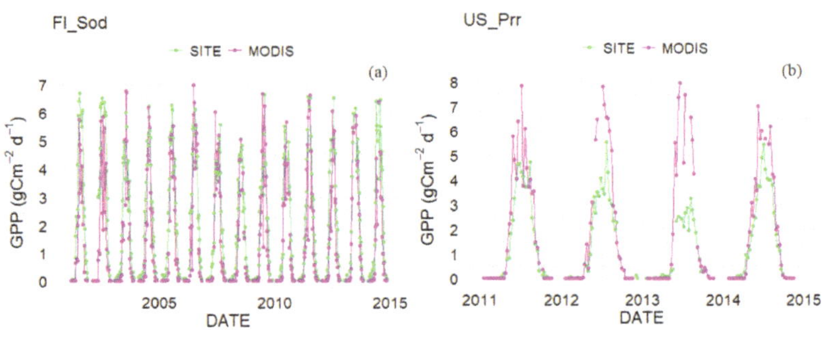

Figure 4. Time series comparison between site-based GPP and MOD17A2H over forest sites, (**a**) and (**b**) refers to FI_Sod and US_Prr.

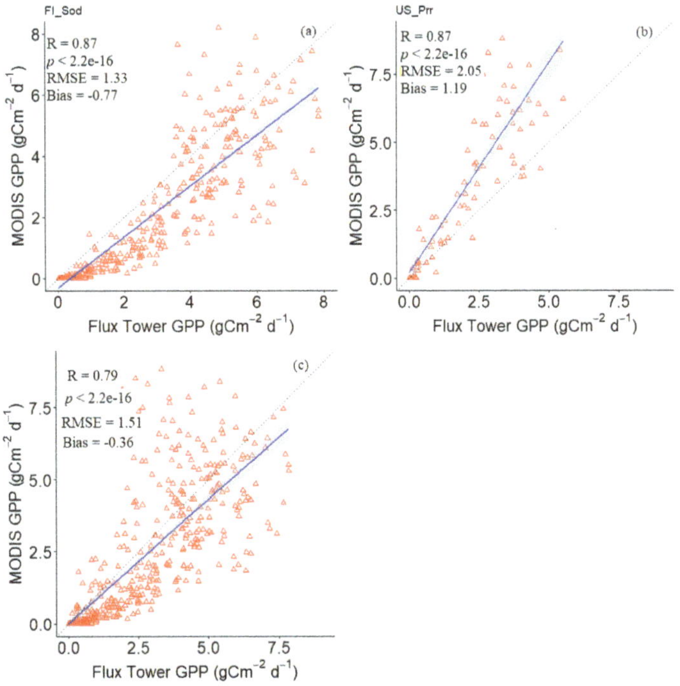

Figure 5. Scatter plots of site-based GPP against MOD17A2H over different forest sites (**a**,**b**), and the overall scatter plots by combining all forest sites (**c**). Flux tower GPP means in situ GPP. Gray belt refers to the confidence interval of 95%.

Figures 6 and 7 present the comparison results over grasslands, shrublands, and permanent snow and ice. MOD17A2H generally underestimates GPP over these land cover types but the degree of underestimation varies with sites, as it is relatively weak for RU_Cok (RMSE = 1.02 g C m^{-2} d^{-1}, Bias = −0.44) but strong for GL_ZaH and SJ_Blv (with the RMSEs of 0.61 and 0.19 g C m^{-2} d^{-1} and the Bias of −0.5 and −0.17, respectively). It is important to remember that the latitudes of GL_ZaH and SJ_Blv are higher than at RU_Cok, which further demonstrates that the accuracy of MOD17A2H is related to the latitude of the sites.

When combining the data points of all the sites (Figure 8), MOD17A2H slightly underestimates GPP, with the bias of −0.32. The overall accuracy of MOD17A2H is reasonable over the Arctic, with RMSE of 1.26 g C m^{-2} d^{-1} and R of 0.8, respectively. These indicators demonstrate that MOD17A2H is able to capture the spatiotemporal variation characteristics of GPP in the Arctic.

From the validation results based on in situ measurements, it is shown that the MOD17A2H generally underestimates GPP over these land cover types. Nevertheless, depending on the location of the sites, it may underestimate or overestimate GPP within each land cover type. This demonstrates that the accuracy of MOD17A2H is also influenced by other factors in addition to land cover types. For instance, the latitude seems to be associated with the accuracy of MOD17A2H given that the accuracy of MOD17A2H tends to be higher over low latitudes but lower over high latitudes (>74° N), which might be due to the actual maximum radiation conversion efficiency (ε_{max}) of vegetations being quite different from the given ε_{max} in high latitude. In fact, the misclassification of a pixel is also responsible for the inconsistency of the accuracies derived from different sites within the same land cover type. Because the classification scheme adopted by IGBP is too general, it cannot reveal the detailed categories carrying out photosynthesis. Moreover,

MOD17A2H is calculated based on the concept of light use efficiency (LUE), which assumes a fixed maximum radiation conversion efficiency (ε_{max}) of each land cover type [19]. This treatment also introduced errors when misclassification occurred. Another cause of the inconsistency is the different degrees of surface heterogeneities within the satellite pixel, which cause the sites to be more or less representative of the satellite pixel. Last but not least, since missing data of in situ may occur unequally during each 8-day period, the temporal representativeness varies across these sites. Despite these uncertainties, the validation results still suggest that the performance of MOD17A2H is better over shrublands and wetlands than it is over forests.

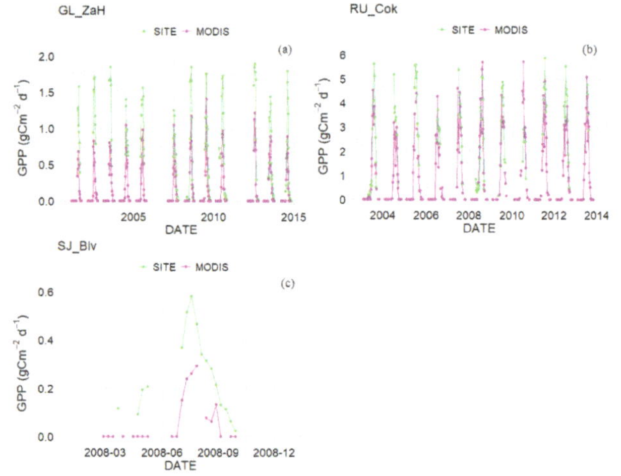

Figure 6. Time series comparison between site-based GPP and MOD17A2H over grasslands (**a**), shrublands (**b**), and permanent snow and ice (**c**) sites.

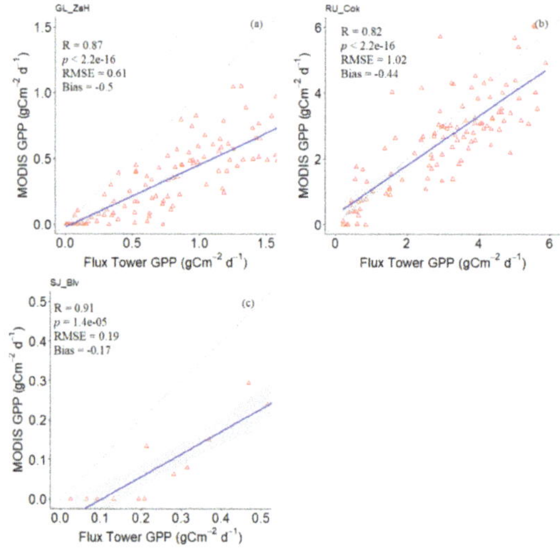

Figure 7. Scatter plots of site-based GPP against MOD17A2H over grasslands (**a**), shrublands (**b**), permanent snow and ice (**c**) sites. Flux tower GPP means in situ GPP. Gray belt refers to the confidence interval of 95%.

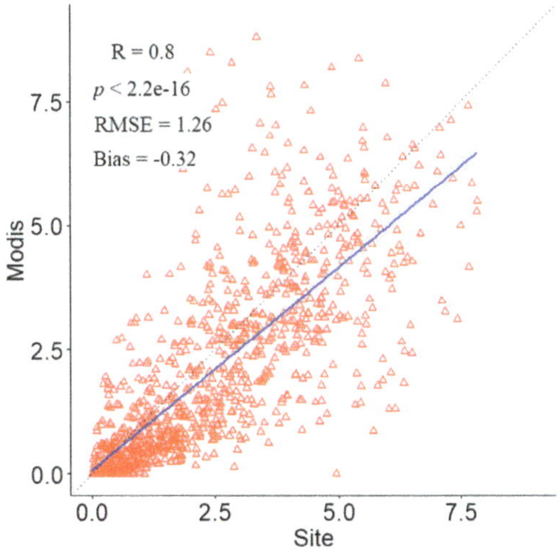

Figure 8. The overall scatterplots of in situ based GPP against MOD17A2H. Flux tower GPP means in situ GPP. Gray belt refers to the confidence interval of 95%.

4.2. Evaluation of the Phenological Characteristics of MOD17A2H

According to Figure 9, MOD17A2H does express the phenological characteristics of GPP in the Arctic, with cross-sites R and RMSE of 0.62 and 8.9 days, respectively. In particular, as shown in Figure 9c, the peak GPP timing DOY of grasslands was the most consistent with that extracted from in situ data (R = 0.86, RMSE = 4 days). By contrast, the discrepancy between MOD17A2H and in situ peak GPP timing over forest sites was the highest (Figure 9b), with R and RMSE of 0.52 and 11.9 days, respectively, which demonstrates that the characteristics of the forests (mostly evergreen needleleaf) GPP, derived from MOD17A2H, are difficult to capture compared with the characteristics of other land cover types. This might be due to two reasons: (1) forests are composed of evergreen needleleaf forests, evergreen broadleaf forests, deciduous needleleaf forests, deciduous broadleaf forests, and mixed forests; therefore, the diversity of forests makes phenological characteristics difficult to capture by satellite observations; and (2) evergreen needleleaf forest is one of the major forest types in the Arctic, which is less sensitive to climate changes according to [46], thus the phenological characteristics of evergreen needleleaf forests are hard to capture.

4.3. Spatial Distribution Characteristics

4.3.1. Spatial Distribution of Annual-Averaged GPP

Figure 10 shows the annual maximum (Figure 10a) and annual-averaged GPP (Figure 10b), derived from MOD17A2H, in the Arctic on a pixel basis. The two metrics generally present similar spatial distribution patterns. GPP is relatively low in the northeast of Canada and the regions surrounding Greenland, with an annual-maximum range of 0 to 1.200 g C m^{-2} d^{-1} and an annual-averaged range of 0 to 0.900 g C m^{-2} d^{-1}. The low GPP over these areas can be explained by the fact that these areas are almost completely covered by grassland and barren land, which have lower GPP (Figure 11c). Therefore, it can be inferred that the spatial distribution of GPP is related to land cover types. Furthermore, the GPP shows a decreasing trend as the latitude increases. This is especially true over the eastern hemisphere of the Arctic, where the annual maximum of GPP drops from 3.300 to 0.300 g C m^{-2} d^{-1} and the annual-averaged of GPP decreases from 2.400 to 0.300 g C m^{-2} d^{-1}.

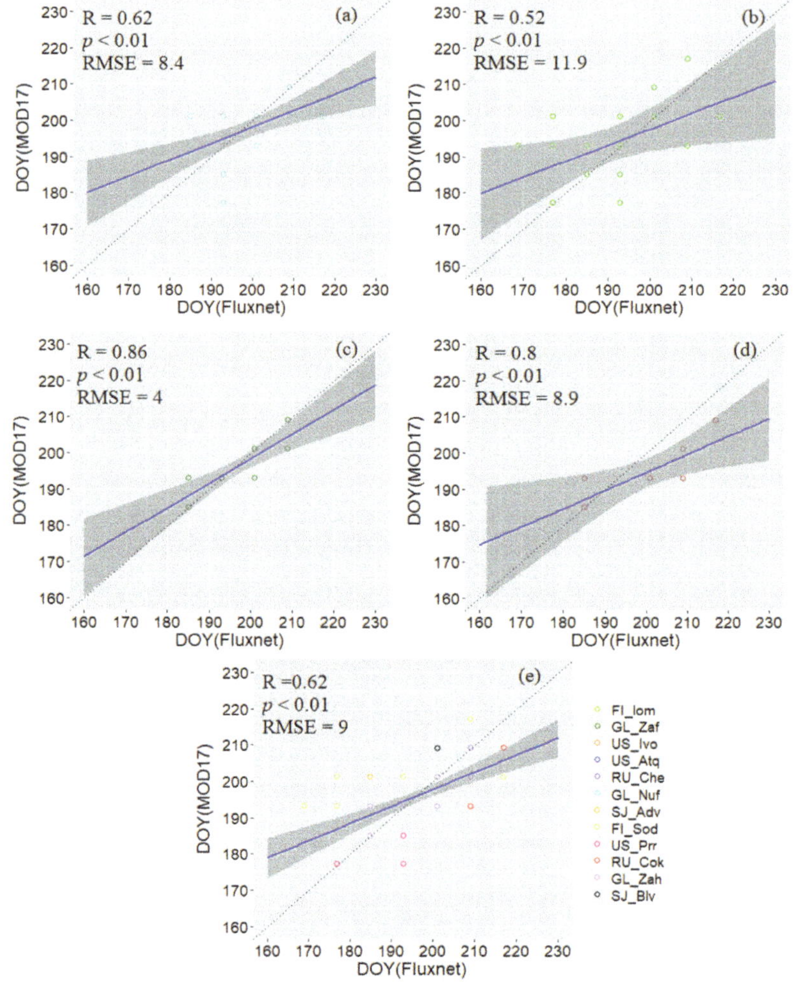

Figure 9. The DOY comparison of the timing of the maximum GPP between the sites and MOD17A2H. (**a–e**) Represent permanent wetlands, forests, grasslands, shrublands, and all sites combined, respectively. The gray belt refers to the confidence interval of 95%.

Figure 11a,b display the distribution of GPP with latitude and elevation, respectively. It can be seen that GPP generally shows a decreasing trend with latitude, which is in line with the results of Gounand et al. [25]. Nevertheless, the sensitivity of GPP to latitude depends on the situation. The results define three groups of latitudes: (1) latitudes less than 62° N and more than 80° N; (2) latitudes higher than 62° N but lower than 66° N; and (3) latitudes between 66° N and 80° N. In the first group, the GPP distribution is sparse because the region located within this scope is quite limited. Thus, GPP shows irregular variation patterns with latitude. In the second group, GPP is relatively stable, indicating the insensitivity of GPP to latitude within this scope. Nevertheless, GPP presents a clear decreasing trend in the third group, demonstrating that GPP is most sensitive to latitude from 66° N to 80° N. Similar to latitude, GPP generally shows a decreasing trend with increasing elevation (Figure 11b). However, the decreasing rates are different depending on the elevation, which is smaller for elevations lower than 700 m, but larger for elevations higher than 700 m.

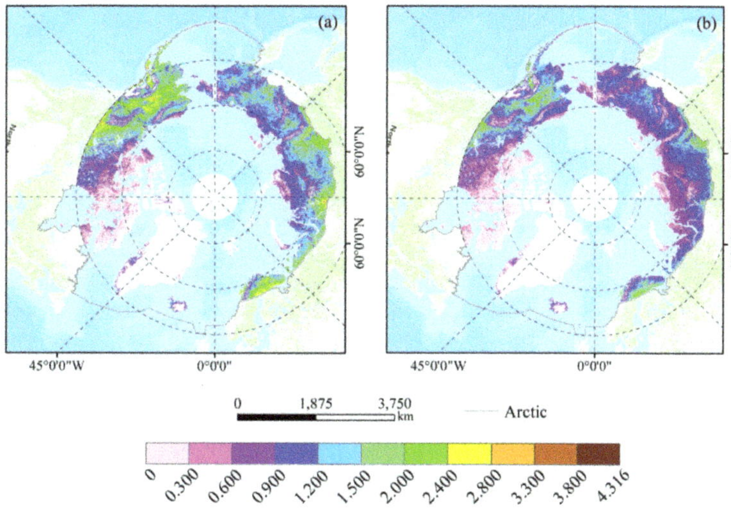

Figure 10. The spatial distribution characteristics of annual-maximum (**a**) and annual-averaged GPP (**b**) over the Arctic. Units of GPP are g C m^{-2} d^{-1}.

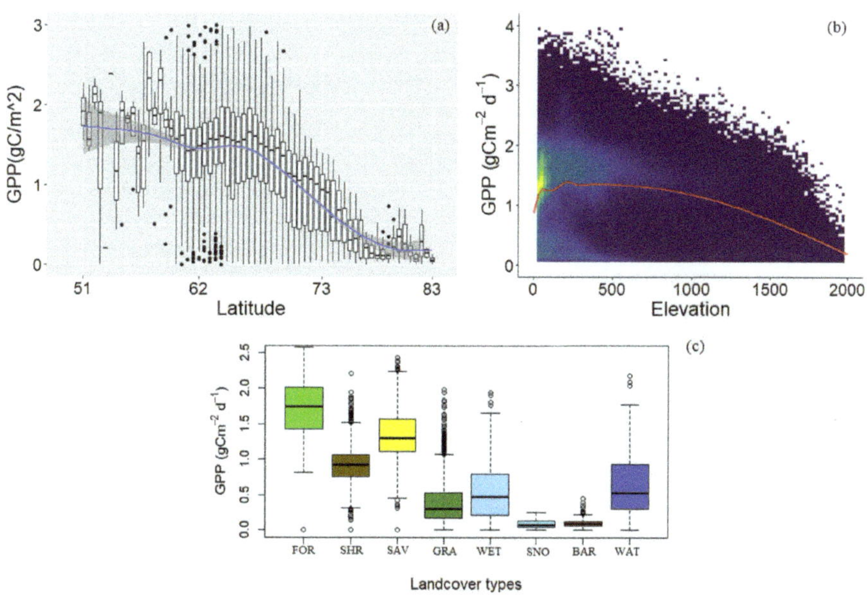

Figure 11. The annual-averaged GPP distribution with latitude (**a**) and elevation (m) (**b**) for the whole Arctic. The gray belt refers to the confidence interval of 95%, the blue line refers to the fit line, and the boxplots denote the distribution of the annual-averaged GPP with the latitude. (**c**) The annual-averaged GPP distribution for the entire study period 2001–2019 over different land cover types. FOR, SHR, SAV, GRA, WET, SNO, BAR, and WAT denote the forests, shrublands, savannas, grasslands, permanent wetlands, permanent snow and ice, barren, and water bodies, respectively.

Figure 11c presents the annual-averaged GPP distribution over different land cover types for the entire study period. It can be seen that all land cover types show considerable interannual variation except for permanent snow and ice as well as barren land (as indicated by the wide distribution of boxplots).

4.3.2. Variation of Monthly GPP

Figure 12 displays the spatial pattern of multiyear (2001–2019) averaged monthly GPP. The phenological cycle of vegetation is clearly shown in the figure. GPP is very low from November to March, with values close to 0. This is because the photosynthesis of vegetation was restricted due to the extremely harsh environment and limited lighting hours. From April to July, with rising temperatures, melting snow and sea ice, and increasing hours of light, the carbon fixation ability of vegetation is stronger, which contributes to the gradual expansion of GPP from the northwest of Canada and the low latitudes of Russia to the entire Arctic. In addition, the mean GPP over the whole Arctic shows a rapid increase during this period, with the values increasing from 0.0859 g C m^{-2} d^{-1} in April to 3.739 g C m^{-2} d^{-1} in July. However, the GPP gradually decreases from August to October as the mean values decrease from 2.215 to 0.0453 g C m^{-2} d^{-1}. This may be attributed to the decrease in temperature, the shortening of day-length, and the senescence of vegetation during this period. The northwestern region of Canada, a small low latitude portion of Russia, and the regions surrounding Iceland have the longest growth period since they are the first to begin and the last to stop photosynthesis. Figure 12 demonstrates that spatial heterogeneity is small during the dormant months (from November to March) and the mid and late summer months (July and August) when GPP is consistently small or large, but large during the transition months (April, May, September, and October) and early summer (June). There are two main reasons for this: the first is the spatial difference in land cover types, the second is the temporal difference in climate conditions [47].

The GPP in the Arctic has a distinct seasonality with the greatest values in July (Figure 13). Throughout the year, most land cover types follow the general seasonality of GPP; they are lowest from January to March, begin to increase in April and reach a maximum in July, decrease from then on and fall back to the lowest values in November. Forest and savannas present the largest GPP from April to August, followed by shrublands, water bodies, permanent wetlands, and grasslands (Figure 13). Permanent snow and ice as well as barren land ranks last. Nevertheless, from September to October, water bodies show slightly larger GPP than other land cover types. These results demonstrate that forests and savannas have a rather high carbon storage capacity during the growing season (from April to August) but water bodies are the biggest contributors to carbon fixation from September to October.

The results of Figure 13 are certainly not anticipated because land cover types such as permanent snow and ice, barren land, and water bodies, which cannot carry out photosynthesis, show considerable GPP. This can be explained by the definition of land cover type of IGBP, which is determined by the dominant land cover of a pixel. Water bodies refer to those pixels that are at least 60% covered by permanent water bodies; barren denotes that at least 60% of the pixel is non-vegetated barren (sand, rock, and soil) areas with less than 10% vegetation; permanent Snow and ice means that at least 60% of the pixel is covered by snow and ice for at least 10 months of the year. Therefore, the misclassified part of a pixel is the source of GPP for these three land cover types. From the results above, it is inferred that only extracting the vegetated pixels, as classified by the land cover, will lead to errors in calculating carbon storage in the Arctic.

Figure 13 also shows that the monthly GPP has the widest distribution in June, indicating that the interannual variation of GPP is the most significant in June. This is understandable since vegetation grew at the fastest speed from May to June (indicated by the largest slope) and thus is more affected by climate change.

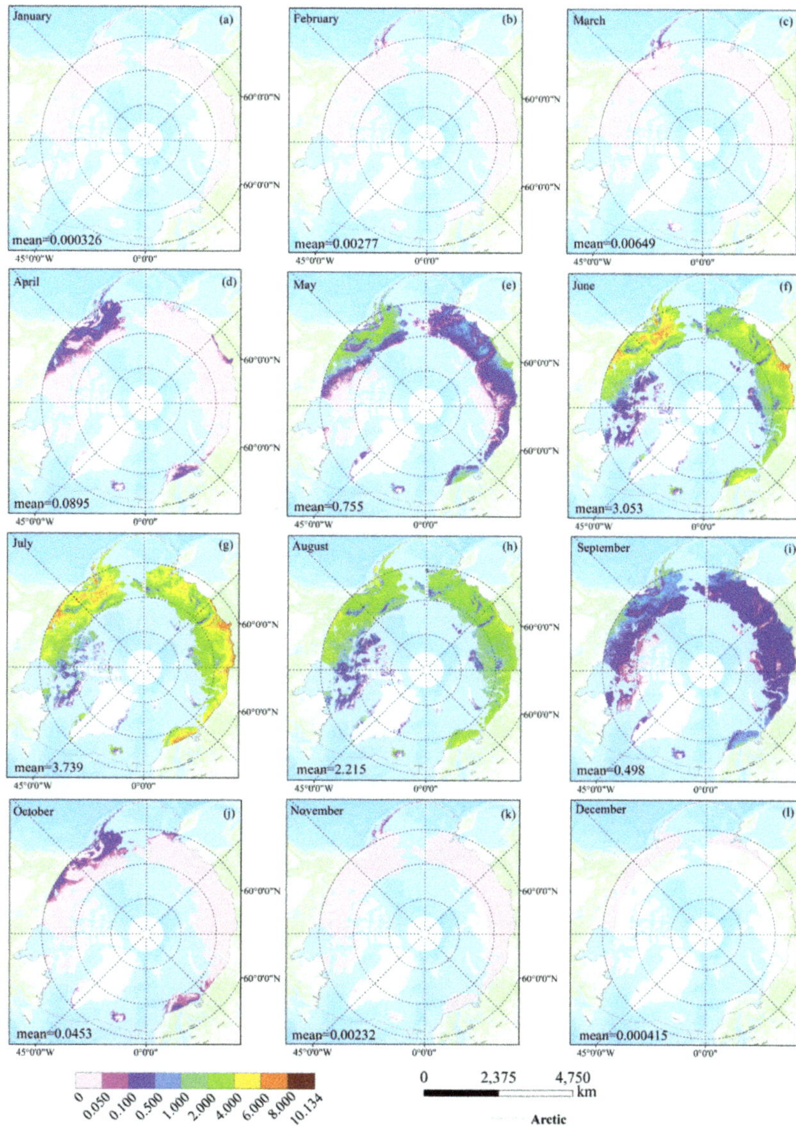

Figure 12. Spatial distribution characteristics of the multiyear averaged monthly GPP over the Arctic, (**a–l**) refers to January to December. The mean values of the whole Arctic are also shown in the figure for each month. Units of GPP are g C m^{-2} d^{-1}.

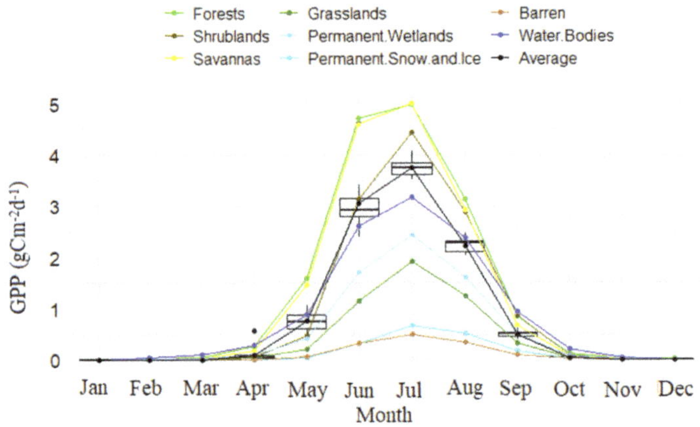

Figure 13. Multiyear averaged monthly GPP distribution over the entire study period for different land cover types (denoted by colored lines) over the Arctic. The black boxplots denote the multiyear monthly GPP distribution of all land cover types.

4.4. Trend Estimates of GPP

The interannual variation trend of GPP is shown on a pixel basis (Figure 14); almost half of the Arctic presents significant positive trends, but the magnitude of trends shows distinct spatial variation. In the northwest of Canada and the latitude lower than 70° N of Russia, the trend of GPP varies significantly, ranging from 0.005 to 0.08 g C m^{-2} $year^{-1}$. By contrast, in the northeast of Canada and the regions surrounding Greenland, the GPP trend varies slightly, ranging from 0 to 0.01 g C m^{-2} $year^{-1}$. The small range of the GPP trend is likely to be associated with grassland and barren land. The former has a relatively small spatial variation, as indicated by the centralized distribution of the interannual trends in GPP. The latter has a very small GPP, which is almost constant over time (Figure 11c).

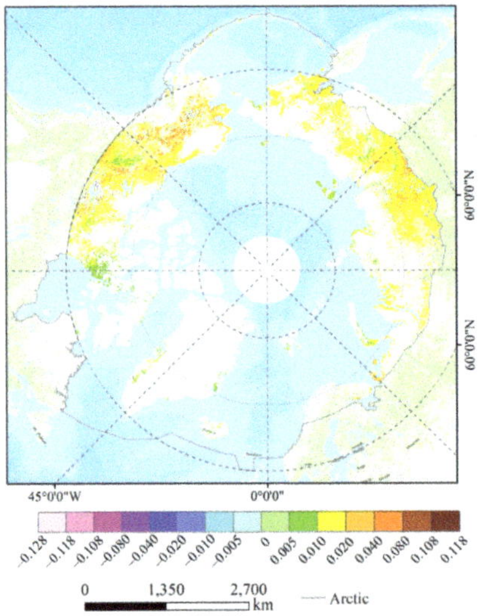

Figure 14. The spatial distribution of interannual trend of GPP (g C m^{-2} $year^{-1}$) over the Arctic (at 97.5% confidence level based on MK test).

As seen from Figure 12, there is almost no vegetation productivity in the Arctic from November to March. Thus, Figure 15 only presents the interannual trends from April to October. It is clear that the interannual trends show significant differences between different months. From April to June, the interannual trends in northwestern Canada gradually increase and reach a maximum in June, with an overall trend of 0.068 g C m^{-2} $year^{-1}$ (Figure 15c). From then on, the interannual trend decreases gradually until October (Figure 15d–g). Furthermore, we also find that the interannual trends in July and August are more spatially heterogeneous than in other months. These results demonstrate that the response of vegetation to climate change is not consistent between different months and over different areas. It is interesting to find that the interannual trend is the most significant in June (Figure 15c) and shows a similar spatial pattern to the overall interannual trend (Figure 14). This demonstrates that the interannual variations of GPP in the Arctic may be dominated by the change of vegetation productivity in June.

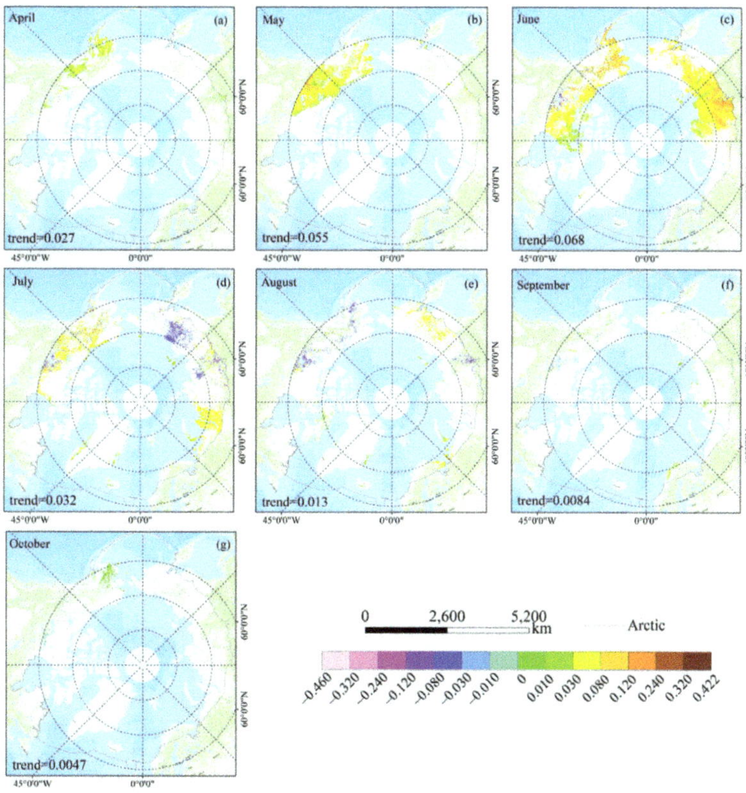

Figure 15. Spatial distribution of the interannual variation trend (g C m^{-2} $year^{-1}$) of monthly GPP over the Arctic, (**a**–**g**) refers to April to October.

The distributions of the interannual trends in GPP with latitude and elevation are shown in Figure 16a,b. The interannual trends first increase at the latitude of 51° N and reach a maximum of 0.018 g C m^{-2} $year^{-1}$ at 57° N. From then on, the interannual trend decreases significantly until 62° N. However, the interannual trend seems to be independent of latitude from 62° N to 66° N. Then a clear decreasing trend can be observed from 66° N until 80° N. Therefore, we can conclude that the interannual trend of GPP is sensitive to the latitude, except in the regions located between 62° N and 66° N, which is similar to the distribution of annual-averaged GPP with the variation of latitude. This is mainly because the regions located between (62° N, 66° N) are in the same climatic zone, namely in the north temperate zone. Figure 16b demonstrates that the interannual trend of GPP generally shows a decreasing trend with increasing elevations. However, the sensitivity is relatively weak at low elevations (<700 m) and significant at larger elevations (>700 m). The low GPP and interannual trend in the regions with relatively higher altitude and latitude Figure 11a,b and Figure 16a,b mainly due to the low precipitation and temperature, which are not conducive to plant growth [48,49]. These results could help us understand the latitude or elevation range in which the change of GPP mainly occurred, providing a basis for understanding the changes of the Arctic GPP.

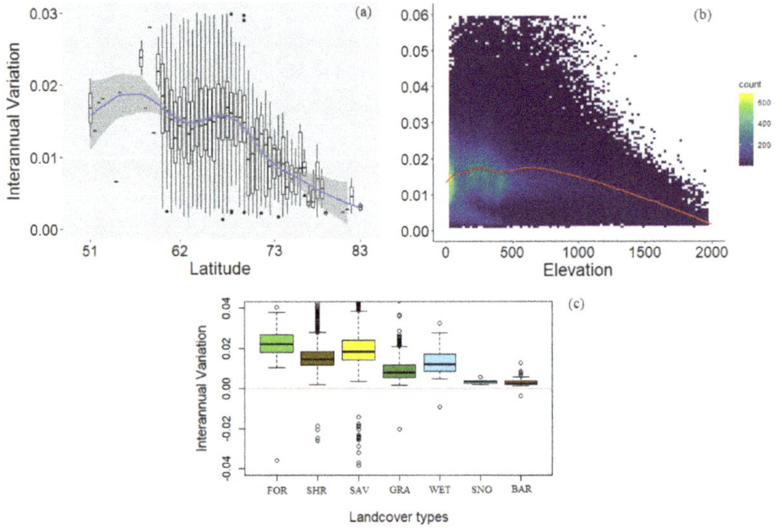

Figure 16. Interannual variation trend of GPP (g C m^{-2} $year^{-1}$) distribution with latitude (**a**), elevation (m) (**b**) over the Arctic. The gray belt refers to the confidence interval of 95%, the blue line refers to the fit line, and the boxplots denote the distribution of the annual-averaged GPP with the latitude. (**c**) Land cover-dependent interannual variation trend of GPP (g C m^{-2} $year^{-1}$) over the Arctic. FOR, SHR, SAV, GRA, WET, SNO, and BAR denote the forests, shrublands, savannas, grasslands, permanent wetlands, permanent snow and ice, barren, and water bodies, respectively.

Figure 16c displays the boxplots of the interannual trends by combining the pixels for each land cover type. It can be seen that almost each land cover type presents positive trends. However, their magnitude, as well as their spatial variation, depend on the land cover types. Forests have the largest interannual trend, followed by savannas and shrublands. The interannual trend of grassland is not as large as we expected and is even slightly smaller than the trend of permanent wetlands. Savannas and forests show the largest spatial variations in interannual trends, as indicated by the most widespread distribution of boxplots. Permanent wetlands and grasslands present the smallest spatial variations when excluding permanent snow and ice as well as barren land. These results demonstrate that the interannual trends of savannas and forests are more influenced by other factors, while those of permanent wetlands and grasslands are less sensitive to other factors.

5. Conclusions

Arctic ecosystems have undergone great changes in the context of climate change. GPP is one of the most crucial indicators of the response of ecosystems to climate change. However, few efforts have been devoted to exploring the spatial variation and phenological characteristics of GPP in the Arctic. In response to this challenge, this study investigated the spatial distribution as well as the seasonal (phenological) and interannual variations of GPP in the Arctic using MOD17A2H. Furthermore, the GPP variation trends with land cover types, latitude, and elevation were also explored. In order to ensure that the results were reliable, the accuracy of MOD17A2H was first evaluated using in situ measurements from FLUXNET.

This study found that MOD17A2H generally underestimates GPP over the land cover types investigated in this study, and its accuracy tends to be higher over low latitude but lower over high latitude. However, the overall accuracy suggests that MOD17A2H is consistent with the FLUXNET 2015 dataset (RMSE = 1.26 g C m^{-2} d^{-1}, R = 0.8, Bias = −0.32), and MOD17A2H can represent the phenological characteristics of GPP (RMSE = 8.9 days, R = 0.62). Based on MOD17A2H, it was demonstrated that the maximum GPP occurred in July. In addition, the spatial distribution of GPP is related to land cover types; for example, forests and savannas have relatively high carbon storage capacity from April to August. By comparing the GPP variation with latitude and elevation, it was shown that GPP generally decreases as the latitude and elevation increase. However, the phenomenon is not evident for latitudes in the range (62° N, 66° N) and elevation lower than 700 m. The overall trend of GPP in the Arctic is greater than zero and is dominated by the variation of vegetation productivity in June. Furthermore, the response to climate change is different across these land cover types; for example, forests are most sensitive to climate warming. The distribution of the interannual trend in GPP across latitudes and elevations is consistent with the changes in GPP as a function of latitude and elevation.

This study is helpful for understanding the spatiotemporal distribution characteristics of GPP over the Arctic as well as the response of ecosystems to climate change. Nevertheless, the results need to be validated with different satellite products. The number of sites used for validation is limited. Therefore, the presented conclusion about the accuracy of MOD17A2H may not be transferable to other regions. Another limitation is that only individual factors such as land cover type, latitude, and elevation were considered in this paper. Other factors, such as air temperature, precipitation, and snow, that are related to vegetation growth status need to be explored further. This will also be our focus in the future.

Author Contributions: Conceptualization, D.M. and J.W.; methodology, D.M. and X.W.; validation, X.L., C.M. and X.W.; formal analysis, D.M. and X.W; investigation, J.W.; resources, X.W.; data curation, D.M.; writing—original draft preparation, D.M. and X.W.; writing—review and editing, D.M., X.W. and X.M.; visualization, J.W.; supervision, C.M. All authors have read and agreed to the published version of the manuscript.

Funding: This work was jointly supported by the National Key Research and Development Program of China (2019YFA0607003), the National Natural Science Foundation of China (Grant No. 42071296 and 41801226), and the Fundamental Research Funds for the Central Universities (lzujbky-2020-72).

Data Availability Statement: Publicly available datasets were analyzed in this study. This data can be found in Google earth engine platform (https://earthengine.google.com/, accessed on 7 February 2021).

Acknowledgments: We gratefully acknowledge the anonymous reviewers and editor of the paper in providing critical and constructive comments, which has helped us to significantly improve the manuscript. We also appreciate the assistance of the all authors of this paper.

Conflicts of Interest: The authors declare no conflict of interest.

References

1. Liu, G.M.; Wu, T.H.; Hu, G.J.; Wu, X.D.; Li, W.P. Permafrost existence is closely associated with soil organic matter preservation: Evidence from relationships among environmental factors and soil carbon in a permafrost boundary area. *Catena* **2021**, *196*, 104894. [CrossRef]
2. Kug, J.S.; Jeong, J.H.; Jang, Y.S.; Kim, B.M.; Folland, C.K.; Min, S.K.; Son, S.W. Two distinct influences of arctic warming on cold winters over North America and East Asia. *Nat. Geosci.* **2015**, *8*, 759–762. [CrossRef]

3. Berra, E.F.; Gaulton, R. Remote sensing of temperate and boreal forest phenology: A review of progress, challenges and opportunities in the intercomparison of in-situ and satellite phenological metrics. *For. Ecol. Manag.* **2021**, *480*, 118663. [CrossRef]
4. Zhu, X.Y.; Pei, Y.Y.; Zheng, Z.P.; Dong, J.W.; Zhang, Y.; Wang, J.B.; Chen, L.J.; Doughty, R.B.; Zhang, G.L.; Xiao, X.M. Underestimates of grassland gross primary production in modis standard products. *Remote Sens.* **2018**, *10*, 1771. [CrossRef]
5. Sun, Z.; Wang, X.; Yamamoto, H.; Tani, H.; Nie, T.; Oppenheimer, M.; Yohe, G. The effects of spatiotemporal patterns of atmospheric co2 concentration on terrestrial gross primary productivity estimation. *Clim. Chang.* **2020**, *163*, 913–930. [CrossRef]
6. Francis, J.A.; Vavrus, S.J. Evidence linking arctic amplification to extreme weather in mid-latitudes. *Geophys. Res. Lett.* **2012**, *39*, 39. [CrossRef]
7. Cohen, J.; Screen, J.A.; Furtado, J.C.; Barlow, M.; Whittleston, D.; Coumou, D.; Francis, J.; Dethloff, K.; Entekhabi, D.; Overland, J.; et al. Recent arctic amplification and extreme mid-latitude weather. *Nat. Geosci.* **2014**, *7*, 627–637. [CrossRef]
8. Screen, J.A.; Simmonds, I. Amplified mid-latitude planetary waves favour particular regional weather extremes. *Nat. Clim. Chang.* **2014**, *4*, 704–709. [CrossRef]
9. Wallace, J.M. Global warming and winter weather. *Science* **2014**, *343*, 969. [CrossRef]
10. Kim, J.S.; Kug, J.S.; Jeong, S.J.; Huntzinger, D.N.; Michalak, A.M.; Schwalm, C.R.; Wei, Y.X.; Schaefer, K. Reduced north american terrestrial primary productivity linked to anomalous arctic warming. *Nat. Geosci.* **2017**, *10*, 572–576. [CrossRef]
11. Wang, S.H.; Zhang, Y.G.; Ju, W.M.; Qiu, B.; Zhang, Z.Y. Tracking the seasonal and inter-annual variations of global gross primary production during last four decades using satellite near-infrared reflectance data. *Sci. Total Environ.* **2021**, *755*, 142569. [CrossRef]
12. Zheng, Y.; Shen, R.Q.; Wang, Y.W.; Li, X.Q.; Liu, S.G.; Liang, S.L.; Chen, J.M.; Ju, W.M.; Zhang, L.; Yuan, W.P. Improved estimate of global gross primary production for reproducing its long-term variation, 1982–2017. *Earth Syst. Sci. Data* **2020**, *12*, 2725–2746. [CrossRef]
13. Myrstener, M.; Gomez-Gener, L.; Rocher-Ros, G.; Giesler, R.; Sponseller, R.A. Nutrients influence seasonal metabolic patterns and total productivity of arctic streams. *Limnol. Oceanogr.* **2021**, *66*, S182–S196. [CrossRef]
14. May, J.L.; Parker, T.; Unger, S.; Oberbauer, S.F. Short term changes in moisture content drive strong changes in normalized difference vegetation index and gross primary productivity in four arctic moss communities. *Remote Sens. Environ.* **2018**, *212*, 114–120. [CrossRef]
15. Ryu, Y.; Baldocchi, D.D.; Kobayashi, H.; van Ingen, C.; Li, J.; Black, T.A.; Beringer, J.; van Gorsel, E.; Knohl, A.; Law, B.E.; et al. Integration of modis land and atmosphere products with a coupled-process model to estimate gross primary productivity and evapotranspiration from 1 km to global scales. *Glob. Biogeochem. Cycles* **2011**, *25*, 4. [CrossRef]
16. Xia, J.Y.; Niu, S.L.; Ciais, P.; Janssens, I.A.; Chen, J.Q.; Ammann, C.; Arain, A.; Blanken, P.D.; Cescatti, A.; Bonal, D.; et al. Joint control of terrestrial gross primary productivity by plant phenology and physiology. *Proc. Natl. Acad. Sci. USA* **2015**, *112*, 2788–2793. [CrossRef] [PubMed]
17. Ma, X.L.; Huete, A.; Yu, Q.; Restrepo-Coupe, N.; Beringer, J.; Hutley, L.B.; Kanniah, K.D.; Cleverly, J.; Eamus, D. Parameterization of an ecosystem light-use-efficiency model for predicting savanna gpp using modis evi. *Remote Sens. Environ.* **2014**, *154*, 253–271. [CrossRef]
18. Heinsch, F.A.; Zhao, M.S.; Running, S.W.; Kimball, J.S.; Nemani, R.R.; Davis, K.J.; Bolstad, P.V.; Cook, B.D.; Desai, A.R.; Ricciuto, D.M.; et al. Evaluation of remote sensing based terrestrial productivity from modis using regional tower eddy flux network observations. *IEEE Trans. Geosci. Remote Sens.* **2006**, *44*, 1908–1925. [CrossRef]
19. Sjostrom, M.; Zhao, M.; Archibald, S.; Arneth, A.; Cappelaere, B.; Falk, U.; de Grandcourt, A.; Hanan, N.; Kergoat, L.; Kutsch, W.; et al. Evaluation of modis gross primary productivity for africa using eddy covariance data. *Remote Sens. Environ.* **2013**, *131*, 275–286. [CrossRef]
20. Kanniah, K.D.; Beringer, J.; Hutley, L.B.; Tapper, N.J.; Zhu, X. Evaluation of collections 4 and 5 of the modis gross primary productivity product and algorithm improvement at a tropical savanna site in northern australia. *Remote Sens. Environ.* **2009**, *113*, 1808–1822. [CrossRef]
21. Wang, J.M.; Sun, R.; Zhang, H.L.; Xiao, Z.Q.; Zhu, A.R.; Wang, M.J.; Yu, T.; Xiang, K.L. New global musyq gpp/npp remote sensing products from 1981 to 2018. *IEEE J. Sel. Top. Appl. Earth Obs. Remote Sens.* **2021**, *14*, 5596–5612. [CrossRef]
22. O'Sullivan, M.; Smith, W.K.; Sitch, S.; Friedlingstein, P.; Arora, V.K.; Haverd, V.; Jain, A.K.; Kato, E.; Kautz, M.; Lombardozzi, D.; et al. Climate-driven variability and trends in plant productivity over recent decades based on three global products. *Glob. Biogeochem. Cycles* **2020**, *34*, 34.
23. Yu, T.; Sun, R.; Xiao, Z.Q.; Zhang, Q.; Liu, G.; Cui, T.X.; Wang, J.M. Estimation of global vegetation productivity from global land surface satellite data. *Remote Sens.* **2018**, *10*, 327. [CrossRef]
24. Tang, X.G.; Li, H.P.; Huang, N.; Li, X.Y.; Xu, X.B.; Ding, Z.; Xie, J. A comprehensive assessment of modis-derived GPP for forest ecosystems using the site-level FLUXNET database. *Environ. Earth Sci.* **2015**, *74*, 5907–5918. [CrossRef]
25. Gounand, I.; Little, C.J.; Harvey, E.; Altermatt, F. Global quantitative synthesis of ecosystem functioning across climatic zones and ecosystem types. *Glob. Ecol. Biogeogr.* **2020**, *29*, 1139–1176. [CrossRef]
26. Fisher, J.I.; Mustard, J.F. Cross-scalar satellite phenology from ground, Landsat, and MODIS data. *Remote Sens. Environ.* **2007**, *109*, 261–273. [CrossRef]
27. Xu, X.J.; Zhou, G.M.; Du, H.Q.; Mao, F.J.; Xu, L.; Li, X.J.; Liu, L.J. Combined modis land surface temperature and greenness data for modeling vegetation phenology, physiology, and gross primary production in terrestrial ecosystems. *Sci. Total Environ.* **2020**, *726*, 137948. [CrossRef] [PubMed]

28. Yang, S.S.; Zhang, J.H.; Zhang, S.; Wang, J.W.; Bai, Y.; Yao, F.M.; Guo, H.D. The potential of remote sensing-based models on global water-use efficiency estimation: An evaluation and intercomparison of an ecosystem model (BESS) and algorithm (MODIS) using site level and upscaled eddy covariance data. *Agr. Forest Meteorol.* **2020**, *287*, 107959. [CrossRef]
29. Reichstein, M.; Falge, E.; Baldocchi, D.; Papale, D.; Aubinet, M.; Berbigier, P.; Bernhofer, C.; Buchmann, N.; Gilmanov, T.; Granier, A.; et al. On the separation of net ecosystem exchange into assimilation and ecosystem respiration: Review and improved algorithm. *Glob. Chang. Biol.* **2005**, *11*, 1424–1439. [CrossRef]
30. Lasslop, G.; Reichstein, M.; Papale, D.; Richardson, A.D.; Arneth, A.; Barr, A.; Stoy, P.; Wohlfahrt, G. Separation of net ecosystem exchange into assimilation and respiration using a light response curve approach: Critical issues and global evaluation. *Glob. Chang. Biol.* **2010**, *16*, 187–208. [CrossRef]
31. Monteith, J.L. Solar radiation and productivity in tropical ecosystems. *J. Appl. Ecol.* **1972**, *9*, 747–766. [CrossRef]
32. Running, S.W.; Zhao, M. User's Guide Daily GPP and Annual NPP (MOD17A2H/A3H) and Year-end Gap-Filled (MOD17A2HGF/A3HGF) Products NASA Earth Observing System MODIS Land Algorithm (For Collection 6). 2019. Available online: https://landweb.modaps.eosdis.nasa.gov/QA_WWW/forPage/user_guide/MOD17UsersGuide2019.pdf (accessed on 15 July 2021).
33. Liang, D.; Zuo, Y.; Huang, L.S.; Zhao, J.L.; Teng, L.; Yang, F. Evaluation of the consistency of modis land cover product (mcd12q1) based on chinese 30 m globeland30 datasets: A case study in Anhui Province, China. *ISPRS Int. J. Geo-Inf.* **2015**, *4*, 2519–2541. [CrossRef]
34. Yamazaki, D.; Ikeshima, D.; Tawatari, R.; Yamaguchi, T.; O'Loughlin, F.; Neal, J.C.; Sampson, C.C.; Kanae, S.; Bates, P.D. A high-accuracy map of global terrain elevations. *Geophys. Res. Lett.* **2017**, *44*, 5844–5853. [CrossRef]
35. McClean, F.; Dawson, R.; Kilsby, C. Implications of using global digital elevation models for flood risk analysis in cities. *Water Resour. Res.* **2020**, *56*, 56. [CrossRef]
36. Uuemaa, E.; Ahi, S.; Montibeller, B.; Muru, M.; Kmoch, A. Vertical accuracy of freely available global digital elevation models (ASTER, AW3D30, MERIT, TanDEM-X, SRTM, and NASADEM). *Remote Sens.* **2020**, *12*, 3482. [CrossRef]
37. Hao, S.R.; Jiang, L.M.; Shi, J.C.; Wang, G.X.; Liu, X.J. Assessment of modis-based fractional snow cover products over the Tibetan plateau. *IEEE J. Sel. Top. Appl. Earth Obs. Remote Sens.* **2019**, *12*, 533–548. [CrossRef]
38. Yang, J.T.; Jiang, L.M.; Menard, C.B.; Luojus, K.; Lemmetyinen, J.; Pulliainen, J. Evaluation of snow products over the Tibetan plateau. *Hydrol. Process.* **2015**, *29*, 3247–3260. [CrossRef]
39. Mir, R.A.; Jain, S.K.; Saraf, A.K.; Goswami, A. Accuracy assessment and trend analysis of modis-derived data on snow-covered areas in the Sutlej basin, Western Himalayas. *Int. J. Remote Sens.* **2015**, *36*, 3837–3858. [CrossRef]
40. Ueyama, M.; Iwata, H.; Harazono, Y.; Euskirchen, E.S.; Oechel, W.C.; Zona, D. Growing season and spatial variations of carbon fluxes of Arctic and Boreal ecosystems in Alaska (USA). *Ecol. Appl.* **2013**, *23*, 1798–1816. [CrossRef]
41. Wu, X.D.; Wen, J.G.; Xiao, Q.; You, D.Q.; Dou, B.; Lin, X.; Hueni, A. Accuracy assessment on modis (v006), glass and Musyq land-surface albedo products: A case study in the heihe river basin, china. *Remote Sens.* **2018**, *10*, 2045. [CrossRef]
42. Li, P.X.; Zhu, W.Q.; Xie, Z.Y. Diverse and divergent influences of phenology on herbaceous aboveground biomass across the tibetan plateau alpine grasslands. *Ecol. Indic.* **2021**, *121*, 107036. [CrossRef]
43. De Castro, H.C.; de Carvalho, O.A.; de Carvalho, O.L.F.; de Bem, P.P.; de Moura, R.D.; de Albuquerque, A.O.; Silva, C.R.; Ferreira, P.H.G.; Guimaraes, R.F.; Gomes, R.A.T. Rice crop detection using lstm, bi-lstm, and machine learning models from sentinel-1 time series. *Remote Sens.* **2020**, *12*, 2655.
44. Yue, S.; Wang, C.Y. The mann-kendall test modified by effective sample size to detect trend in serially correlated hydrological series. *Water Resour. Manag.* **2004**, *18*, 201–218. [CrossRef]
45. Gorelick, N.; Hancher, M.; Dixon, M.; Ilyushchenko, S.; Thau, D.; Moore, R. Google Earth engine: Planetary-scale geospatial analysis for everyone. *Remote Sens. Environ.* **2017**, *202*, 18–27. [CrossRef]
46. Lu, L.; Guo, H.; Kuenzer, C.; Klein, I.; Zhang, L.; Li, X. Analyzing phenological changes with remote sensing data in Central Asia. *IOP Conf. Ser. Earth Environ. Sci.* **2014**, *17*, 012005. [CrossRef]
47. Korner, C. The use of 'altitude' in ecological research. *Trends Ecol. Evol.* **2007**, *22*, 569–574. [CrossRef]
48. Kato, T.; Tang, Y.H. Spatial variability and major controlling factors of co2 sink strength in asian terrestrial ecosystems: Evidence from eddy covariance data. *Glob. Chang. Biol.* **2008**, *14*, 2333–2348. [CrossRef]
49. Yu, G.R.; Zhu, X.J.; Fu, Y.L.; He, H.L.; Wang, Q.F.; Wen, X.F.; Li, X.R.; Zhang, L.M.; Zhang, L.; Su, W.; et al. Spatial patterns and climate drivers of carbon fluxes in terrestrial ecosystems of China. *Glob. Chang. Biol.* **2013**, *19*, 798–810. [CrossRef]

remote sensing

MDPI

Article

Remote Sensing-Based Quantification of the Summer Maize Yield Gap Induced by Suboptimum Sowing Dates over North China Plain

Sha Zhang [1], Yun Bai [1,2,*] and Jiahua Zhang [1,2]

[1] Research Center for Remote Sensing Information and Digital Earth, College of Computer Science and Technology, Qingdao University, Qingdao 266071, China; zhangsha@radi.ac.cn (S.Z.); zhangjh@radi.ac.cn (J.Z.)
[2] Key Laboratory of Digital Earth Science, Aerospace Information Research Institute, Chinese Academy of Sciences, Beijing 100094, China
* Correspondence: baiyun@qdu.edu.cn

Abstract: Estimating yield potential (Yp) and quantifying the contribution of suboptimum field managements to the yield gap (Yg) of crops are important for improving crop yield effectively. However, achieving this goal on a regional scale remains difficult because of challenges in collecting field management information. In this study, we retrieved crop management information (i.e., emerging stage information and a surrogate of sowing date (SDT)) from a remote sensing (RS) vegetation index time series. Then, we developed a new approach to quantify maize Yp, total Yg, and the suboptimum SDT-induced Yg (Yg_0) using a process-based RS-driven crop yield model for maize (PRYM–Maize), which was developed in our previous study. PRYM–Maize and the newly developed method were used over the North China Plain (NCP) to estimate Ya, Yp, Yg, and Yg_0 of summer maize. Results showed that PRYM–Maize outputs reasonable estimates for maize yield over the NCP, with correlations and root mean standard deviation of 0.49 ± 0.24 and 0.88 ± 0.14 t hm^{-2}, respectively, for modeled annual maize yields versus the reference value for each year over the period 2010 to 2015 on a city level. Yp estimated using our new method can reasonably capture the spatial variations in site-level estimates from crop growth models in previous literature. The mean annual regional Yp of 2010–2015 was estimated to be 11.99 t hm^{-2}, and a Yg value of 5.4 t hm^{-2} was found between Yp and Ya on a regional scale. An estimated 29–42% of regional Yg in each year (2010–2015) was induced by suboptimum SDT. Results also show that not all Yg_0 was persistent over time. Future studies using high spatial-resolution RS images to disaggregate Yg_0 into persistent and non-persistent components on a small scale are required to increase maize yield over the NCP.

Keywords: remote sensing; crop sowing date; development stage; yield gap; yield potential; process-based model

Citation: Zhang, S.; Bai, Y.; Zhang, J. Remote Sensing-Based Quantification of the Summer Maize Yield Gap Induced by Suboptimum Sowing Dates over North China Plain. *Remote Sens.* **2021**, *13*, 3582. https://doi.org/10.3390/rs13183582

Academic Editors: David M. Johnson and Jose Moreno

Received: 7 June 2021
Accepted: 5 September 2021
Published: 8 September 2021

1. Introduction

Yield potential (Yp) is the upper limit of the yield of a specific crop type within a given domain and is limited by only the local heat and light resources [1]. Narrowing the gap (yield gap, Yg) between Yp and on-farm yield (Ya) is critical for increasing food production. Yg is caused by multiple factors, but not all could be controlled [2,3]. Factors contributing to Yg were categorized into either persistent (field managements, terrain, and soil quality) or non-persistent factors (adverse climate, insect attack, and other non-management factors) by Lobell, et al. [4]. The persistent factor, field management, could be controlled in the field and has made considerable contributions to Yp, as presented in previous studies [2,5,6]. Therefore, the knowledge of Yp and the contributions of persistent factors to Yg can provide useful information for improving crop yield.

Sowing date (SDT) is one of the important management factors that affect crop yield [7–9]. A few economic inputs are required to optimize SDT on a farm. Several methods may be available for exploring Yp and quantifying the contribution of this factor to Yg. These methods could be generally categorized into two types: model simulation and field experiments. The latter method provides an estimate of Yp and other levels of yield under controlled experimental conditions, thus it can quantify the effect of SDT on Yg [10]. However, for analysis over a wide spatial or temporal range, such a method is time-costing and uneconomical. In this case, the field experiment generally served as a means of acquiring data for calibrating and validating crop growth models (CGMs) [11–13]. Model simulation is favored for its low cost and high efficiency [1,14]. CGMs such as the agricultural production system simulator (APSIM) [15], crop environment resource synthesis (CERES) models [16], and CSM-IXIM [17], after being calibrated, can produce reliable estimates of crop Yp and Ya and can also simulate crop yield under controlled conditions [11,18,19]. The knowledge of the contribution of management factors, including SDT, to Yg could be revealed by comparing simulated values of crop yield under different management scenarios [6,20]. However, quantifying the contribution of SDT to Yp is not feasible over a broad region using CGMs owing to the inaccessibility of spatiotemporally continuous field management information and the sparse spatial distribution of accessible meteorological sites at present. Simulations and analyses with CGMs in existing literature were merely performed on meteorological-site levels or within a small region and provided limited information for understanding Yp and Yg in space [6,20].

The use of remote sensing (RS)-based methods may be an alternative way to address the above issues, as the use of RS data can make a model less dependent on meteorological data and management information and show better performances in regional simulations [21,22]. Unlike CGMs, the key factor, leaf area index (LAI), for simulating photosynthesis rate RS-based models is remotely sensed rather than simulated by the model [23], which makes RS-based models more applicable to regional applications. An RS-based land process model has long been used to simulate ecosystem productivity and is also useful for simulating crop yield [23,24]. SDT, which is also an essential input of RS-based models to map crop yield, can be obtained by analyzing the RS vegetation indices (VI) time series [25–27]. Hence, we can use RS data to study the contribution of suboptimum SDT to Yg over a broad region. Vegetation parameters retrieved from RS data reflect the actual growing condition of crops and seem to be useful for simulating only Ya. However, spatiotemporal variations in pixel-level Ya predicted using an RS method are potentially useful for quantifying Yp [28] and understanding the contribution of different factors to Yg [2,29]. Assuming that potential yield is realized on a local scale, pixel-level Yp could be computed as the high percentile (95th or 99th) of yield distribution of surrounding pixels [28,30]. To avoid this assumption, Lobell [28] suggested using a hybrid method that estimates the real Yp by fitting the Yp derived from the above RS-based method to the estimate of a calibrated CGM. However, the hybrid approach may in turn be restricted by CGMs' high input data requirement over a broad region. In addition, such a method may fail when Yg and factors contributing to Yg varied significantly and irregularly in space. Therefore, an RS-based method to quantify regional-scale Yp, Yg, and the effect of SDT on crop yield, without the need for CGMs, is needed.

Multiple types of satellite data are available for modeling crop yield and Ygs. High-spatial-resolution data, such as Landsat and SPOT, that have relatively long revisit times have been generally used to develop empirical relationships between crop yield and spectral indices at a specific developmental stage of the crop [29,31–33]. These empirical methods are limited to quantify Yp and Yg for regions with lower levels of field management [28]. Besides, these methods do not explicitly consider the effect of SDT on crop yield, being not able to quantify the SDT to Yg. The two Sentinel-2 satellites (Sentinel-2A and B) can provide 10 to 20 m resolution multispectral data with a revisit period of 4–5 days and are useful for driving a process-based model to estimate crop yield, which requires time-continuous inputs [34]. However, Sentinel-2B data were only available from 2018 when the satellite was launched; hence, using process-based methods with Sentinel-2 data to reproduce Yg and Yp from earlier years is impossible. Although the widely used

MODIS data have a coarser spatial resolution than Landsat, SPOT, and Sentinel-2, it can provide time-continuous images with more available pixels in time and are available from 2000 to present. The data have also been used to estimate crop yield with process-based models [23,35].

In this study, we aim to develop a new RS approach that uses MODIS data to quantify Yp and the contribution of suboptimum SDT to Yg of summer maize over the North China Plain (NCP), with three main objectives: (1) To develop a novel RS-based method driven by MODIS data to simulate Yp and Yg at a large spatial scale; (2) to assess the reliability of the developed RS-based method in simulating Ya and Yp over the NCP; and (3) to quantify Yg and the contribution of suboptimum SDT to the Yg of summer maize over the NCP in the period 2010 to 2015.

2. Materials and Methods

2.1. Study Region

We quantified the Ya, Yp, Yg, and the contribution of suboptimum SDT to Yg of summer maize over the NCP. The study region in this study spans five provincial-level administrative divisions of China (Beijing, Tianjin, Hebei, Shandong, and Henan) and covers 42 prefecture and two provincial-level cities of China (110.98° E–122.71° E, 32.27° N–41.06° N). In the period 2010 to 2015, the accumulated effective temperature (or growing degree days, GDD, with a base temperature of 8 °C) during the maize growing season (May to September) and annual precipitation shows a significant spatial gradient with mean values of 1950 °C d year^{-1} and 720 mm year^{-1}, respectively, for the entire region. The NCP is one of the major cultivation areas for maize in China, and it is approximated to produce 30% of the total maize production of the country [36]. We determined the study region based on administrative boundaries (Figure 1). The region does not correspond to the exact domain of the plain, but it does cover the major maize cultivation areas.

2.2. Data

2.2.1. Remotely Sensed Crop Information

The remotely sensed crop information we used in this study includes maize distribution maps and vegetation indices (VIs) data. We used annual maize cultivation areas over the study region in the period 2010 to 2015, which were estimated by Xun, et al. [37] and have users' accuracies greater than 80%. The remotely sensed VIs were required in the RS crop model to simulate the actual yield. We used the Moderate Resolution Imaging Spectroradiometer (MODIS) normalized difference vegetation index (NDVI), enhanced vegetation index (EVI), and LAI in this study. NDVI and EVI data with 1 km and 16-day resolution, retrieved from MOD13A2 and MYD13A2 products (available through https://earthdata.nasa.gov/, accessed in 3 May 2020), were used. As the study area is large enough and the typical maize planted region is usually larger than 1 km^2, the used images with 1 km resolution could capture the main spatial characteristics of maize in the study area. The retrieved data were preprocessed in terms of the following procedures before use to remove unreliable data and noise:

- Quality control: Pixel values with snow or cloud cover in the retrieved data were replaced by linearly interpolating the nearest available pixels in time.
- Time-series filter: TIMESAT3.3 was used to filter the retrieved VI data, with the required parameters were set, as shown in Section 2.3.5.
- Temporal interpolation: Eight-day VI data were linearly interpolated to daily data.

Figure 1. The study region (North China Plain) and reference region (Rongcheng county and Dingxing county). The reference region is a county, where most of the croplands were cultivated with summer maize, and we obtained a reference LAI time-series curve by averaging the LAI time series of all maize pixels retrieved from MODIS products in the reference region (Section 2.3.4). The reference LAI time series was an important factor for computing the Yp of maize (Section 2.3.4).

The LAI of maize was not directly retrieved from the MODIS product as crop LAI was reported to be significantly underestimated by the MODIS product. In this study, maize LAI was calculated using empirical equations, calibrated by Bai, et al. [38], in terms of EVI. This method was calibrated with samples collected from the US, Europe, and China. Thus,

$$LAI = \begin{cases} 24.805EVI^2 - 15.444EVI + 2.382 & \text{before EVI peaking} \\ 9.249EVI - 1.236 & \text{after EVI peaking} \end{cases} \quad (1)$$

where LAI was calculated in two ways during one growing season, and we used a quadratic equation before EVI peaking and a linear equation after it.

All the computations involving gridded MODIS data were carried out using the GDAL (the Geospatial Data Abstraction Library) package under the Python2.7 environment.

2.2.2. Meteorological Data

Gridded daily meteorological data retrieved from the ERA-Interim reanalysis (ERA) dataset and multi-site data of the China Meteorological Administration (CMA) were required to drive the RS crop yield model. Surface net radiation (Rn), vapor pressure deficit (VPD), and air temperature (T) were directly retrieved from the ERA-Interim dataset (https://www.ecmwf.int/en/forecasts/datasets/reanalysis-datasets/era-interim, accessed in May and June, 2020). Precipitation (Pr) and global solar radiation (Rg) were retrieved from CMA multi-site data (http://data.cma.cn/, accessed in June 2020). Site-scale CMA data were spatially interpolated to a raster dataset using inverse distance weighted method provided by ArcGIS software (v10.1), and key parameters required were set as follows:

- Output cell size = 5 km;
- Power = 2;
- Search radius = variable; and
- Search radius settings: Number of points = 12; maximum distances = 100 km.

Rather than directly retrieving from the site-scale observations, daily solar radiation (Rg) data were calculated in terms of the interpolated site-observed daily air temperature range (T_R) and daily sun hours (Hr_S) because sites observing Rg is too sparse to be interpolated. We referred to Chen, et al. [39] to calculate daily Rg, such that

$$Rg/R_0 = a \times \ln(T_R) + b \times \left(Hr_S/Hr_{day}\right)^c + d \tag{2}$$

where R_0 denotes the extra-terrestrial solar radiation; Hr_{day} denotes the number of daytime hours; and a, b, c, and d are empirical coefficients. The average of values for each coefficient across multiple sites over China was used, such that a = 0.04, b = 0.48, c = 0.83, and d = 0.11. ERA-Interim datasets provide gridded global meteorological variables from 1981 to present in multiple temporal and spatial resolutions. The temporal and spatial resolutions of gridded data retrieved from ERA-Interim were 12 h and 0.125 arc-degree. The daily value of each variable is the sum (for Pr) or average (for variables except for Pr) of two 12-h values in one day.

2.2.3. Reference Maize Yield

Prefecture-level statistics of maize yield in the period 2010 to 2015 reported by the National Bureau of Statistics of China were used to validate Ya simulated by a process-based and RS-driven crop yield model for maize (PRYM–Maize). These data were retrieved from the statistical yearbook of Shandong Province, Hebei Province, Henan Province, Beijing, and Tianjin for 2010–2015. When validating the simulations against statistical values, simulated pixel-level yields were averaged by prefecture-level districts (Figure 1).

2.3. *Methods*

2.3.1. Quantifying Yg and the Contribution of SDT to Yg

An RS process-based crop yield model (Section 2.3.2 driven by MODIS data was used to simulate Ya, limiting the potential yield by only SDT (Yp_0) over the NCP. Yp was computed on the basis of the modeled Yp_0 (see Section 2.3.4. Two Ygs and the contribution of suboptimum SDT to Yg were calculated as follows:

$$\mathrm{Yg} = \mathrm{Yp} - \mathrm{Ya} \tag{3}$$

$$\mathrm{Yg}_0 = \mathrm{Yp} - \mathrm{Yp}_0 \tag{4}$$

$$\mathrm{CYg}_0 = \mathrm{Yg}_0/\mathrm{Yg} \tag{5}$$

where Yg or Yg_0 is the gap between Yp and Ya or potential yield limited by suboptimum SDT, and CYg_0 is the contribution of Yg_0 to Yg.

When mapping the mean annual Yg and CYg0 based on MODIS 1 km resolution Ya and Yg_0, the two 1 km data were aggregated to 5 km-resolution rasters. Most 1 km pixels over the NCP were not continuously cropped with maize in time (Supplementary Figure S2), whereas most 5 km grids have more than one 1 km pixels continuously cropped with maize in 2010–2015. A 5 km grid represents all maize fields located on the grid. However, we noted that the aggregation may reduce the effects of random factors (uncertainties in RS data or meteorological inputs) on yield, as discussed in Section 4.1.

2.3.2. A Brief Overview of PRYM–Maize

A process-based RS crop yield model, PRYM–Maize, which was developed in our previous study and running on a daily step, was used to simulate maize yield in conjunction with MODIS data in this study [35]. PRYM–Maize consists of two basic modules: water balance and productivity modules. The water balance module (WBM) follows the RS and water balance-based Penman-Monteith model version 2 (RS-WBPM2) [38], which was modified from Bai, et al. [40], to simulate water stress based on MODIS VIs and meteorological data. The productivity module consists of multiple sub-processes, photosynthesis, grain conversion, and respirations. The ecosystem-level photosynthesis rate was scaled from the leaf-level value based on a two-leaf canopy model and the MODIS-based LAI (Equation (1))). A full description of the model can be found in Zhang, Bai, Zhang and Shahzad [35] or Supplementary Text S1.

2.3.3. Modifications to PRYM–Maize

In PRYM–Maize, we simulated the growth of maize grain on a daily step as a function of development stage (DVS) and daily net primary productivity (NPP_{daily}), as presented in Supplementary Text S1.3. The DVS index is required to compute the proportions of photosynthesis-produced dry matters that were allocated to different organs, namely, root, stem, leaf, and grain. In the previous version of PRYM–Maize, the DVS is calculated as a function of GDDs, as also presented in other CGMs [41–43]. However, GDDs are vulnerable to uncertainties in input air temperature data. Alternatively, GPP was potentially useful to indicate the crop DVS because it is more directly related to crop growth than are GDDs [24]. In addition, RS information represented by GPP may partly eliminate the error in input air temperature data. Huang, Ryu, Jiang, Kimm, Kim, Kang and Shim [24] proposed using normalized accumulated GPP (GPP_{nor_acc}), estimated using an RS-based model, to indicate the DVS of crop and found that the two factors were tightly correlated. This method was successfully used in simulating rice grain-filling across three flux sites over the Korean Peninsula [24]. In this study, we do not use GDDs but adopt GPP_{nor_acc} to compute the DVS. GPP_{nor_acc} was designed to vary from 0 to 1 in Huang, Ryu, Jiang, Kimm, Kim, Kang and Shim [24] as crops grow accompanied by accumulating GPP, and GPP_{nor} = 0 and 1 correspond to emerging and maturing stages of crops. Such that

$$GPP^{(t_0)}_{nor\_acc} = \frac{\sum_{t=EM}^{t_0} GPP^{(t)}}{GPP_{acc}} \quad (6)$$

$$GPP_{acc} = \sum_{t=EM}^{HV} GPP^{(t)} \quad (7)$$

where GPP_{acc} denotes the accumulated GPP along the entire growing season; $GPP^{(t_0)}_{nor\_acc}$ denotes normalized accumulated GPP for day t_0 and is equivalent to DVS; and EM and HV denote the emerging and harvest date, respectively. Pixel-level EM and HV were retrieved from the RS NDVI time series (Section 2.3.5).

DVS varies from 0 to 2, indicating the development of crops from emerging to maturing stages; therefore, we estimated DVS by scaling GPP_{nor_acc} using 2, as follows:

$$DVS^{(t_0)} = 2 \times GPP^{(t_0)}_{nor\_acc} \quad (8)$$

2.3.4. Simulating Potential LAI and Yp

Temporal variations in LAI of crops fully depend on simulations in CGMs; thus, it is feasible to simulate Yp using a calibrated crop model by removing all field management limits for crop growth in the model. However, not all management factors in an RS-based yield model can be adjusted to the optimum because RS-LAI, as a critical input of the RS-based model, does include environmental stress on crop growth. However, it is still possible to use an RS-based model to simulate Yp if the stress on crop growth represented by actual LAI is removed. Here, we define the LAI time series, representing no environmental stress on crops and leading to potential yielding, as potential LAI (LAI_p).

RS-LAI derived from the MODIS VI (Equation (1)) may fail in representing the magnitude of LAI_p but could capture the variation trend of LAI_p in time because the DVS of crops, including maize, primarily depends on the effective accumulative temperature rather than field managements, as represented by existing CGMs [41,42,44]. Figure 2 is a scatter plot, with 45 data pairs, for maize yield vs. peak LAI observed during the growing season (PLAI). These data were collected from 21 published and 8 unpublished super-high yield experiments over China. These experiments attempted to optimize all field managements, and the growth of maize was primarily limited by air temperature and solar radiation. Under optimum field managements, maize yield positively correlates with PLAI. When PLAI is greater than 6.6, maize yield tends to saturate at a high level, ~20 t hm^{-2}, and the correlations between PLAI and maize yield become insignificant (Figure 2). This implies that a PLAI greater than 6.6 may be adequate for maize yielding.

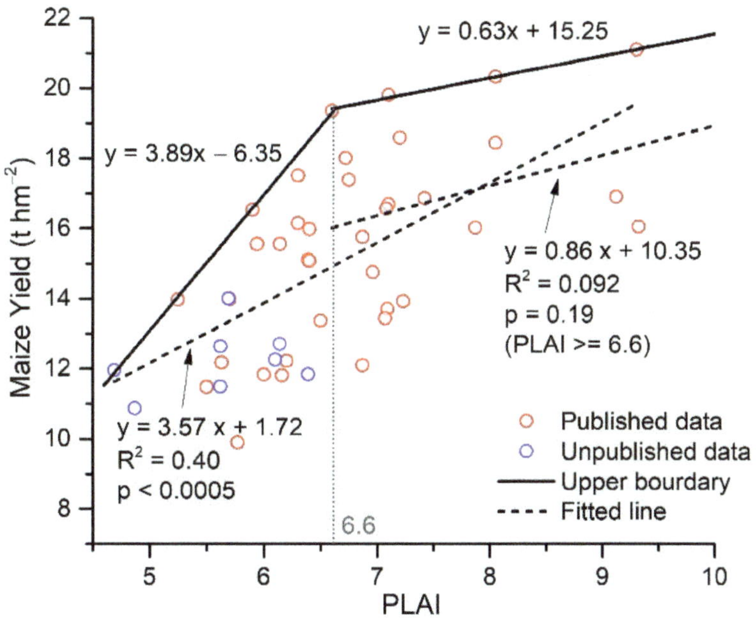

Figure 2. Maize yield against growing season maximum leaf area index (LAI_{max}) over China. Red circles denote data retrieved from 21 published [45–65] and 8 unpublished trials. For one trial involving multiple treatments or maize spices, the data pair from the highest yield treatment of each species was picked up. Blue circles denote recent field trails that have not been published yet.

Based on the above analyses, we proposed using the following steps to calculate LAI_p and Yp:

1. Derive a reference curve ($C_{RF} = \left\{LAI_{RF}^{(0)}, LAI_{RF}^{(2)}, \ldots, LAI_{RF}^{(J_{HV})}\right\}$) from the LAI time series within the reference region (Figure 1), where J_{HV} denotes the number of days from EM to HV, LAI_{RF} denotes a reference LAI value of a given day. C_{RF} has a PLAI greater than 6.6, representing the roughly temporal variation in LAI_p. We obtain C_{RF} using two procedures: (i) Average the LAI time series of all maize pixels retrieved

from MODIS VI-based LAI (Equation (1)) by date in 2010 within the reference region (Figure 2) to obtain an LAI curve ($C_0 = \left\{LAI_0^{(0)}, LAI_0^{(2)}, \ldots, LAI_0^{(J_{HV})}\right\}$) for maize; (ii) calculate C_{RF} from C_0 as $C_{RF} = PLAI_{opt} \times \frac{C_0 - \min(C_0)}{\max(C_0) - \min(C_0)}$, where $PLAI_{opt}$ denotes the optimum PLAI for maize potentially yielding and is expected to be greater than 6.6. In this study, we made a conservative estimate for $PLAI_{opt}$ and set it to 7.5, and this estimate was supported by Liu, Hou, Xie, Ming, Wang, Xu, Liu, Yang and Li [50], who reported PLAI = 7.53 of the maize plants, achieving the highest maize yield (22.5 t hm^{-2}) record in China.

2. Derive LAI_p for each pixel by fitting the time series of RS-LAI ($C_{RS} = \left\{LAI_{RS}^{(0)}, LAI_{RS}^{(2)}, \ldots, LAI_{RS}^{(J'_{HV})}\right\}$) of the pixel to C_{RF} using linear regression analysis, as shown in Figure 3, where J'_{HV} denotes the number of days from EM to HV for a given pixel. In other words, we fitted $C_{RF} = k \times C_{RS} + b$ for each pixel to obtain k and b and then calculated LAI_p as $k \times C_{RS} + b$. If $J'_{HV} \neq J_{HV}$, C_{RF} was linearly stretched in the timeline to match the time span of C_{RS}.
3. Run the RS process-based crop yield model to simulate Yp_0 with input LAI substituted by LAI_p, $f_N(N) = 1$, and $g_{sm,2000} = 0.017$ m s^{-1}.
4. We referred to the method of Lobell [28], which assumed optimal field management (SDT in this study) can be achieved within a given domain with similar climate conditions, to compute Yp at each pixel as the 95th of Yp_0 values of surrounding pixels. These surrounding pixels were selected using two criteria: (i) Within a 50 km buffer around the central pixel and (ii) differences in accumulated GDDs and total solar radiation during the growing season (June–September) between the surroundings and the center pixel were less than 50 °C d and 50 MJ, respectively. The buffer size here referred to the size of a zone represented by a meteorological site in Wart, et al. [66].

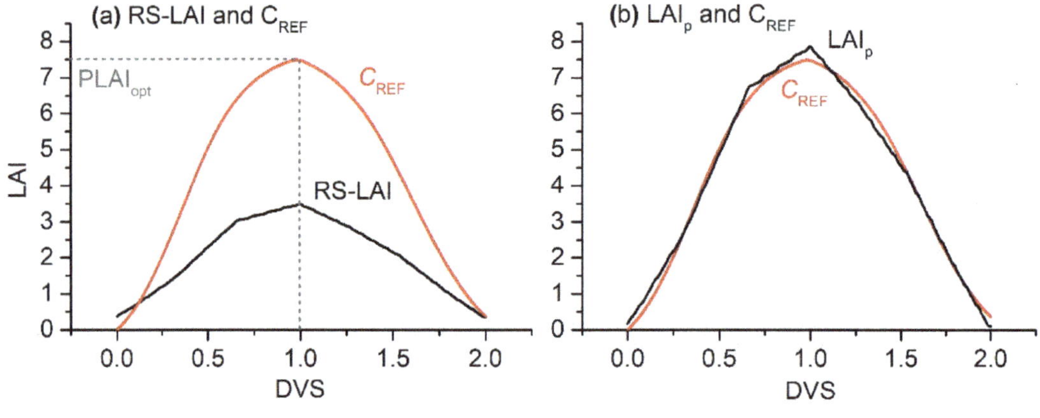

Figure 3. An example of deriving LAI_p from RS-LAI and C_{REF}. LAI_p on the right panel (**b**) was derived by fitting the RS-LAI on the left panel (**a**) to C_{REF} using the least-square method.

2.3.5. Extracting Crop Phenology Using RS Data

Emerging (EM) and harvest (HV) dates of each pixel in each year were retrieved from the 8-day and 1 km time-series NDVI data using TIMESAT3.3. The Savitzky–Golay filter was used to denoise the NDVI time series. The EM and HV dates were estimated on the basis of the denoised NDVI time series, and they corresponded to the start and end of the season (SOS and EOS, respectively) defined in TIMESAT. The key parameters required were set as follows:

- Window size = 64 days,
- Start of season method = relative amplitude, and
- Season start/end value = 0.1/0.2,

where "window size" means the total nearest days that are used to denoise the current data; "relative amplitude" indicates that the SOS or EOS was estimated as the time when NDVI increases or decreases to a given proportion, as specified by the season start or end value, of the relative amplitude of NDVI time series during a specific growing season.

3. Results

3.1. Modeled Ya

Pixel-level Ya in 2010–2015 over the NCP was simulated using the modified PRYM–Maize. Results are shown in Figure 4. In each year, modeled prefecture-level yields had a root mean standard deviation (RMSD) value of 0.88 $\pm$ 0.14 t hm^{-2} and an R value of 0.49 $\pm$ 0.24 with the reference value. A pooled analysis for modeled yield vs. reference yield in 2010–2015 shows modeled yield by PRYM–Maize has an R (RMSD) of 0.45 (0.87 t hm^{-2}). The performance of PRYM–Maize is improved when modeled and reference prefecture-level yields are averaged in time (Figure 5b). Uncertainties in modeled yield are significantly reduced, and the RMSD of mean annual modeled yield vs. the reference value is 0.66 t hm^{-2}, which is less than that in any one year (Figure 4).

The average of R (RMSD) values for modeled yield vs. reference value in each year is 0.49 (0.86 t hm^{-2}). The model achieves the best simulations in 2011 with an R value of 0.66. Generally, R values achieved by PRYM–Maize in this study are not high, and one important reason for this is that the prefecture-level maize yield values in 2010–2015 are in a narrow range (4–9 t hm^{-2}) and show relatively small dynamics over space, as represented by both reference (standard deviation (STD) = 0.88 t hm^{-2} for reference yield) and modeled yield (Figures 4 and 5a). Therefore, these results demonstrate that PRYM–Maize outputs reasonable estimates for the Ya of summer maize over the NCP.

3.2. Modeled Yp

The Yp of summer maize over the NCP was estimated using the RS-based method presented in Section 2.3.4. Yp estimated for each year in the period 2010 to 2015 was averaged over time (Figure 6a). Results show that Yp estimated in this study has a tight correlation (R = 0.81, RMSD = 0.87 t hm^{-2}) with that estimated using a calibrated APSIM-Maize model at 10 agricultural meteorological (AM) sites from existing studies (Figure 6c). The result demonstrates that the performance of the developed approach in estimating Yp is comparable to that of a calibrated CGM. However, Yp values from the two methods were not the same (Figure 6c). Regardless of the differences in formulations between the two methods, Yp in this study represented the period 2010 to 2015, which is different from the data we used for comparisons in Figure 6.

Our results show that the mean annual Yp in the period 2010 to 2015 ranged between 9 and 16 t hm^{-2} over the study region with a regional value of 11.99 t hm^{-2}. Mean annual Yp over the study region generally increased from southwest to northeast. The northeast of Shandong Province had the highest Yp, whereas the lowest Yp appeared in the southeast of Henan province.

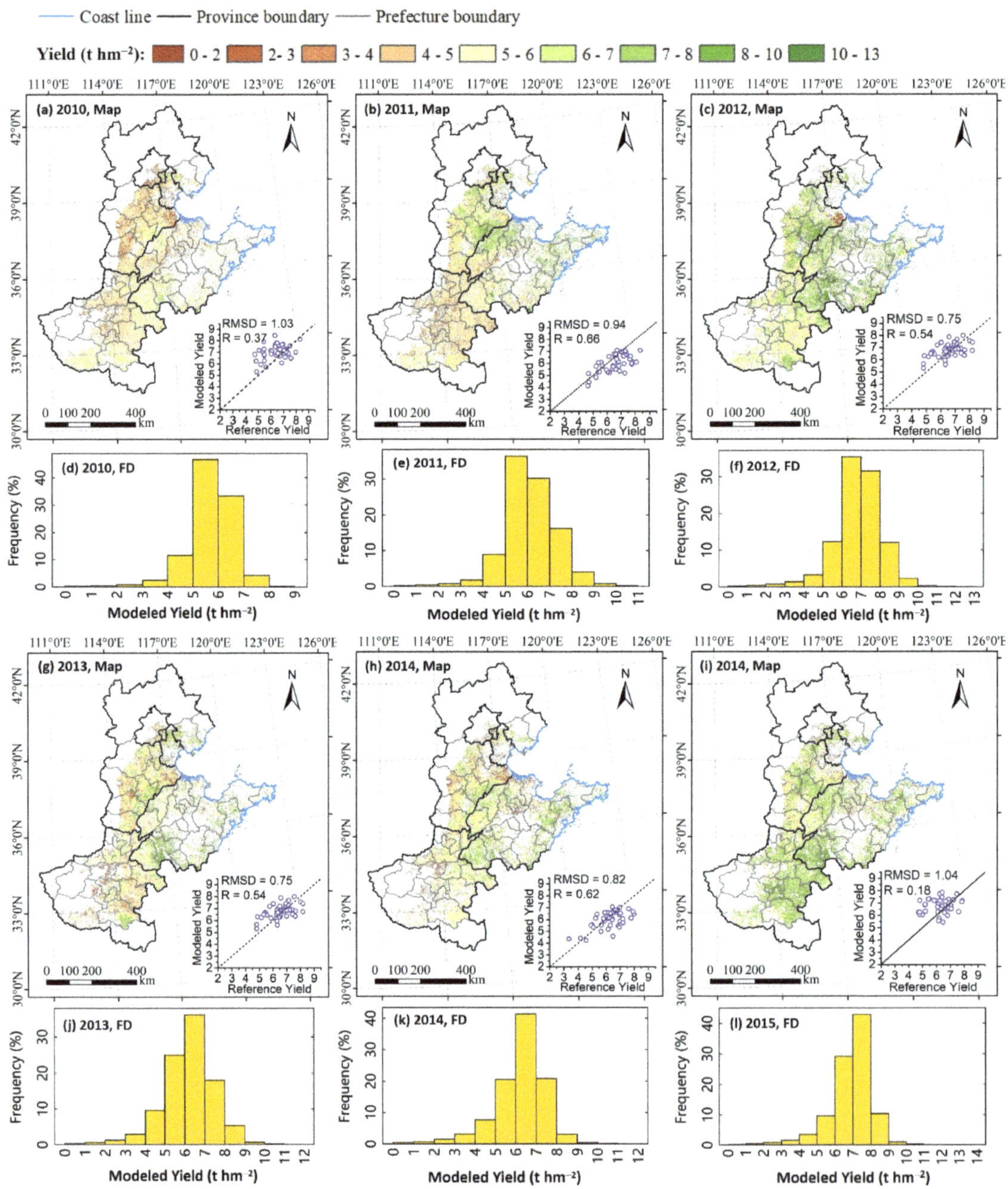

Figure 4. Maps of modeled Ya of summer maize in 2010 (**a**), 2011 (**b**), 2012 (**c**), 2013 (**g**), 2014 (**h**), and 2015 (**i**); and the pixel-level frequency distribution (FD) of pixel level values in 2010 (**d**), 2011 (**e**), 2012 (**f**), 2013 (**j**), 2014 (**k**), and 2015 (**l**). The scatter plot represents a comparison between modeled yield and reference yield on a prefecture-level, the modeled yield on a prefecture-level is the average of yield values of all pixels within a prefecture-level district, and each scatter plot has 44 samples; the solid line in each scatter plot represents the "1:1 line." The value ranges in the legend are right-closed and left-open.

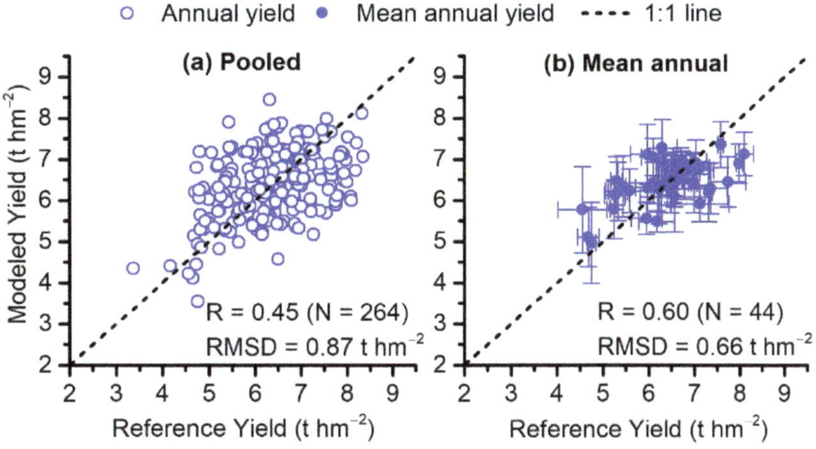

Figure 5. Modeled maize yields vs. reference values on a prefecture-level for pooled data (**a**) and mean annual data (**b**) in the period 2010 to 2015; R, N, and RMSD denote correlation, sample size, and root mean standard deviation, respectively; the error bar represents the standard deviation of multi-year data.

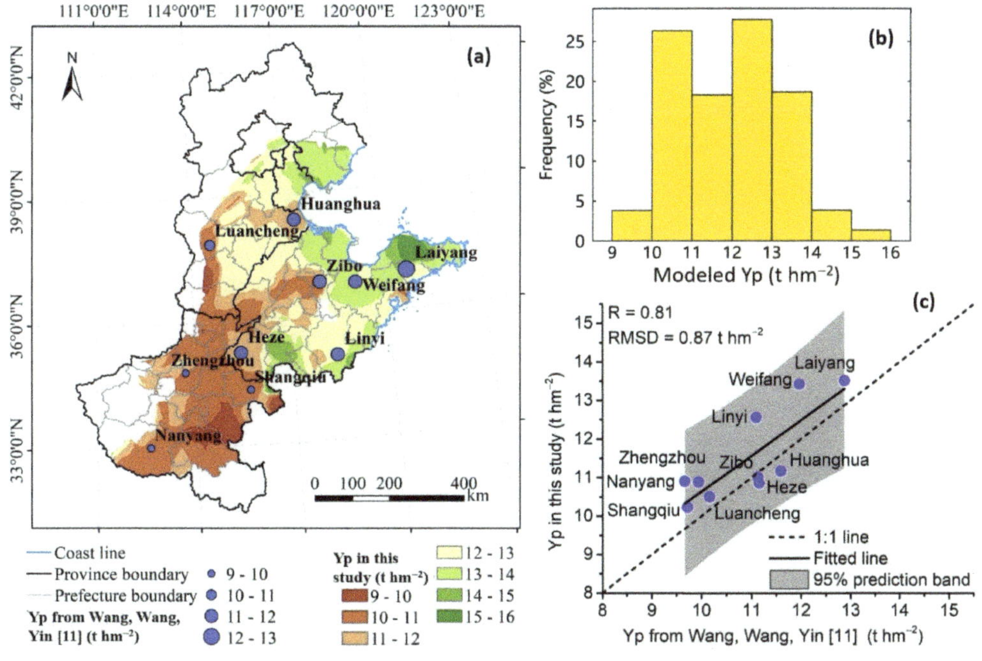

Figure 6. The map (**a**) and frequency distribution (**b**) of mean annual modeled Yp of summer maize over the period 2010 to 2015, and the modeled Yp of 10 agricultural meteorological (AM) sites vs. the simulations of corresponding sites from Wang, Wang, Yin, Feng and Zhang [11] (**c**). The value ranges in the legend of panel (**a**) are right-closed and left-open. Yp data from Wang, Wang, Yin, Feng and Zhang [11] represent the average of annual Yp in the period 1982 to 2005 for Linyi, 1982 to 2008 for Zibo and Laiyang, and 1982 to 2009 for the remaining sites.

3.3. *Yg and the Contribution of Suboptimum SDT to Yg*

Modeled regional Ya and potential yield limited by suboptimal SDT (Yp_0) and Yp in the period 2010 to 2015 are presented in Table 1 and Figure 7a. The ratio of Yg to Yp (Yg/Yp), the ratio of Ya to Yp (Ya/Yp), Yg caused by suboptimal SDT (Yg_0), and the contribution of suboptimal SDT to Yg (CYg_0) computed on the basis of Ya, Yp_0, and Yp are also presented in Table 1 and Figure 7b. The spatial variations in annual mean Ya, Yg, Yg/Yp, Yp_0, Yg_0, and CYg_0 are shown in Figure 8. Results show that large gaps (Ygs) remained between Ya and Yp on a regional scale or at a specific location over the NCP. Most areas of the study region had Yg values greater than 3 t hm^{-2}, and high Yp values were mainly distributed in the north and northeast (Figure 8b). Yg of approximately 99% of the study areas accounted for more than 30% of local Yp (Figure 8c). Annual regional Yg of summer maize in NCP ranged in 4.9–6.4 t hm^{-2} with a mean value of 5.4 t hm^{-2}, accounting for approximately 45% of the mean annual regional Yp (Table 1 and Figure 8). As shown in Table 1 and Figure 7, considerable proportions of Yg were induced by suboptimum SDT. An estimated 80% of the study areas, Yg_0 ranged from 1 to 4 t hm^{-2} (Figure 8e). Yg_0 also contributed to more than 20% of Yg in ~85% of the study areas. Regional Yg_0 in each year ranged from 1.4 to 2.2 t hm^{-2}, accounting for 29–42% of the annual regional Yg. The annual average of regional Yg_0 contributes to 35% of annual averaged regional Yg in 2010–2015 (Table 1). The analyses above demonstrate that large Yg remained in summer maize croplands over the NCP, and the values of Yg varied in space and could be considerably reduced by optimizing the SDT.

Table 1. Ya, potential yield limited by suboptimal SDT (Yp_0), Yp, Yg, ratio of Yg to Yp (Yg/Yp), ratio of Ya to Yp (Ya/Yp), Yg caused by suboptimal SDT (Yg_0), and contribution of suboptimal SDT to Yg (CYg_0) in the period 2010 to 2015 for the entire NCP [a].

Year	Ya	Yp	Yg	Yg/Yp	Ya/Yp	Yp_0	Yg_0	CYg_0
2010	5.8	10.8	4.9	45%	55%	9.4	1.4	29%
2011	6.1	11.4	5.3	46%	54%	9.2	2.2	42%
2012	7.0	12.7	5.7	45%	55%	10.7	2.0	35%
2013	6.4	11.3	4.9	43%	57%	9.6	1.7	35%
2014	6.3	11.2	4.8	43%	57%	9.4	1.8	38%
2015	7.1	13.4	6.4	48%	52%	11.3	2.1	33%
Average	6.5	11.8	5.4	45%	55%	9.9	1.9	35%

[a] Values of Ya, Yp, and Yp_0 were computed on the basis of the actual distribution of maize cultivation areas in each year, and Yg/Yp, Ya/Yp, and CYg_0 were computed using the regional statistics.

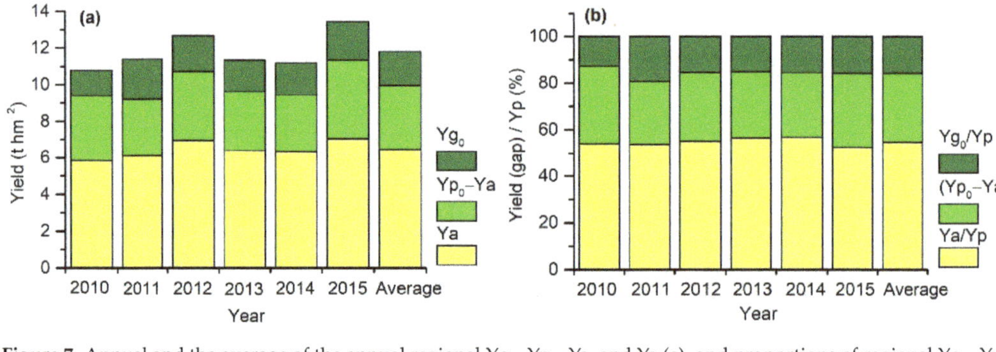

Figure 7. Annual and the average of the annual regional Yg_0, Yp_0−Ya, and Ya (**a**), and proportions of regional Yg_0, Yp_0−Ya, and Ya in regional Yp (**b**) in the period 2010 to 2015.

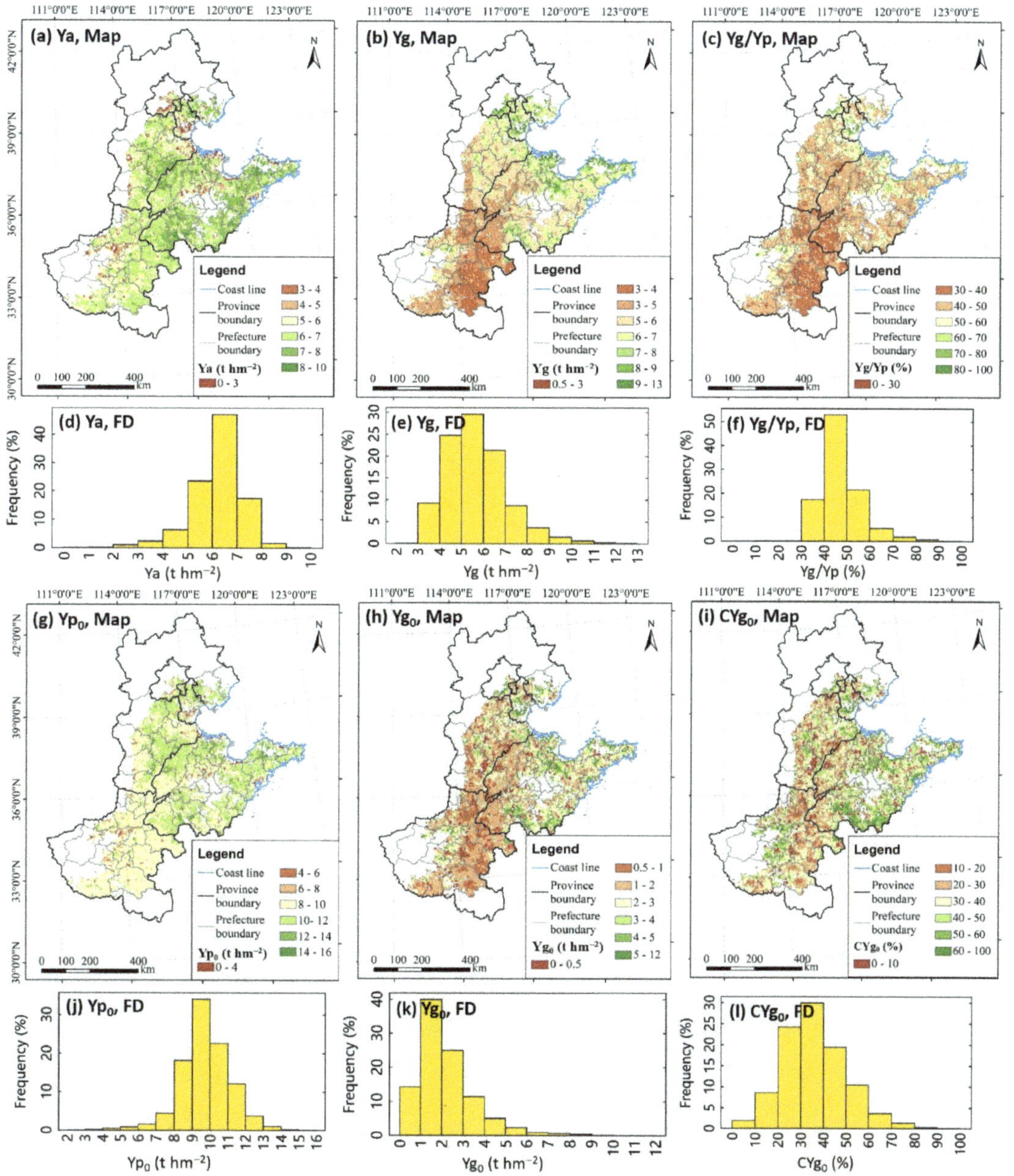

Figure 8. Maps of simulated mean annual Ya (**a**), Yg (**b**), the ratio of Yg to Yp (Yg/Yp) (**c**), potential yield limited by suboptimum SDT (Yp_0) (**g**), Yg induced by suboptimum SDT(Yg_0) (**h**), and the contribution of suboptimum SDT to Yg (CYg_0) (**i**), with a 5 km resolution in 2010–2015; and the frequency distribution (FD) of pixel-level values of Ya (**d**), Yg (**e**), Yg/Yp (**f**), Yp_0 (**j**), Yg_0 (**k**), and CYg_0 (**l**). The value ranges in the legends are right-closed and left-open. Annual 1 km Ya, Yp_0, and Yp were aggregated to 5 km and then averaged in time to derive the mean annual Ya, Yp_0, and Yp. Yg and Yg/Yp were derived from the 5 km Ya and Yp, and Yg_0 and CYg_0 were derived from the 5 km Yp_0, Yg, and Yp. Each 5 km grid represents all 1 km cropland pixels inside the grid.

4. Discussion

4.1. Yg_0 Is Partly Persistent

It is infeasible to perfectly optimize the summer maize SDT across the entire study region to fully eliminate Yg_0 because the optimum SDT depends on weather conditions during the growing season and the growing season of summer maize generally lasts ~100 days, but predicting weather conditions over such a long term (~100 days) precisely is not feasible at present. During the growing season, some unfavorable weather conditions (e.g., shifting of heat, radiation, and precipitation among different development stages) may cause SDT optimization to fail. Therefore, the contribution of suboptimum SDT, as a result of unfavored weather conditions, to Yg is non-persistent. To separate the persistent factors (primarily the knowledge and management skills of farmers) affecting SDT from the non-persistent will help understand the likelihood of reducing Yg by optimizing SDT.

Here, we adopted the method (Supplementary Text S2) of Farmaha, Lobell, Boone, Cassman, Yang and Grassini [2] to assess persistent Yg_0 based on both 1 km Yg_0 and 5 km Yg_0 time series (Supplementary Figure S3). The 1 km result was derived on the basis of the Yg_0 time series in croplands continuously cropped with maize (Supplementary Figure S2). We showed only the results of persistent factor percentage (PFP) in Yg based on "small Yg group (SYg)," PFP_{SYg}. Figure S3a,b show the percentage of persistent values in Yg_0 based on 1 km Yg_0 and 5 km Yg_0, respectively. The 1 km result covers a smaller spatial extent, and a pixel-by-pixel comparison between the two results was performed over the overlapped region (Figure S3c). The spatial variations in the PFP of the two results were moderately correlated with each other with a correlation coefficient (R) of 0.59. However, the 5 km result shows an overall overestimation of PFP. Regional values of PFP of the 1 and 5 km results over the overlapped regions are 59% and 69%, respectively. The reason for the higher PFP estimated by 5 km Yg_0 may be that some spatial dynamics in Yg_0 were eliminated in 5 km Yg_0. Nevertheless, both panels (Figure S3a,b) indicate non-negligible non-persistent components in Yg_0 and significant variations in persistent Yg_0 over space. Smaller percentages of persistent Yg_0 are found in the south of the NCP. Figure S3 presents two examples of assessing persistent Yg_0, implying that further studies are required to reveal the impact of climates on SDT and that assessing persistent Yg_0 within smaller regions using high-resolution RS data is necessary to understand the likelihood of narrowing Yg_0 on a local scale.

4.2. Yp Derived from Ya and Yp_0

Ya derived from remotely sensed data is also useful for quantifying Yp and Yg [1,28]. Pixel-level Yp could be estimated as a high percentile (95th or 99th) of pixel-level Ya values within a small region around the pixel under investigation. However, as previously mentioned, this method assumes optimum field managements are achieved in some croplands (or pixels) within the domain under investigation. A novel approach avoiding the assumption in estimating Yp based on satellite data was proposed in this study, and the new method only assumes optimum SDT is achieved in on-farm managements. The "Potential yield" derived from Ya is equivalent to "potential farmers' yield (Yp_f)," as defined in Liu, Yang, Lin, Hubbard, Lv and Wang [6]. In the maize belt of the US, where farmers' management skills were maintained at a high level, Yp_f was close to Yp; however, in other regions, where crop growth was strongly stressed, Yp was poorly represented by Yp_f [1]. We investigated the differences between Yp_f and Yp derived using the new method proposed in this study (Supplementary Figure S4). Modeled Yp_f values were significantly smaller than modeled Yp. The regional-scale mean annual Yp_f is 8.7 t hm^{-2}, which is significantly less than the Yp estimated in this study as well as previous studies [11,67,68], whereas our method produced a result closer to previous estimates. This implies that large gaps exist between farmers' potential yield and the Yp of summer maize over the NCP. The newly developed method in this study provides a more reliable approach to estimating Yp and can improve the understanding of the Yg of summer maize in the study region.

4.3. Limits of the Method in This Study

PRYM–Maize is proved to perform reasonably well in reproducing regional crop yield, having comparable or even better performances than models in recent studies in terms of RMSD [2,69]. PRYM–Maize was then used to develop a new method in this study to quantify Yp, Yp_0, Yg, and Yg_0. This new method produced a Yp magnitude similar to that produced by the calibrated CGM (Section 3.2), and the spatial pattern in Yp over the NCP simulated using our model was also closed to the results of Li [68], who reproduced regional Yp across the NCP using CGM simulations at multiple meteorological sites. This new method can also be used in other regions, where farmers' potential yields are far below the potential levels. However, one should be careful with the value of $PLAI_{opt}$. We used $PLAI_{opt} = 7.5$ in this study, and this value was a conservative estimate for $PLAI_{opt}$ and was obtained by analyzing historical field trials over China. However, the value of $PLAI_{opt}$ may be reduced in higher latitude or altitude regions, where low temperatures and heat dominate the growth of maize.

The accuracy of an RS-based approach to estimate crop yield highly depends on the input RS data. The accuracy of Ya estimates over the NCP in this study was degraded by mixed pixels. There was an overall underestimation of the Ya over Shandong Province, where many pixels were mixed with plastic greenhouses. The plastic greenhouse is widely used across Shandong, and it weakens the vegetation information [70], reducing the yield estimated from mixed pixels. Future studies are required to resolve such issues. Using higher-resolution images may be feasible, but the temporal resolution of such data such as Landsat, Sentinel-2, and SPOT may become a new limitation. Alternative approaches, such as pixel downscaling, can also be useful for addressing the above issue. For example, we can merge RS data with coarse spatial resolution and high temporal resolution (e.g., MODIS), high-spatial-resolution panchromatic product [71], or other bands [72] to obtain high-spatial- and high-temporal-resolution data to drive the RS-based crop yield model, thereby reducing the effect of non-vegetation information. In addition, the pixel change detection method [73] is available for further removing the bad pixels or outliers in yield or Ygs produced by the RS-based model. We should consider these approaches in our future work to improve the quantification of regional crop yield and Ygs.

The WBM is a critical part of PRYM–Maize for crop photosynthesis modeling in the context of climate change in the future. Extreme climate events (e.g., drought and heatwave) have great impacts on crop water status and thus crop yield [74,75]. Thus, with the elevating intensification of these extreme climate events [76,77], water availability estimated using the WBM of PRYM–Maize will play a more important role in quantifying crop yield. The WBM consists of evapotranspiration (loss of water of croplands) and soil water balance processes; hence, water availability can, in turn, affect the water balance process through its impact on evapotranspiration [38,40]. Evapotranspiration modeling in the current PRYM–Maize does not explicitly include the effect of extreme climate events. Therefore, in future work, the improvement of PRYM–Maize with regard to better characterizing the water status of crops during extreme climate events will be required.

5. Conclusions

The knowledge of how field managements contribute to Yg can help improve crop yield. In this study, we modified a process-based RS crop yield model for simulating Ya and proposed a new approach based on the modified model to quantifying Yp and the contribution of suboptimum SDT to Yg over a broad region. The above methods were used to estimate Ya, Yp, Yg, Yp_0, Yg_0, and CYg_0 of summer maize over the NCP in the period 2010 to 2015. We have the following conclusions:

1. The modified RS crop yield model reasonably estimated the Ya of summer maize over the NCP, but the model's accuracy was limited by input RS data.
2. Modeled Yp showed close relationships with site-level results given by CGMs in previous studies, which demonstrated that the proposed RS-based approach to estimate Yp was effective in modeling Yp over the NCP.

3. Large gaps, Ygs, remained between Ya and Yp of summer maize over the NCP and suboptimum SDT, which considerably contributed to Yg; regional Yg over the NCP in the period 2010 to 2015 was 5.0 t hm^{-2}, and the Yg, which accounted for suboptimum SDT (Yg_0), was approximately 41% of Yg. However, not all Yg_0 could be filled by optimizing SDT because Yg_0 was also affected by non-persistent factors. Thus, studies on small regions with higher-resolution RS data are required to decompose the persistent portion from Yg_0.

PRYM–Maize's robust performance under extreme weather conditions, such as drought and heatwaves, will need to be improved in the future. In addition, it is necessary to improve the performance of the RS-based method in estimating Yp within a specific region in conjunction with finer-resolution data or a pixel downscaling method.

Supplementary Materials: The following are available online at https://www.mdpi.com/2072-4292/13/18/3582/s1, Figure S1: The diagram for calculating Small yield gap (SYg) and Large yield gap (LYg) in ranking and non-ranking rasters, Figure S2: 1-km pixels that were continuously cropped with summer maize in 2010–2015 over the NCP, Figure S3: Persistent factor percentage (PFP) based on 1-km Yg_0 (a) and 5-km Yg_0 (b), and a comparison between 1-km PFP and 5-km PFP over space (c). The value intervals in the legends of panel (a) and (b) are right-closed and left-open. PFPSYg denotes the PFP value calculated in terms of Yg of croplands grouped in small Yg as defined in Farmaha, Lobell, Boone [25] (Reference [25] is cited in the supplementary materials) or illustrated in Supplementary Text S2. In this study, PFP was calculated for each pixel using surrounding pixels within a buffer of 50 km. But not all pixels within the buffer were used, only pixels meet the criteria (see Section 2.3.4—Step 4) for computing Yp from Yp_0 were kept. Figure S4: Modeled yield potential (Yp) vs. modeled farmers' yield penitential (Ypf), Table S1: Values of coefficients for calculating maize respiration and dry matter allocation.

Author Contributions: S.Z.: Methodology, writing—original draft preparation, writing—review and editing, investigation, formal analysis; Y.B.: Conceptualization, methodology, data curation, writing—review and editing; J.Z.: Writing—review and editing, supervision, project administration. All authors have read and agreed to the published version of the manuscript.

Funding: This research was funded by the Shandong Provincial Natural Science Foundation (grant number ZR2020QD016), National Natural Science Foundation of China (grant number 41901342, 31571565, 31671585), "Taishan Scholar" Project of Shandong Province (grant number TSXZ201712), Key Basic Research Project of Shandong Natural Science Foundation of China (grant number ZR2017ZB0422), National Key Research and Development Program of China (grant number 2016YFD0300101), and CAS Strategic Priority Research Program (grant number XDA19030402).

Acknowledgments: The authors would like to show great appreciation to three anonymous reviewers for their valuable comments, the editor of "Remote Sensing" and the guest editor of "Remote Sensing of Land Surface Phenology special issue" for offering us this opportunity to submit our manuscript, and Rufino O for the English language editing.

Conflicts of Interest: The authors declare no conflict of interest.

References

1. van Ittersum, M.K.; Cassman, K.G.; Grassini, P.; Wolf, J.; Tittonell, P.; Hochman, Z. Yield gap analysis with local to global relevance—A review. *Field Crop Res.* **2013**, *143*, 4–17. [CrossRef]
2. Farmaha, B.S.; Lobell, D.B.; Boone, K.E.; Cassman, K.G.; Yang, H.S.; Grassini, P. Contribution of persistent factors to yield gaps in high-yield irrigated maize. *Field Crop Res.* **2016**, *186*, 124–132. [CrossRef]
3. Di Mauro, G.; Cipriotti, P.A.; Gallo, S.; Rotundo, J.L. Environmental and management variables explain soybean yield gap variability in Central Argentina. *Eur. J. Agron.* **2018**, *99*, 186–194. [CrossRef]
4. Lobell, D.B.; Ortiz-Monasterio, J.I.; Falcon, W.P. Yield uncertainty at the field scale evaluated with multi-year satellite data. *Agric. Syst.* **2007**, *92*, 76–90. [CrossRef]
5. Gambin, B.L.; Coyos, T.; Di Mauro, G.; Borrás, L.; Garibaldi, L.A. Exploring genotype, management, and environmental variables influencing grain yield of late-sown maize in central Argentina. *Agric. Syst.* **2016**, *146*, 11–19. [CrossRef]
6. Liu, Z.; Yang, X.; Lin, X.; Hubbard, K.G.; Lv, S.; Wang, J. Maize yield gaps caused by non-controllable, agronomic, and socioeconomic factors in a changing climate of Northeast China. *Sci. Total Environ.* **2016**, *541*, 756–764. [CrossRef]

7. Wang, Y.; Zhang, Y.; Zhang, R.; Li, J.; Zhang, M.; Zhou, S.; Wang, Z. Reduced irrigation increases the water use efficiency and productivity of winter wheat-summer maize rotation on the North China Plain. *Sci. Total Environ.* **2018**, *618*, 112–120. [CrossRef] [PubMed]
8. Cui, X.; Xie, W. Adapting Agriculture to Climate Change through Growing Season Adjustments: Evidence from Corn in China. *Am. J. Agric. Econ.* **2021**, 1–24. [CrossRef]
9. de Souza Nóia Júnior, R.; Sentelhas, P.C. Soybean-maize succession in Brazil: Impacts of sowing dates on climate variability, yields and economic profitability. *Eur. J. Agron.* **2019**, *103*, 140–151. [CrossRef]
10. Zhang, X.; Wang, S.; Sun, H.; Chen, S.; Shao, L.; Liu, X. Contribution of cultivar, fertilizer and weather to yield variation of winter wheat over three decades: A case study in the North China Plain. *Eur. J. Agron.* **2013**, *50*, 52–59. [CrossRef]
11. Wang, J.; Wang, E.; Yin, H.; Feng, L.; Zhang, J. Declining yield potential and shrinking yield gaps of maize in the North China Plain. *Agric. For. Meteorol.* **2014**, *195–196*, 89–101. [CrossRef]
12. Sun, H.; Zhang, X.; Wang, E.; Chen, S.; Shao, L. Quantifying the impact of irrigation on groundwater reserve and crop production—A case study in the North China Plain. *Eur. J. Agron.* **2015**, *70*, 48–56. [CrossRef]
13. Mohammadi-Ahmadmahmoudi, E.; Deihimfard, R.; Noori, O. Yield gap analysis simulated for sugar beet-growing areas in water-limited environments. *Eur. J. Agron.* **2020**, *113*, 125988. [CrossRef]
14. Lobell, D.B.; Cassman, K.G.; Field, C.B. Crop Yield Gaps: Their Importance, Magnitudes, and Causes. *Annu. Rev. Environ. Resour.* **2009**, *34*, 179–204. [CrossRef]
15. Keating, B.A.; Carberry, P.S.; Hammer, G.L.; Probert, M.E.; Robertson, M.J.; Holzworth, D.; Huth, N.I.; Jng, H.; Meinke, H.; Hochman, Z. An overview of APSIM, a model designed for farming systems simulation. *Eur. J. Agron.* **2003**, *18*, 267–288. [CrossRef]
16. Basso, B.; Liu, L.; Ritchie, J.T. A Comprehensive Review of the CERES-Wheat, -Maize and -Rice Models' Performances. *Adv. Agron.* **2016**, *136*, 27–132. [CrossRef]
17. Lizaso, J.I.; Boote, K.J.; Jones, J.W.; Porter, C.H.; Echarte, L.; Westgate, M.E.; Sonohat, G. CSM-IXIM: A New Maize Simulation Model for DSSAT Version 4.5. *Agron. J.* **2011**, *103*, 766–779. [CrossRef]
18. Amarasingha, R.P.R.K.; Suriyagoda, L.D.B.; Marambe, B.; Gaydon, D.S.; Galagedara, L.W.; Punyawardena, R.; Silva, G.L.L.P.; Nidumolu, U.; Howden, M. Simulation of crop and water productivity for rice (*Oryza sativa* L.) using APSIM under diverse agro-climatic conditions and water management techniques in Sri Lanka. *Agric. Water Manag.* **2015**, *160*, 132–143. [CrossRef]
19. Kwesiga, J.; Grotelüschen, K.; Senthilkumar, K.; Neuhoff, D.; Döring, T.F.; Becker, M. Rice Yield Gaps in Smallholder Systems of the Kilombero Floodplain in Tanzania. *Agronomy* **2020**, *10*, 1135. [CrossRef]
20. Devkota, K.P.; McDonald, A.J.; Khadka, A.; Khadka, L.; Paudel, G.; Devkota, M. Decomposing maize yield gaps differentiates entry points for intensification in the rainfed mid-hills of Nepal. *Field Crop. Res.* **2015**, *179*, 81–94. [CrossRef]
21. Gilardelli, C.; Stella, T.; Confalonieri, R.; Ranghetti, L.; Campos-Taberner, M.; García-Haro, F.J.; Boschetti, M. Downscaling rice yield simulation at sub-field scale using remotely sensed LAI data. *Eur. J. Agron.* **2019**, *103*, 108–116. [CrossRef]
22. Huang, J.; Tian, L.; Liang, S.; Ma, H.; Becker-Reshef, I.; Huang, Y.; Su, W.; Zhang, X.; Zhu, D.; Wu, W. Improving winter wheat yield estimation by assimilation of the leaf area index from Landsat TM and MODIS data into the WOFOST model. *Agric. For. Meteorol.* **2015**, *204*, 106–121. [CrossRef]
23. Wang, P.; Sun, R.; Zhang, J.; Zhou, Y.; Xie, D.; Zhu, Q. Yield estimation of winter wheat in the North China Plain using the remote-sensing–photosynthesis–yield estimation for crops (RS–P–YEC) model. *Int. J. Remote Sens.* **2011**, *32*, 6335–6348. [CrossRef]
24. Huang, Y.; Ryu, Y.; Jiang, C.; Kimm, H.; Kim, S.; Kang, M.; Shim, K. BESS-Rice: A remote sensing derived and biophysical process-based rice productivity simulation model. *Agric. For. Meteorol.* **2018**, *256–257*, 253–269. [CrossRef]
25. Chang, Q.; Zhang, J.; Jiao, W.; Yao, F. A Comparative Analysis of the NDVIg and NDVI3g in Monitoring Vegetation Phenology Changes in the Northern Hemisphere. *Geocart. Internat.* **2016**, *33*, 1–20. [CrossRef]
26. Lobell, D.B.; Ortiz-Monasterio, J.I.; Sibley, A.M.; Sohu, V.S. Satellite detection of earlier wheat sowing in India and implications for yield trends. *Agric. Syst.* **2013**, *115*, 137–143. [CrossRef]
27. Ji, Z.; Pan, Y.; Zhu, X.; Wang, J.; Li, Q. Prediction of Crop Yield Using Phenological Information Extracted from Remote Sensing Vegetation Index. *Sensors* **2021**, *21*, 1406. [CrossRef]
28. Lobell, D.B. The use of satellite data for crop yield gap analysis. *Field Crop Res.* **2013**, *143*, 56–64. [CrossRef]
29. Zhao, Y.; Chen, X.; Cui, Z.; Lobell, D.B. Using satellite remote sensing to understand maize yield gaps in the North China Plain. *Field Crop Res.* **2015**, *183*, 31–42. [CrossRef]
30. Lobell, D.B.; Ortiz-Monasterio, J.I.; Lee Addams, C.; Asner, G.P. Soil, climate, and management impacts on regional wheat productivity in Mexico from remote sensing. *Agric. For. Meteorol.* **2002**, *114*, 31–43. [CrossRef]
31. Lobell, D.B.; Thau, D.; Seifert, C.; Engle, E.; Little, B. A scalable satellite-based crop yield mapper. *Remote Sens. Environ.* **2015**, *164*, 324–333. [CrossRef]
32. Dehkordi, P.A.; Nehbandani, A.; Hassanpour-bourkheili, S.; Kamkar, B. Yield Gap Analysis Using Remote Sensing and Modelling Approaches: Wheat in the Northwest of Iran. *Int. J. Plant Prod.* **2020**, *14*, 443–452. [CrossRef]
33. Deines, J.M.; Patel, R.; Liang, S.-Z.; Dado, W.; Lobell, D.B. A million kernels of truth: Insights into scalable satellite maize yield mapping and yield gap analysis from an extensive ground dataset in the US Corn Belt. *Remote Sens. Environ.* **2021**, *253*, 112174. [CrossRef]

34. He, L.; Mostovoy, G. Cotton Yield Estimate Using Sentinel-2 Data and an Ecosystem Model over the Southern US. *Remote Sens.* **2019**, *11*, 2000. [CrossRef]
35. Zhang, S.; Bai, Y.; Zhang, J.-H.; Shahzad, A. Developing a process–based and remote sensing driven crop yield model for maize (PRYM–Maize) and its validation over the Northeast China Plain. *J. Integr. Agric.* **2020**, *20*, 408–423. [CrossRef]
36. Liu, Z. *The Yield Gaps and Constraint Factors of Spring Maize in Northeast China*; China Agricultural University: Beijing, China, 2013.
37. Xun, L.; Wang, P.; Li, L.; Wang, L.; Kong, Q. Identifying crop planting areas using Fourier-transformed feature of time series MODIS leaf area index and sparse-representation-based classification in the North China Plain. *Int. J. Remote Sens.* **2019**, *40*, 2034–2052. [CrossRef]
38. Bai, Y.; Zhang, J.; Zhang, S.; Yao, F.; Magliulo, V. A remote sensing-based two-leaf canopy conductance model: Global optimization and applications in modeling gross primary productivity and evapotranspiration of crops. *Remote Sens. Environ.* **2018**, *215*, 411–437. [CrossRef]
39. Chen, R.; Ersi, K.; Yang, J.; Lu, S.; Zhao, W. Validation of five global radiation models with measured daily data in China. *Energy Convers. Manag.* **2004**, *45*, 1759–1769. [CrossRef]
40. Bai, Y.; Zhang, J.; Zhang, S.; Koju, U.A.; Yao, F.; Igbawua, T. Using precipitation, vertical root distribution and satellite-retrieved vegetation information to parameterize water stress in a Penman-Monteith approach to evapotranspiration modelling under Mediterranean climate. *J. Adv. Model. Earth Syst.* **2017**, *9*, 168–192. [CrossRef]
41. Supit, I.; Hooijer, A.A.; Van Diepen, C.A. *System Description of the WOFOST 6.0 Crop Simulation Model Implemente in CGMS.*; Volume 1: Theory and Algorithms; Joint Research Centre, European Commission: Luxembourg, 1994.
42. Peng, B.; Guan, K.; Chen, M.; Lawrence, D.M.; Pokhrel, Y.; Suyker, A.; Arkebauer, T.; Lu, Y. Improving maize growth processes in the community land model: Implementation and evaluation. *Agric. For. Meteorol.* **2018**, *250–251*, 64–89. [CrossRef]
43. Shawon, A.R.; Ko, J.; Ha, B.; Jeong, S.; Kim, D.K.; Kim, H.-Y. Assessment of a Proximal Sensing-integrated Crop Model for Simulation of Soybean Growth and Yield. *Remote Sens.* **2020**, *12*, 410. [CrossRef]
44. Osborne, T.; Gornall, J.; Hooker, J.; Williams, K.; Wiltshire, A.; Betts, R.; Wheeler, T. JULES-crop: A parametrisation of crops in the Joint UK Land Environment Simulator. *Geosci. Model Dev.* **2015**, *8*, 1139–1155. [CrossRef]
45. Wang, Y. *Study on Population Quality and Individual Physiology Function of Super High-yielding Maize (Zea mays L.)*; Shandong Agricultural University: Tai'an, China, 2008.
46. Yang, G.; Xin, L.; Chenglian, W.; Xiangning, L. Study on Effects of Plant Densities on the Yield and the Related Characters of Maize Hybrids. *Acta Agric. Boreali-Occident. Sin.* **2006**, *15*, 57–60.
47. Jing, L. *Study on Population Quality Indices for High or Super High-Yield of Maize*; Yangzhou University: Yangzhou, China, 2011.
48. Chu, G.; Zhang, J. Effects of Nitrogen Application on Photosynthetic Characteristics, Yield and Nitrogen Use Efficiency in Drip Irrigation of Super High-yield Spring Maize. *J. Maize Sci.* **2016**, *24*, 130–136. [CrossRef]
49. Huang, Z. *Studies on Photosynthetic and Nutrient Physiological Characteristics of Super-High Yield Summer Maize Hybrids*; Shandong Agricultural University: Tai'an, China, 2007.
50. Liu, G.; Hou, P.; Xie, R.; Ming, B.; Wang, K.; Xu, W.; Liu, W.; Yang, Y.; Li, S. Canopy characteristics of high-yield maize with yield potential of 22.5 Mg ha^{-1}. *Field Crop Res.* **2017**, *213*, 221–230. [CrossRef]
51. Wang, J. *Characteristics on Canopy Vertical Structures and Agronomic Regulation of Super-High Yield of Spring Maize*; Inner Mongolia Agricultural University: Huhehot, China, 2009.
52. Liu, W.; Lv, P.; Su, K.; Yang, J.; Zhang, J.; Dong, S.; Liu, P.; Sun, Q. Effects of planting density on the grain yield and source-sink characteristics of summer maize. *Chin. J. Appl. Ecol.* **2010**, *21*, 1737–1743.
53. Hu, W. *A Study on Characteristics of Radiation and Photosynthesis in Canopy of Super High-Yield Summer Maize*; Henan Agricultural University: Zhengzhou, China, 2012.
54. Cao, Y. *Study on the Activity of Photosynthesis Enzymes and Protective Enzymes in Super High Yield Corns and Common Corns*; Jilin Agricultural University: Changchun, China, 2008.
55. Jin, X.; Du, X.; Liu, J.; Cheng, F.; Cui, Y. Physiological Characters of the Summer Maize Population with High Yield in the North Areas of the Yellow River, Huai and Hai Rivers Plain. *J. Maize Sci.* **2012**, *20*, 79–83. [CrossRef]
56. Wang, Z. *Structural and Functional Properties of Canopy and Root of Super High Yield Spring Maize & Agronomic Water Saving Compensatory Mechanism*; Inner Mongolia Agricultural University: Huhehot, China, 2009.
57. Chang, J.; Zhang, G.; Li, Y.; Yan, L.; Li, C. Study on Growth of Super-high-yield Summer Maize in the Ecological Area of the Yellow River, Huai and Hai Rivers. *J. Maize Sci.* **2011**, *19*, 75–79. [CrossRef]
58. Yang, D.; Zhao, W.; Qin, D.; Liu, F.; Zhang, Q.; Guan, Y.; Yang, K. Yield and Canopy Structure of Maize under the Condition of High Yield Cultivation. *J. Maize Sci.* **2016**, *24*, 129–135. [CrossRef]
59. Bao, Y. *Study on Canopy Structure and Photosynthesis Character of Super-High-Yield Maize*; Jilin Agricultural University: Changchun, China, 2006.
60. Zhang, S.; Wang, Y.; Qi, T.; Cheng, L.; Xu, G.; Su, Y.; Qin, Y.; Li, Y. Study on Cultivated Technology for Super High Yield of Summer Maize in Huanghuaihai Region. *Chin. Agric. Sci. Bull.* **2009**, *25*, 130–133.
61. Zhang, Y.; Yang, H.; Gao, J.; Zhang, R.; Wang, Z.; Xu, S.; Fan, X.; Yang, S. Study on Canopy Structure and Physiological Characteristics of Super-High Yield Spring Maize. *Sci. Agric. Sin.* **2011**, *44*, 4367–4376. [CrossRef]
62. Ma, X.; Wang, Q.; Qian, C.; Ke, F.; Wang, C. Canopy Characteris tics of Super-high Yielding Maize Under Different Nitrogen Application. *J. Maize Sci.* **2008**, *16*, 158–162. [CrossRef]

63. Li, X.; Tang, Q.; Li, D.; Li, W.; Li, H.; Cai, Q. Effects of Different Plant Densities on the Photosynthetic-Physiological Characters and Yield Traits in Spring Maize Grown on Super-High Yielding Paddy Field. *Acta Agric. Boreali-Sin.* **2011**, *26*, 174–180.
64. Yang, J.; Gao, H.; Liu, P.; Li, G.; Dong, S.; Zhang, J.; Wang, J. Effects of Planting Density and Row Spacing on Canopy Apparent Photosynthesis of High-Yield Summer Corn. *Acta Agron. Sin.* **2010**, *36*, 1226–1233. [CrossRef]
65. Wu, Z. *Creation High-Yield Maize Canopy Structure and Micro-Environmental Factors*; Jilin Agricultural University: Changchun, China, 2002.
66. Wart, J.V.; Kersebaum, K.C.; Peng, S.; Milner, M.; Cassman, K.G. Estimating crop yield potential at regional to national scales. *Field Crop Res.* **2013**, *143*, 34–43. [CrossRef]
67. Meng, Q.; Hou, P.; Wu, L.; Chen, X.; Cui, Z.; Zhang, F. Understanding production potentials and yield gaps in intensive maize production in China. *Field Crop Res.* **2013**, *143*, 91–97. [CrossRef]
68. Li, K. *Yield Gap Analysis Focused on Winter Wheat and Summer Maize Rotation in the North China Plain*; China Agricultural University: Beijing, China, 2014.
69. Jin, Z.; Azzari, G.; Lobell, D.B. Improving the accuracy of satellite-based high-resolution yield estimation: A test of multiple scalable approaches. *Agric. For. Meteorol.* **2017**, *247*, 207–220. [CrossRef]
70. Yang, D.; Chen, J.; Zhou, Y.; Chen, X.; Chen, X.; Cao, X. Mapping plastic greenhouse with medium spatial resolution satellite data: Development of a new spectral index. *Int. J. Photogramm. Remote Sens.* **2017**, *128*, 47–60. [CrossRef]
71. Li, J.; Luo, J.; Ming, D.; Shen, Z. A new method for merging IKONOS panchromatic and multispectral image data. In Proceedings of the International Geoscience and Remote Sensing Symposium, Seoul, Korea, 29 July 2005; pp. 3916–3919.
72. Hilker, T.; Wulder, M.A.; Coops, N.C.; Seitz, N.; White, J.C.; Gao, F.; Masek, J.G.; Stenhouse, G. Generation of dense time series synthetic Landsat data through data blending with MODIS using a spatial and temporal adaptive reflectance fusion model. *Remote Sens. Environ.* **2009**, *113*, 1988–1999. [CrossRef]
73. Celestre, R.; Rosenberger, M.; Notni, G. A novel algorithm for bad pixel detection and correction to improve quality and stability of geometric measurements. *J. Phys. Conf. Ser.* **2016**, *772*, 012002. [CrossRef]
74. Lobell, D.B.; Burke, M.B. On the use of statistical models to predict crop yield responses to climate change. *Agric. For. Meteorol.* **2010**, *150*, 1443–1452. [CrossRef]
75. Challinor, A.J.; Watson, J.; Lobell, D.B.; Howden, S.M.; Smith, D.R.; Chhetri, N. A meta-analysis of crop yield under climate change and adaptation. *Nat. Clim. Chang.* **2014**, *4*, 287–291. [CrossRef]
76. Wang, Y.; Song, Q.; Du, Y.; Wang, J.; Zhou, J.; Du, Z.; Li, T. A random forest model to predict heatstroke occurrence for heatwave in China. *Sci. Total Environ.* **2019**, *650*, 3048–3053. [CrossRef] [PubMed]
77. Han, X.; Wu, J.; Zhou, H.; Liu, L.; Yang, J.; Shen, Q.; Wu, J. Intensification of historical drought over China based on a multi-model drought index. *Int. J. Climatol.* **2020**, *40*, 5407–5419. [CrossRef]

remote sensing

Article

Quantification of Urban Heat Island-Induced Contribution to Advance in Spring Phenology: A Case Study in Hangzhou, China

Yingying Ji [1,2], Jiaxin Jin [1,3,*], Wenfeng Zhan [2,4], Fengsheng Guo [1] and Tao Yan [1]

[1] Hydrology and Water Resources College, Hohai University, Nanjing 211100, China; 1709040214@hhu.edu.cn (Y.J.); 181309010013@hhu.edu.cn (F.G.); 191309010014@hhu.edu.cn (T.Y.)
[2] Jiangsu Provincial Key Laboratory of Geographic Information Science and Technology, International Institute for Earth System Science, Nanjing University, Nanjing 210023, China; zhanwenfeng@nju.edu.cn
[3] National Earth System Science Data Center, National Science & Technology Infrastructure of China, Beijing 100101, China
[4] Jiangsu Center for Collaborative Innovation in Geographical Information Resource Development and Application, Nanjing 210023, China
* Correspondence: jiaxinking@hhu.edu.cn; Tel.: +86-15951871334

Abstract: Plant phenology is one of the key regulators of ecosystem processes, which are sensitive to environmental change. The acceleration of urbanization in recent years has produced substantial impacts on vegetation phenology over urban areas, such as the local warming induced by the urban heat island effect. However, quantitative contributions of the difference of land surface temperature (LST) between urban and rural (ΔLST) and other factors to the difference of spring phenology (i.e., the start of growing season, SOS) between urban and rural (ΔSOS) were rarely reported. Therefore, the objective of this study is to explore impacts of urbanization on SOS and distinguish corresponding contributions. Using Hangzhou, a typical subtropical metropolis, as the study area, vegetation index-based phenology data (MCD12Q2 and MYD13Q1 EVI) and land surface temperature data (MYD11A2 LST) from 2006–2018 were adopted to analyze the urban–rural gradient in phenology characteristics through buffers. Furthermore, we exploratively quantified the contributions of the ΔLST to the ΔSOS based on a temperature contribution separation model. We found that there was a negative coupling between SOS and LST in over 90% of the vegetated areas in Hangzhou. At the sample-point scale, SOS was weakly, but significantly, negatively correlated with LST at the daytime ($R^2 = 0.2$ and $p < 0.01$ in rural; $R^2 = 0.14$ and $p < 0.05$ in urban) rather than that at nighttime. Besides, the ΔSOS dominated by the ΔLST contributed more than 70% of the total ΔSOS. We hope this study could help to deepen the understanding of responses of urban ecosystem to intensive human activities.

Keywords: plant phenology; land surface temperature; urban heat island effect; contribution; Hangzhou

Citation: Ji, Y.; Jin, J.; Zhan, W.; Guo, F.; Yan, T. Quantification of Urban Heat Island-Induced Contribution to Advance in Spring Phenology: A Case Study in Hangzhou, China. *Remote Sens.* **2021**, *13*, 3684. https://doi.org/10.3390/rs13183684

Academic Editor: Alfredo Huete

Received: 11 August 2021
Accepted: 13 September 2021
Published: 15 September 2021

1. Introduction

Plant phenology is the time of a certain growth event in the growth cycle, such as germination, branching, leafing, flowering, fruiting, defoliation and dormancy [1–4]. It directly or indirectly regulates several processes of plant growth, such as carbon and water cycle, playing a crucial role in the earth system [5,6]. Adapting to seasonal changes of the environment, plants show a growth rhythm, which is sensitive to environmental change [7,8]. As one of the most critical factors affecting plant phenology, an increase in temperature can promote the activity of enzymes, thereby prolonging plant development. Specifically, an increase in spring temperature promotes the release of plant dormancy in spring, and generally extends the growth cycle of plants [9–14].

Urbanization is an important feature of world development today, and it is one of the main causes of global environmental change in the 21st century. The acceleration of urbanization in recent years has produced substantial impacts on plant phenology over both urban areas and their rural surroundings [15–19]. This is mostly associated to the local warming effect induced by the urban heat island effect, which resulted from the increase in impervious surface percentage and anthropogenic heat emissions [20–23]. Moreover, it is as well as through the fertilization effect induced by the increase in the concentration of carbon dioxide (CO_2), nitrogen oxides (NO_x), and other atmospheric trace gases over urban surfaces [24–26]. These changes affect urban environments that plants depend on, and have impacts on the growth of plants, thereby changing the plant phenology [27,28].

At present, many studies have paid attention to impacts of urbanization on the change of plant phenology [27–32]. There are two methods to explore the impacts above: the historical comparison method and the urban–rural comparison method. The historical comparison method compared the phenology before and after urbanization, which was mainly for fast-developing cities [31]. However, due to the difficulty of obtaining long time series data, the historical comparison method is greatly restricted. The urban–rural comparison method used the data of the urban and the rural at the same time to explore the impact of urbanization on phenology, which is a method of changing space for time. The second one has been widely used, because of the great advantages in large-scale observations of remote sensing data [15–19,32]. Meng et al. investigated the urban and rural phenology of the of 85 giant cities in the continental United States from 2001 to 2014, and the results showed that the start of growing season (SOS) in the urban was 6 days earlier than that in the rural [33]. Wohlfahrt et al. found that with the acceleration of urbanization, the SOS advanced and the senescence delayed in the urban areas where the temperature rises [34]. Hu et al. used the Enhanced Vegetation Index (EVI) to explore the spatio-temporal changes of plant phenology and its response to land surface temperature (LST) in Northeast China [35]. The results showed that the LST was significantly negatively correlated with the SOS. Recently, most current studies focused on varieties of plant phenology and influences of temperature on plant phenology under urbanization, but did not quantitatively evaluate the contribution of the temperature differences to the phenological differences between the urban and the rural. That is, the quantitative contribution of the local warming induced by the urban heat island effect (the difference of LST between urban and rural, ΔLST) to the difference of spring phenology (SOS) between urban and rural (ΔSOS) was less understood in past research.

With the development of statistical methods, it was possible to distinguish the influence of different factors. Li et al. used a statistical method to quantify the contribution of cooling and water supply to the yield benefits due to irrigation. They found that 16% of irrigation yield increase was due to irrigation cooling while the rest (84%) is due to water supply and other factors [36]. Besides, Zhao et al. also used a statistical method to quantifying the impacts of urbanization on vegetation growth. They found that the growth enhancement offset about 40% of direct loss of vegetation productivity caused by replacing productive vegetated surfaces with nonproductive impervious surfaces [16,30]. Based on the studies above, a statistical model was used to carry out this study.

Therefore, the objective of this study is to explore impacts of urbanization on SOS and exploratively distinguish contributions of local warming induced by the urban heat island effect (ΔLST) and other factors to the difference of spring phenology between urban and rural (ΔSOS). Hangzhou, a typical subtropical metropolis, was selected as the study area. Specifically, the spatial differences and inter-annual changes of the phenology in the urban and the rural were compared through a gradient analysis method using satellite-based phenology and LST data from 2006 to 2018. Then, the coupling relationship between phenology and temperature were investigated. After that, taking typical forest grid cells in the urban and the rural areas of Hangzhou as test samples, the local SOS was extracted using a remote sensing vegetation index from 2006 to 2018, and the difference of responses of SOS to LST between the urban and rural was explored. Finally, we exploratively distinguish quantitative contributions of the ΔLST and other factors to the ΔSOS.

2. Materials and Methods

2.1. Study Area

Hangzhou (118°21′–120°31′E, 29°11′–30°33′N) is the capital city of Zhejiang Province, whose GDP ranks among the top 10 in China, with 8.133 million in urban population, 2.227 million in rural population and an urbanization rate of 78.5% in 2019. It is located in the north of Zhejiang Province with a subtropical humid monsoon climate. As for temperature, it is lowest in January (average of 3–5 °C) and highest in July (average of 28–29 °C) with an annual average of 15.3–17 °C. The extreme maximum and minimum temperatures in Hangzhou reached 42.9 °C (31 July 1971) and −15 °C (5 January 1977). For precipitation, the annual average is 1100–1600 mm with rainy days of 130–160 days/year. There are two rainy seasons throughout the year. The first is the plum flood season from May to June, with an average rainfall of 350–500 mm, accounting for 25–31% of the year. The second rainy season is the typhoon rainy season from August to September, with an average rainfall of 120–220 mm, accounting for 8–13%. The forest coverage rate is over 64.77% (about 10,900 km^2), dominated by evergreen broad-leaved forests and deciduous broad-leaved forests. In this study, the land cover map with 10 m-spatial resolution in 2017 from Gong Peng Research Group of Tsinghua University was aggregated to pixels of 500 m × 500 m to extract forest areas for the following analysis. To assure both a certain level of homogeneity in the land cover type and an adequate number of pixels for a meaningful analysis, only the pixels of 500 m where the forest type was over 75% were included in this study. Besides, the multi-temporal dataset of global urban boundaries of 2018 was used to divide the scope of the urban and the rural of Hangzhou Figure 1. This dataset is derived from the Global Artificial Impervious Area-GAIA, released by Gong Peng Research Group of Tsinghua University [37]. Then, 5 test areas of deciduous broad-leaved forest were selected in the urban and the rural (i.e., 10 km away from the urban area) of Hangzhou, respectively. In each test area, 2 sample points, a total of 20 sample points, were extracted (Figure 1). Besides, the Google map, latitude and longitude of the test areas in the urban and rural were showed in Table 1.

Table 1. The Google map, latitude and longitude of the test areas in the urban and rural.

U1	U2	U3

U4	**U5**	**No: Latitude, Longitude**
		U1: 30.36051N, 120.19516E U2: 30.35946N, 120.17856E U3: 30.18648N, 120.30131E U4: 30.18959N, 120.42429E U5: 30.42698N, 120.28390E

Table 1. *Cont.*

R1	R2	R3
		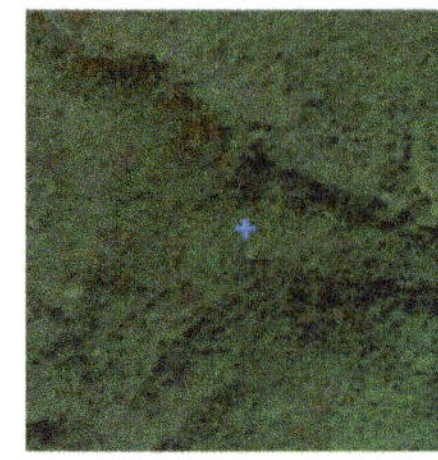
R4	**R5**	**No: Latitude, Longitude**
		R1: 30.23112N, 119.35119E R2: 30.05648N, 119.51766E R3: 29.72447N, 119.35588E R4: 29.97818N, 119.27411E R5: 30.24293N, 119.33563E

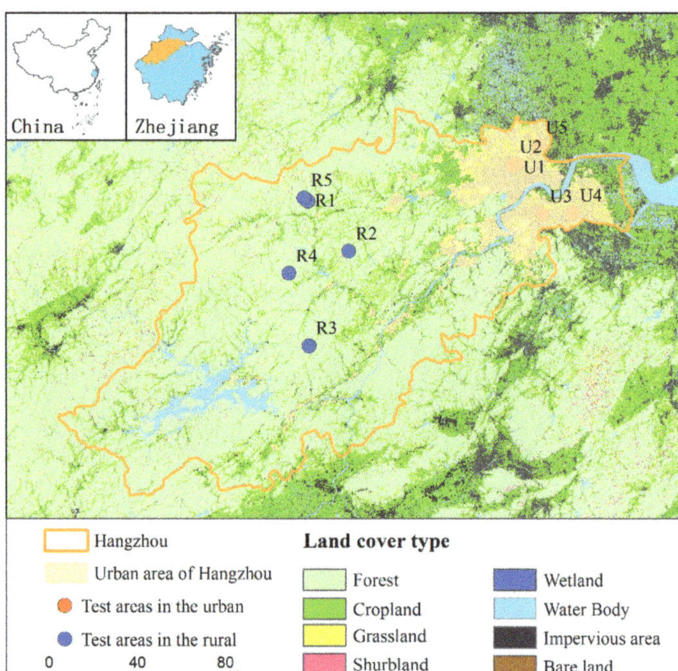

Figure 1. Spatial distribution of Hangzhou and the location of forest test areas. The small maps in the upper right corner show the location of Hangzhou in Zhejiang Province and the location of Zhejiang Province in China.

2.2. Remote Sensing Data

2.2.1. Land Surface Temperature

MODIS MYD11A2 LST product was used in this study, including the LST during the daytime and the nighttime, with a spatial resolution of 1000 m and a temporal resolution of 8 days. The LST of Aqua satellite is observed at 1:30 and 13:30 local solar time, the lowest and highest temperature of the day, which is more representative than that of Terra (monitored at 10:30 and 22:30) in the study of urban heat island. Therefore, the LST data of the Aqua satellite were used in this study [38,39]. The MODIS Reprojection Tool (MRT) was used to process the original images of LST, and so did the following data. Then, they were extracted in light of the study area and resampled to 500 m to be consistent with the phenology data. In addition, to explore the quantitative contributions of the ΔLST to the ΔSOS under urbanization, we collected 1000-m spatial resolution LST data of daytime and nighttime from 2006 to 2018 according to the coordinates of forest sample points.

2.2.2. MODIS Phenology

This paper tried to explore the difference of plant phenology between the urban and the rural of Hangzhou and its temporal and spatial patterns. The SOS from the MCD12Q2 (i.e., a MODIS phenology dataset) across the study area during 2006–2018 was used in this study. The spatial and time resolution of the dataset is 500 m and 1 year, respectively. The phenological events are derived from time series of MODIS 2-band Enhanced Vegetation Index (EVI2), which are fitted by QA/QC-weighted penalized cubic smoothing spline. The phenology data in some high-latitude regions and some semi-arid and arid environments exhibiting low-amplitude EVI2 variation, having uncertainty, while the data of Hangzhou in mid-latitude regions is relatively stable [40]. Besides, these data have been validated with field observations [41] and have been widely used and approved in previous studies [17,18,42].

2.2.3. Enhanced Vegetation Index

The contributions of the ΔLST to the ΔSOS were used to explore at a finer spatial resolution. Here, the Aqua MODIS 16-day EVI data (MYD13Q1, 250 m × 250 m), which matched the LST data from the same satellite (Aqua) to reduce uncertainties, was used to extract the phenology. Previous studies have suggested that EVI could accurately reflect the growth status of vegetation, and effectively extract phenology at both regional and local scales [43–47]. In this study, the asymmetric Gaussian function, which is widely used in curve fitting and phenological extraction, while a 20% threshold was used to fit the EVI curve and extract the SOS for each selected forest sample point from 2006 to 2018 [6,7].

Notably, both the MODIS phenology product and EVI-derived phenology were used in this study. On the one hand, to explore the changes in plant phenology caused by the urban heat island effect, we needed to focus on the forest, which is severely affected by urbanization. At this time, the plant phenology data with a resolution of 500 m cannot satisfy the demand, so the EVI data with a resolution of 250 m at a smaller scale was used to improve the resolution and reduce the influence of mixed pixels. On the other hand, MCD12Q2 uses EVI2 data to extract plant phenology, while EVI2 data lacks the blue band, different from EVI [48]. Therefore, the more widely used and robust EVI data was used for extracting more accurate plant phenology. Considering the reasons above, we selected EVI-based SOS for the research of sample-scale.

Besides, the SOS extracted from MCD12Q2 phenology data and EVI data showed a linear correlation ($R^2 = 0.54$) at the forest sample area in Hangzhou Figure 2. First, the original data and methods to extract SOS were different. The one used the EVI data and the method of asymmetric Gaussian function with a 20% threshold, while the other used the EVI2 data and the method of QA/QC-weighted penalized cubic smoothing spline. Although the correlation of SOS derived from MCD12Q2 and EVI was not satisfied, the relationship was statistically significant ($p < 0.05$), which could support the consistency between them. Second, in this study, we aimed to explore the quantitative contributions of the difference of land surface temperature between urban and rural and other factors to the difference of spring phenology (ΔSOS) under urbanization. The relative ΔSOS was effective instead of the absolute value of SOS. Therefore, despite the difference in the absolute SOS of the two data, they had a statistically significant linear relationship and there was also a certain relationship between the ΔSOS. Besides, the RMSE of the two data was 5 days, which was smaller than the average ΔSOS of >9 days. In summary, the two data of SOS had a certain consistency in this study.

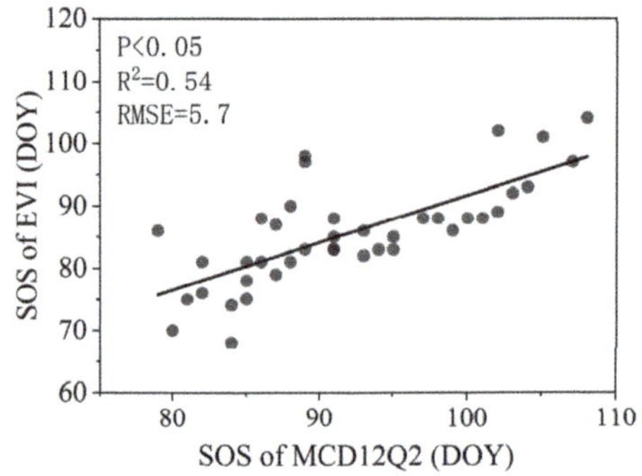

Figure 2. Correlation of SOS between MCD12Q2 and EVI. Black line denotes linear regression lines. The DOY denotes the day of year.

2.3. Temperature Contribution Separation Model

The rapid development of urbanization dramatically changes the environments which terrestrial ecosystem depended on. Compared with the rural surroundings, there are differences in temperature, photoperiod and atmosphere conditions, having a certain impact on plant phenology. A large number of studies have shown that the acceleration of urbanization in recent years produced substantial impacts on plant phenology over both urban areas and their rural surroundings [15–18,33]. Therefore, in order to distinguish the contributions of ΔLST and other factors between urban and rural to the difference of spring phenology (ΔSOS), we followed the statistical method of quantifying the contributions of cooling and water supply to the yield benefits due to irrigation of Li et al. [36], establishing a temperature contribution separation model based on the laboratory of the rural and urban of Hangzhou.

Firstly, we performed regression analysis on SOS and average LST during the daytime in spring (February, March, April) from 2006 to 2018 in the rural and the urban Equations (1) and (2), respectively. Secondly, Equations (3)–(5) were used to distinguish the contributions of the ΔLST and other factors to the ΔSOS:

$$SOS_{rural} = f_{rural}(T_{rural}) \tag{1}$$

$$SOS_{urban} = f_{urban}(T_{urban}) \tag{2}$$

$$T_{contribution} = f_{rural}(T_{rural}) - f_{rural}(T_{urban}) \tag{3}$$

$$Other_{contribution} = f_{rural}(T_{urban}) - f_{urban}(T_{urban}) \tag{4}$$

$$T_{percent} = T_{contribution} / (T_{contribution} + Other_{contribution}) \tag{5}$$

where the subscripts of *rural* and *urban* denote the corresponding parameters of the rural and the urban, respectively; the *SOS* denotes the start of the growing season; the *T* denotes the average LST during the daytime in spring; the *f* denotes the regression relationship between LST and SOS. The $T_{contribution}$ and $Other_{contribution}$ denote the contributions of the ΔLST and other factors to the ΔSOS, respectively; and the $T_{percent}$ denotes the percentage of the contribution of the ΔLST.

The temperature contribution separation model was shown in Figure 3. (1) In the figure, the blue and red lines represent the regression relationships between SOS and LST in the rural and the urban Equations (1) and (2), respectively. Points A and D represent the average LST and SOS of the sample points in the same year of the rural and urban, respectively. Line B-D is the difference of the SOS between the rural and the urban. (2) We supposed that the rural surroundings were heated to reach the LST of the urban in the same year. Then, the SOS of the rural (point A) in that year moved to point C according to Equation (3). At this time, the two points A and C were the phenological state only in different LST, and line B-C was the phenological difference only when the LST rose Equation (3), which refers to the influence of the urban heat island effect (ΔLST). (3) However, in the same year, the average LST and SOS of the sample point in the urban was located at point D. There was a phenological difference C-D Equation (4) from point C, which refers to the influence of other environmental factors in the urban except temperature. (4) Furthermore, the percentage of the contribution of the ΔLST to the ΔSOS was calculated by Equation (5).

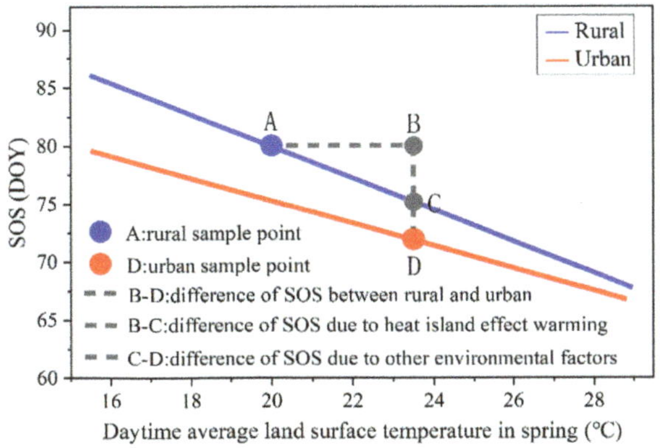

Figure 3. Temperature contribution separation model. The red/blue solid lines denote the linear regression line of the urban/rural sample point data. Points A and D denote the data of sample points in the rural and the urban in the same year, points B and C denote the predicted values of the model, and gray dotted lines denote the contributions of different factors. The DOY denotes the day of year.

2.4. Statistical Analysis

In this study, the multi-year average phenology of the SOS of 2006–2018 were calculated to compare the differences of phenology between that in the urban and the rural of Hangzhou, which were extracted from the MCD12Q2 phenology dataset. Besides, a buffer analysis method was adopted to compare the difference of SOS in urban–rural gradient and its relationship with LST in more detail [7,28]. First, the urban boundary of Hangzhou in 2018 was derived from the global urban boundary dataset [12]. Second, the circular buffer zones outside the urban boundary were drawn every 2 km, which were 2, 4, 6, 8, 10, 12, 14, 16, 18 and 20 km, respectively. Finally, the average of SOS and LST in each buffer zone were calculated to explore the relationship between SOS and LST with the distance away from the urban.

Furthermore, the two-factor combination mapping is a very intuitive visualization method that can express the coupling relationship between two variables. The two factors that are incomparable numerically can be compared in a hierarchical manner, and the different levels of the two factors are matched in pairs to different combinations, representing the different coupling relationships of the two factors. In this study, to explore the coupling relationship between SOS and LST, the natural breakpoint classification method was used to divide the daily average LST and SOS in spring of 2018 into three levels: low, medium and high. Additionally, we used the two-factor combination mapping method to display the different coupling relationships, including low LST-low SOS, low LST-medium SOS, low LST-high SOS, medium LST-low SOS, medium LST-medium SOS, medium LST-high SOS, high LST-low SOS, high LST-medium SOS, and high LST-high SOS.

3. Results

3.1. Spatial Patterns of the LST and SOS in Hangzhou

The average LST in spring of 2018 was utilized to explore the difference of LST between that in the urban and the rural Figure 4. The results showed that there was a significant spatial heterogeneity in the LST of Hangzhou, showing a gradient of high in the urban and low in the rural. In terms of spatial distribution, the area with a LST greater than 23.5 °C accounted for 10.9%, mainly distributed in the urban area; 16.6%, 32.9% and 26.6% of the area with a LST of 21.5–23.5, 19.5–21.5 and 17.5–19.5 °C distributed in the middle area of Hangzhou, respectively. The area with a LST less than 17.5 °C (accounting for 13%) was mainly distributed in the northern and southern edges of Hangzhou. Moreover, as shown in the inset chart in Figure 4, the LST followed a generally decreasing urban–rural gradient, showing a significant urban heat island effect that the LST was highest (24 °C) in the urban and lowest (18.6 °C) in the rural. In the range of 0–6 km from the urban, the LST decreased fast (0.72 °C/km), and the decrease tended to slow down (0.08 °C/km) after 6 km.

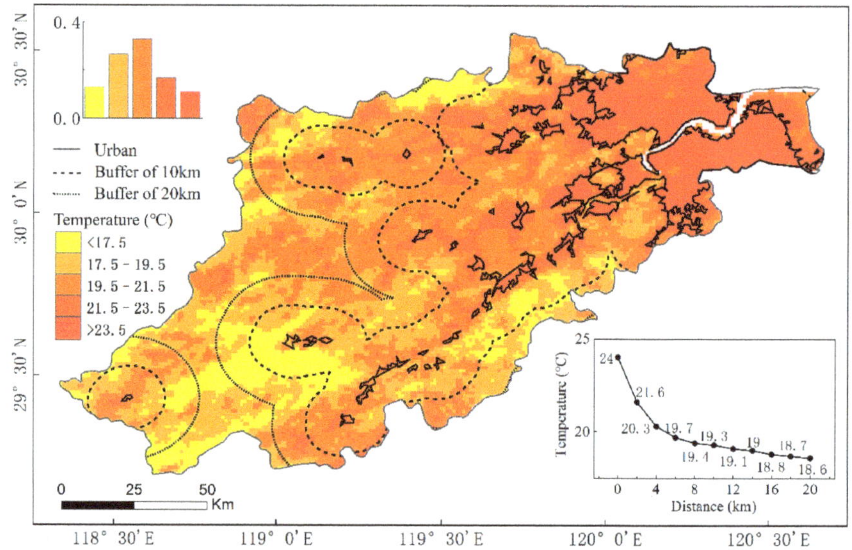

Figure 4. Spatial distribution of land surface temperature in spring of 2018 in Hangzhou. The histogram denotes the frequency distribution of difference temperature. The inset graph denotes the variation trend of temperature in the buffer zone at different distances from urban. The solid lines denote the urban boundaries, and the wide and short dotted lines denote the 10 km and 20 km buffers of the urban boundaries, respectively.

As shown in Figure 5, the annual average SOS of Hangzhou from 2006 to 2018 showed a significant spatial heterogeneity. The SOS was earlier in the urban and later in the rural. In terms of spatial distribution, the area with the SOS less than 76 day of year (termed DOY) accounted for 8.9%, mainly distributed in the urban and the area within 2 km from the urban, located in the east and south of Hangzhou. About 23.1%, 31.1% and 26.6% of the area with SOS of 76–83, 83–88 and 88–94 DOY were distributed in the middle area of Hangzhou, respectively. The area with SOS more than 94 DOY (accounting for 10.3%) mainly distributed in the northern edges of Hangzhou, in the mountainous areas with higher elevations. Besides, as shown in the inset chart in Figure 5, the SOS followed a generally increasing urban–rural gradient, that is, from urban (79 DOY) to rural (87 DOY), the SOS was continuously delayed by 8 days. In the range of 0–6 km from the urban, the SOS increased fast (1.02 days/km), and the increase tended to slow down (0.14 days/km) after 6 km.

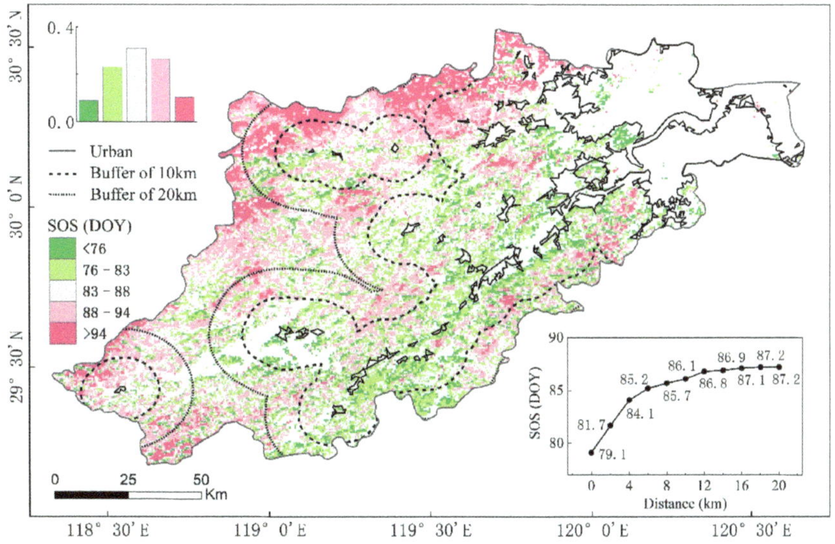

Figure 5. Spatial distribution of annual average of the start of the growing season (SOS) from 2006 to 2018 in Hangzhou. The histogram denotes the frequency distribution of difference SOS. The inset graph denotes the variation trend of SOS in the buffer zone at different distances from urban. The DOY denotes the day of year. The solid lines denote the urban boundaries, and the wide and short dotted lines denote the 10 km and 20 km buffers of the urban boundaries, respectively.

We further explored the spatial difference of SOS between that in the urban and the rural of Hangzhou from 2006 to 2018, and we found that the results from each year had little significant fluctuations. Therefore, in order to avoid information redundancy and excessively long images, we displayed the results every 4 years (2006, 2010, 2014 and 2018) Figure 6. The results showed that although the absolute value of SOS varied from year to year, the spatial differentiation of SOS yearly was consistent with the annual average SOS from 2006 to 2018 in Figure 5. They both showed a significant spatial heterogeneity that the SOS was smaller in the eastern and southern area of Hangzhou and larger in the northern marginal area. As shown in the inset chart in Figure 6, the SOS of the urban was 9, 9, 6 and 6 days earlier than the rural in 2006, 2010, 2014 and 2018, respectively. In addition, the SOS followed a generally increasing urban–rural gradient. In 2006, 2010, 2014 and 2018, the SOS increased fast (1.25, 1.05, 0.83 and 0.93 days/km, respectively) within the range of 0–6 km from the urban, while it tended to be stable (0.07, 0.21, 0.05 and 0.04 days/km, respectively) after 6 km.

Combined with the analysis of the average LST in spring of 2018 and SOS across Hangzhou above, the difference in LST and SOS presented an opposite state and change trend. That is, the LST tended to be higher in the urban and lower in the rural, while SOS tended to be earlier in the urban and later in the rural. The LST followed a generally decreasing urban–rural gradient, while the opposite occurred for SOS, but both the LST and SOS varied greatly within the range of 0–6 km and then tended to be stable. In summary, the LST and SOS showed a negative correlation, and the coupling relationship between them would be further explored below.

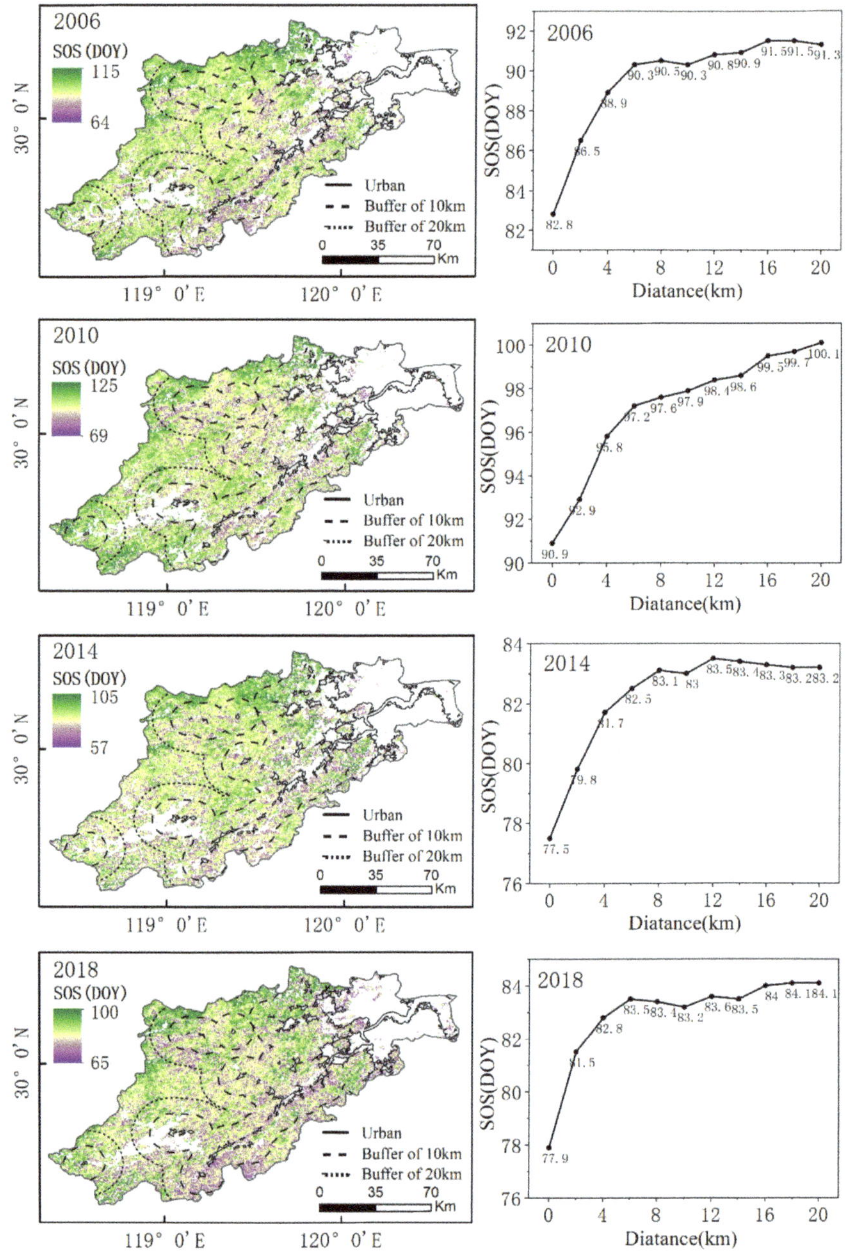

Figure 6. Spatial distribution of the start of the growing season (SOS) in Hangzhou (left charts), and variation trend of SOS in the buffer zone at different distances from urban (right charts) in 2006, 2010, 2014, and 2018. The black line denotes the urban boundary, dark gray denotes the buffer of 10 km, and light grey denotes the buffer of 20 km. The DOY denotes the day of year. The solid lines denote the urban boundaries, and the wide and short dotted lines denote the 10 km and 20 km buffers of the urban boundaries, respectively.

3.2. Relationship between LST and SOS in Hangzhou

The results above showed that the spatial distribution and urban–rural gradient of the average LST in spring and SOS showed an opposite trend. Therefore, it could be inferred that there was negative correlation between LST and SOS. Figure 7 shows the coupling relationship between LST and SOS, and the 94.6% of the area conformed to the inference above. The LST and SOS showed a significant negative correlation accounted for 53.9%, more than a half of the total area. Among them, the low LST-high SOS (accounting for 6.2%) mainly distributed in the northern edge of Hangzhou, the outer rural farthest from urban. The medium LST–medium SOS (45.3%) was mainly distributed in the middle part of Hangzhou, with a moderate distance from urban. The high LST–low SOS (2.4%) was mainly distributed in urban and within 2 km from urban. At the same time, there was 40.7% of the area that LST and SOS showed a weaker negative correlation, including low LST–medium SOS (10.9%), medium LST–low SOS (12.2%), medium LST–high SOS (10.1%), and high LST–medium SOS (7.5%), mainly distributed in the area between urban and outer rural. Besides, there was 5.4% of the area that exhibited a contrary relationship to the inference. That is, the LST and SOS showed a significantly positive correlations, which were low LST–low SOS (3.3%) and high LST–high SOS (2.1%). It might be related to the threshold of LST and SOS for the classification. In addition, in areas with low, moderate and high LST, 83%, 67% and 82% of the SOS has a medium-high, moderate and medium-low distribution, respectively. Overall, the results above confirmed the inference that spring LST was negatively correlated with SOS.

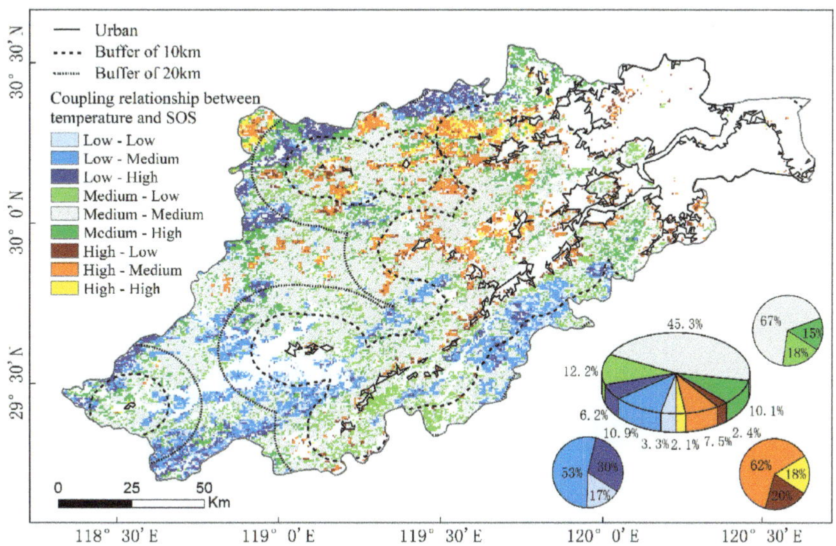

Figure 7. Coupling relationship between land surface temperature in spring and the start of the growing season (SOS) of 2018 in Hangzhou. The 3D pie chart denotes the percentage of each coupling relationship, and the flat pie charts denote the percentage of SOS by level at different temperatures. The solid lines denote the urban boundaries, and the wide and short dotted lines denote the 10 km and 20 km buffers of the urban boundaries, respectively.

In order to further verify the inference above, the annual average SOS of 2006–2018 and its change trend at different levels of LST was calculated to explore the relationship between LST and SOS. As shown in Figure 8a, the annual average SOS at low, medium, and high LST were 88.9, 85.8, and 85.0 DOY, respectively. The SOS continently decreased with the increase of the LST. That is, the spring phenology continued to advance. As shown in Figure 8b, the change trend of SOS at low, medium, and high LST were −5.1, −3.9, and

−2 days/10 years, respectively. The downward trend of SOS decreased with the increase of LST. That is, the rate of advancement of phenology continuously slowed down. In general, the inference established that there was a negative correlation between SOS and LST, and it developed in a consistent direction, showing a trend of convergence.

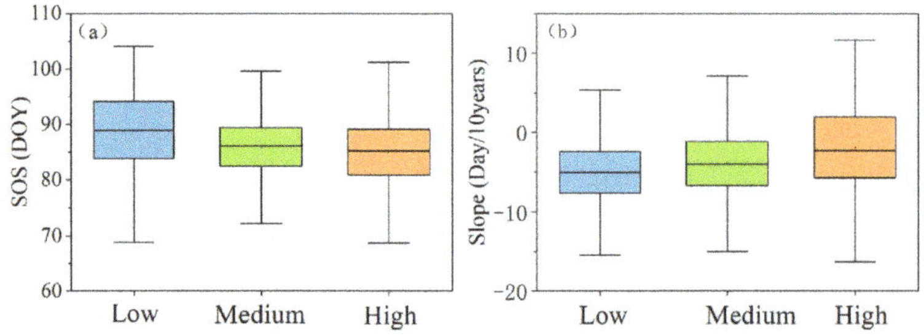

Figure 8. Distribution of (**a**) the annual average of the start of growing season (SOS) of 2006–2018 and (**b**) its interannual variation trend under different temperatures. Slope denotes the slope of the linear regression line between SOS and year. Low, Medium and High denote different temperatures according to the natural breakpoint classification method. The DOY denotes the day of year. In the box charts, the box denote the values of median, lower quartile (Q1) and upper quartile (Q3), respectively; the error bars denote the values of Q1 − 1.5(Q3 − Q1) and Q3 + 1.5(Q3 − Q1), respectively.

3.3. Relationship between LST and SOS at Sample Points

The relationship between LST and SOS at the sample point scale was further employed Figure 9. The LST in the urban of Hangzhou was significantly higher than that in the rural, exhibiting a significant urban heat island effect. In terms of the LST during the daytime in spring, the difference between that in the urban (23.0 ± 1.2 °C) and the rural (19.7 ± 1.0 °C) was significant, reaching 3.3 ± 1.0 °C. From 2006 to 2018, the difference continued to increase, with an average annual increase of 0.2 °C ($p < 0.01$). For the LST during the nighttime in spring, the difference between that in the urban (8.2 ± 1.0 °C) and the rural (7.0 ± 0.7 °C) reached 1.2 ± 0.3 °C. Compared with the daytime, the urban heat island effect was weaker at night, reduced by 2.1 °C. Integrating the temperature during the daytime and nighttime Figure 9c, the difference of the daily average temperature between that in the urban (15.6 ± 0.8 °C) and the rural (13.3 ± 0.8 °C) reached 2.3 ± 0.5 °C, and it continued to increase from 2006 to 2018 (Slope = 0.09 °C/years, $p < 0.01$). To further compare the difference of SOS between that in the urban and the rural Figure 9d, the SOS of urban (79.3 ± 5.6 DOY) was 7.4 ± 2.7 days earlier than that in the rural (86.7 ± 5.0 DOY). It indicated that the plant phenology changed significantly under different environmental backgrounds in the urban and the rural. Besides, it was worth noting that, except 2012 (1.5 days) and 2015 (12.9 days), the difference of SOS between that in the urban and the rural was relatively stable from 2006 to 2018, with a difference of 7.5 ± 1.6 days, and it was consistent with the difference of 6–9 days in Hangzhou Figure 6.

As shown in Figure 10, the LST during the daytime rather than nighttime showed a statistically significantly negative correlation with SOS both in the urban and the rural. In terms of the LST during the daytime Figure 10a, the sample points of LST–SOS in the urban and the rural distributed significantly separately, with LST of 19–28 °C and 16–23 °C, and SOS of 40–100 DOY and 50–110 DOY in the urban and the rural, respectively. Besides, the SOS in the rural was statistically significantly negatively correlated with the LST (Slope = −1.64 days/°C; $p < 0.01$). The correlation in urban had a degree of significance (Slope = −1.01 days/°C, $p < 0.05$), and the change trend of the SOS with LST increasing in the urban was smaller than that in the rural. For the LST during the nighttime Figure 10b, the sample points of LST-SOS in the rural and the urban had poor separation and high similarity. There was less statistically significant correlation between SOS and LST ($p > 0.05$), indicating that the LST during the nighttime had little effect on SOS. However, the fitted trends all showed that the SOS decreased with the increasing LST, which was consistent with the response of SOS to LST in Figure 8.

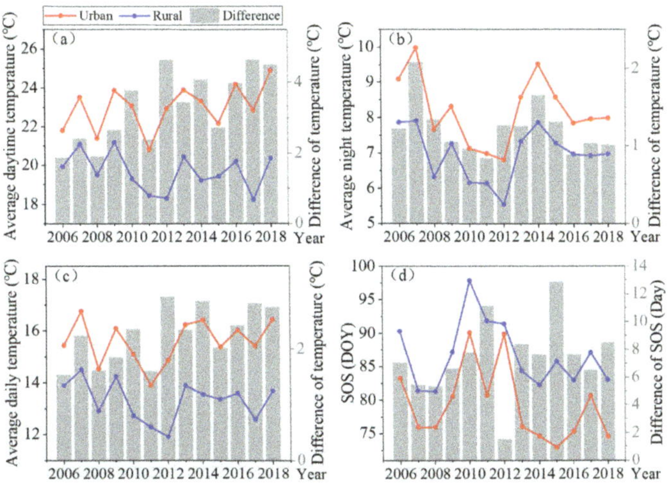

Figure 9. The interannual changes and differences of (**a**) the daytime land surface temperature in spring, (**b**) the nighttime land surface temperature in spring, (**c**) the daily average temperature, and (**d**) the start of growing season (SOS) in the urban and the rural from 2006 to 2018. The red/blue dotted lines denote the interannual changes, and the gray bars denote the differences. The DOY denotes the day of year.

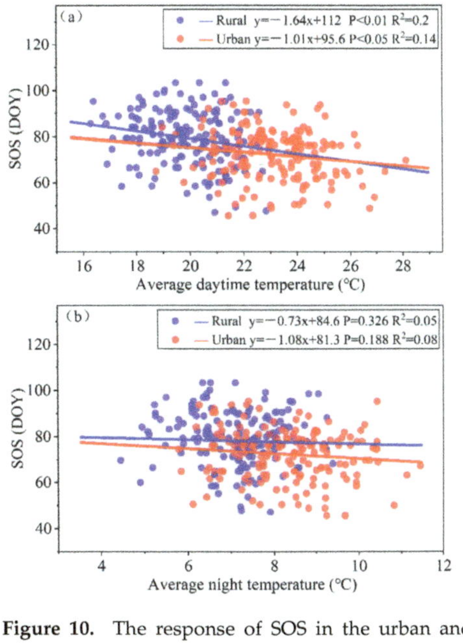

Figure 10. The response of SOS in the urban and the rural to (**a**) daytime temperature and (**b**) nighttime temperature. The SOS means the start of the growing season. The red/blue solid points denote urban/rural sample data, the red/blue solid lines denote linear regression lines, and the regression equation and significance are shown in the illustration. The DOY denotes the day of year.

3.4. Relative Contributions of ΔLST to ΔSOS at Sample Points

The temperature contribution separation model was utilized to explore the contributions of the ΔLST to the ΔSOS under urbanization. The difference of SOS under urbanization between predicted by the model (7.3 ± 1.3 days) and observed through data (7.4 ± 2.7 days) was −0.16 ± 1.4 days Figure 11a. For the results predicted by the model Figure 11b, the ΔLST played a significant role in the advance of SOS under urbanization and the proportion of its contributions was relatively stable, despite the interannual fluctuations. Besides, we found that the ΔSOS dominated by the ΔLST contributed 72 ± 13.3% (5.3 ± 1.7 of 7.3 ± 1.3 days) to the difference of SOS between that in the urban and the rural Figure 11a. The advance of 28 ± 13.3% (1.9 ± 0.7 days) of SOS under urbanization was dominated by other factors such as photoperiod and air pollution. Overall, the local warming effect induced by the urban heat island effect produced substantial impacts on plant phenology under urbanization, but the impact of other factors also cannot be ignored.

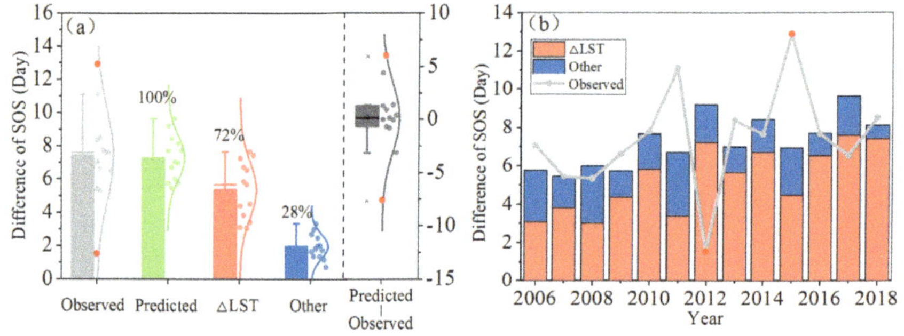

Figure 11. (**a**) The contributions of the difference of land surface temperature between urban and rural (ΔLST) and other factors to the difference of plant phenology under urbanization and (**b**) their interannual variations from 2006 to 2018. The SOS means the start of the growing season. Quantile chart denotes the distribution of SOS differences from 2006 to 2018. Solid red points represent abnormal values. The DOY denotes the day of year.

4. Discussion

Plant phenology is one of the key regulators of ecosystem processes, which is sensitive to environmental change. The acceleration of urbanization in recent years has produced substantial impacts on vegetation phenology over urban areas, such as the local warming induced by the urban heat island effect. This study explored impacts of urbanization on SOS and distinguished contributions of ΔLST and other factors to ΔSOS based on a temperature contribution separation model. We found that the SOS was negatively correlated with the daytime LST in spring, and the ΔSOS dominated by the ΔLST and other factors contributed 72% and 28% to the ΔSOS, respectively. Previous studies showed that there were lots of aspects besides temperature were different between urban and rural under urbanization, which had a certain impact on plant phenology. (1) The land cover changes under urbanization changed the soil properties extremely in the urban, affecting the relationship between plants, water and nutrients [49]. (2) There were high concentration of greenhouse gases (such as CO_2) and major pollutants (such as NO, NO_2, CO, SO_2 and particulates with a diameter of 10µm or less) in the urban, resulting from the emissions produced by factories and automobiles [50]. In this regard, many studies showed that pollutants in the urban environment could cause the advance or delay of plant phenology [51–53]. (3) Due to the increasing artificial light caused by the human activities at night in the urban, the growth of plants was seriously influenced [54,55]. Therefore, there were various differences between the urban areas and their rural surroundings, not only in terms of temperature, but also in other aspects that affect plant phenology.

The LST in spring from February to April was selected to explore the relationship between LST and SOS in this study. Previous studies showed that meteorological parameters such as temperature and precipitation in a period of time before the phenological event were important determinants affecting the occurrence of phenology. The period of time is significant to the study of the relationship between phenology and climate [56,57], which called preseason duration. Polgar et al. found that the temperature in late winter and spring or preseason temperature played an important role in the occurrence of SOS [58]. Zhou et al. and Jia et al. found that the LST showed a statistically significant correlation with SOS ($p < 0.05$, $R^2 > 0.8$) [7,15]. In addition, different temperature indicators were used to explore its effect on the SOS, such as daily maximum temperature and diurnal temperature difference. (1) Piao et al. showed that the SOS was more sensitive to the preseason daily maximum temperature in the northern hemisphere, and 68% of the European Union and 83% of the United States had a preseason duration of 0–3 months [59]. (2) The results of Huang et al. showed that 77.2% of the northern hemisphere, the SOS had the strongest correlation with the average diurnal temperature difference in the preseason period of 1–3 months [60]. In general, the results above indicated that the temperature in spring was relevant to SOS with the preseason duration of 0–3 months.

At the same time, there were some limitations in this study. (1) As shown in Figure 10, the LST during the daytime showed a statistically significantly negative correlation with SOS both in the urban ($p < 0.05$) and the rural ($p < 0.01$), while the R^2 was low. The low R^2 might be related to the impurity of the data where existed many mixed pixels in the urban sample. To explore the changes in plant phenology caused by the urban heat island effect, we needed to focus on forest where is severely affected by urbanization. However, the MODIS EVI data with a resolution of 250 m was the data with long time series, the highest resolution and we could currently obtain. Therefore, the problem of mixed pixels inevitably existed in the urban, which affected the correlation between SOS and LST. Although the R^2 was relatively low, they were both statistically significant ($p < 0.05$), which was meaningful in a certain degree. In the future, data with higher resolutions should be used to reduce the uncertainty caused by the data and make results more reliable. (2) In this study, all types of forest in Hangzhou were used to explore the impact of urbanization on plant phenology. Previous studies found that the phenology and the response to urbanization varied in the different types of vegetation. However, due to the limitation of the accuracy of data, we could only exclude shrub, farmland and grassland. Relatively uniform and stable forests were extracted as the study object to weaken the impact of different vegetation types to a certain extent. Reliably, previous studies utilized all vegetation types for the study and got reliable results, having a certain significance [7,33]. In future research, we hope to more finely distinguish the vegetation types and improve the accuracy of the study, in order to obtain more reliable results. (3) As mentioned above, there were lots of aspects besides temperature that were different between urban and rural under urbanization, which had a certain impact on plant phenology. Although other factors were taken into consideration in this study, no specific data analysis was carried out on other factors such as greenhouse gases. In future studies, a more quantitative and detailed discussion on the effects of other factors on vegetation phenology should be advanced. With the development of high-quality data and online data processing platforms (e.g., Google Earth Engine), it is feasible to apply the methodology of our present work to other study areas even to a global scale.

5. Conclusions

Focused on Hangzhou, a typical subtropical metropolis, to investigate the impact of urbanization on plant phenology. Vegetation index-based phenology data and land surface temperature data were adopted to analyze the urban–rural gradient in phenology characteristics and the contributions of ΔLST to the ΔSOS under urbanization. We found that there was a negative coupling between SOS and LST in over 90% of the vegetated areas in Hangzhou. At the sample-point scale, SOS was weakly, but significantly, negatively correlated with LST at the daytime ($R^2 = 0.2$ and $p < 0.01$ in rural; $R^2 = 0.14$ and $p < 0.05$ in urban) rather than that at nighttime. Besides, the ΔSOS dominated by the ΔLST contributed more than 70% of the total ΔSOS. We consider that the achievements of this study will provide quantitative evidence for the impact of urbanization on the plant phenology, and help to deepen the understanding of urban ecosystem adaptation under intensive human activities.

Author Contributions: Conceptualization, J.J.; data curation, Y.J.; formal analysis, Y.J.; investigation, Y.J.; methodology, Y.J. and J.J.; project administration, J.J.; resources, J.J.; supervision, J.J.; validation, Y.J.; visualization, Y.J.; writing—original draft, Y.J.; writing—review and editing, J.J., W.Z., F.G. and T.Y. All authors have read and agreed to the published version of the manuscript.

Funding: This research was supported in part by National Natural Science Foundation of China (grant No. 41971374), Fundamental Research Funds for the Central Universities (grant No. B200202016) and Jiangsu Provincial Natural Science Foundation (grant No. BK20180009).

Institutional Review Board Statement: Not applicable.

Informed Consent Statement: Not applicable.

Data Availability Statement: The study did not report any data.

Acknowledgments: We are grateful to NASA's Land Processes Distributed Active Archive Center (LP DAAC) and Gong Peng Research Group of Tsinghua University for providing the MODIS products and the multi-temporal dataset Global Urban Boundary, respectively. Additionally, we express our gratitude to anonymous reviewers and editors for their professional comments and suggestions.

Conflicts of Interest: The authors declare no conflict of interest.

References

1. Badeck, F.W.; Bondeau, A.; Bottcher, K.; Doktor, D.; Lucht, W.; Schaber, J.; Sitch, S. Responses of spring phenology to climate change. *New Phytol.* **2004**, *162*, 295–309. [CrossRef]
2. Chen, X.; Wang, D.; Chen, J.; Wang, C.; Shen, M. The mixed pixel effect in land surface phenology: A simulation study. *Remote Sens. Environ.* **2018**, *211*, 338–344. [CrossRef]
3. White, M.A.; de Beurs, K.M.; Didan, K.; Inouye, D.W.; Richardson, A.D.; Jensen, O.P.; O'Keefe, J.; Zhang, G.; Nemani, R.R.; van Leeuwen, W.J.D.; et al. Intercomparison, interpretation, and assessment of spring phenology in North America estimated from remote sensing for 1982–2006. *Glob. Chang. Biol.* **2009**, *15*, 2335–2359. [CrossRef]
4. Linderholm, H.W. Growing season changes in the last century. *Agric. For. Meteorol.* **2006**, *137*, 1–14. [CrossRef]
5. Menzel, A. Phenology: Its importance to the global change community. *Clim. Chang.* **2002**, *54*, 379–385. [CrossRef]
6. Richardson, A.D.; Black, T.A.; Ciais, P.; Delbart, N.; Friedl, M.A.; Gobron, N.; Hollinger, D.Y.; Carriage, W.L.; Longdoz, B.; Luyssaert, S.; et al. Influence of spring and autumn phenological transitions on forest ecosystem productivity. *Philos. Trans. R. Soc. B-Biol. Sci.* **2010**, *365*, 3227–3246. [CrossRef]
7. Zhou, D.; Zhao, S.; Zhang, L.; Liu, S. Remotely sensed assessment of urbanization effects on vegetation phenology in China's 32 major cities. *Remote Sens. Environ.* **2016**, *176*, 272–281. [CrossRef]
8. Wang, L.; Chen, H.; Li, Q.; Yu, W. Research advances in plant phenology and climate. *Acta Ecol. Sin.* **2010**, *30*, 447–454.
9. Richardson, A.D.; Keenan, T.F.; Migliavacca, M.; Ryu, Y.; Sunny Day, O.; Toomey, M. Climate change, phenology, and phenological control of vegetation feedbacks to the climate system. *Agric. For. Meteorol.* **2013**, *169*, 156–173. [CrossRef]
10. Cong, N.; Piao, S.; Chen, A.; Wang, X.; Lin, X.; Chen, S.; He, S.; Zhou, G.; Zhang, X. Spring vegetation green-up date in China inferred from SPOT NDVI data: A multiple model analysis. *Agric. For. Meteorol.* **2012**, *165*, 104–113. [CrossRef]
11. Aono, Y.; Kazui, K. Phenological data series of cherry tree flowering in Kyoto, Japan, and its application to reconstruction of springtime temperatures since the 9th century. *Int. J. Climatol.* **2008**, *28*, 905–914. [CrossRef]
12. Delbart, N.; Picard, G.; le Toan, T.; Kergoat, L.; Quegan, S.; Woodward, I.; Dye, D.; Fedotova, V. Spring phenology in boreal Eurasia over a nearly century time scale. *Glob. Chang. Biol.* **2008**, *14*, 603–614. [CrossRef]
13. Keenan, T.F.; Gray, J.; Friedl, M.A.; Toomey, M.; Bohrer, G.; Hollinger, D.Y.; Munger, J.W.; O'Keefe, J.; Schmid, H.P.; SueWing, I. Net carbon uptake has increased through warming-induced changes in temperate forest phenology. *Nat. Clim. Chang.* **2014**, *4*, 598–604. [CrossRef]
14. Zeng, H.; Jia, G.; Epstein, H. Recent changes in phenology over the northern high latitudes detected from multi-satellite data. *Environ. Res. Lett.* **2011**, *6*, 045508. [CrossRef]
15. Jia, W.; Zhao, S.; Zhang, X.; Liu, S.; Henebry, G.M.; Liu, L. Urbanization imprint on land surface phenology: The urban–rural gradient analysis for Chinese cities. *Glob. Chang. Biol.* **2011**, *27*, 2895–2904. [CrossRef]

16. Jia, W.; Zhao, S.; Liu, S. Vegetation growth enhancement in urban environments of the Conterminous United States. *Glob. Chang. Biol.* **2018**, *24*, 4084–4094. [CrossRef]
17. Li, X.; Zhou, Y.; Asrar, G.R.; Mao, J.; Li, X.; Li, W. Response of vegetation phenology to urbanization in the conterminous United States. *Glob. Chang. Biol.* **2017**, *23*, 2818–2830. [CrossRef]
18. Ren, Q.; He, C.; Huang, Q.; Zhou, Y. Urbanization impacts on vegetation phenology in China. *Remote Sens.* **2018**, *10*, 1905. [CrossRef]
19. Li, D.; Stucky, B.J.; Deck, J.; Kiss, B.; Guralnick, R.P. The effect of urbanization on plant phenology depends on regional temperature. *Nat. Ecol. Evol.* **2019**, *3*, 1661–1667. [CrossRef]
20. Arnfield, A.J. Two decades of urban climate research: A review of turbulence, exchanges of energy and water, and the urban heat island. *Int. J. Climatol.* **2003**, *23*, 1–26. [CrossRef]
21. Clinton, N.; Gong, P. MODIS detected surface urban heat islands and sinks: Global locations and controls. *Remote Sens. Environ.* **2013**, *134*, 294–304. [CrossRef]
22. Oke, T.R. The energetic basis of the urban heat island. *Q. J. R. Meteorol. Soc.* **1982**, *108*, 1–24. [CrossRef]
23. He, C.; Shi, P.; Chen, J.; Xu, X. Process and mechanism of urbanization in Beijing area. *Acta Geogr. Sin.* **2002**, *57*, 363–371. [CrossRef]
24. Wang, S.; Ju, W.; Penuelas, J.; Cescatti, A.; Zhou, Y.; Fu, Y.; Huete, A.; Liu, M.; Zhang, Y. Urban-rural gradients reveal joint control of elevated CO_2 and temperature on extended photosynthetic seasons. *Nat. Ecol. Evol.* **2019**, *3*, 1076–1085. [CrossRef] [PubMed]
25. Franklin, K.A.; Whitelam, G.C. Light signals, phytochromes and cross-talk with other environmental cues. *J. Exp. Bot.* **2004**, *55*, 271–276. [CrossRef]
26. Mimet, A.; Pellissier, V.; Quenol, H.; Aguejdad, R.; Dubreuil, V.; Rink, F. Urbanization induces early flowering: Evidence from Platanus acerifolia and Prunus cerasus. *Int. J. Biometeorol.* **2009**, *53*, 287–298. [CrossRef]
27. Jeong, S.J.; Park, H.; Ho, C.H.; Kim, J. Impact of urbanization on spring and autumn phenology of deciduous trees in the Seoul Capital Area, South Korea. *Int. J. Biometeorol.* **2019**, *63*, 627–637. [CrossRef]
28. Ding, H.; Xu, L.; Elmore, A.J.; Shi, Y. Vegetation phenology influenced by rapid urbanization of The Yangtze Delta region. *Remote Sens.* **2020**, *12*, 1783. [CrossRef]
29. Tian, J.; Zhu, X.; Wu, J.; Shen, M.; Chen, J. Coarse-resolution satellite images overestimate urbanization effects on vegetation spring phenology. *Remote Sens.* **2020**, *12*, 117. [CrossRef]
30. Zhao, S.; Liu, S.; Zhou, D. Prevalent vegetation growth enhancement in urban environment. *Proc. Natl. Acad. Sci. USA* **2016**, *113*, 6313–6318. [CrossRef]
31. Landsberg, H.E. Atmospheric changes in a growing community (the Columbia, Maryland Experience). *Urban Ecol.* **1979**, *4*, 53–81. [CrossRef]
32. Petterson, J.T.; Stoffel, T.L. Analysis of urban-rural radiation data from St. Louis Missouri. *J. Appl. Meteorol.* **1980**, *19*, 275–283. [CrossRef]
33. Meng, L.; Mao, J.; Zhou, Y.; Richardson, A.D.; Lee, X.H.; Thornton, P.E.; Ricciuto, D.M.; Li, X.; Dai, Y.; Shi, X.; et al. Urban warming advances spring phenology but reduces the response of phenology to temperature in the conterminous United States. *Proc. Natl. Acad. Sci. USA* **2020**, *117*, 4228–4233. [CrossRef]
34. Wohlfahrt, G.; Tomelleri, E.; Hammerle, A. The urban imprint on plant phenology. *Nat. Ecol. Evol.* **2019**, *3*, 1669–1674. [CrossRef]
35. Hu, Z.; Dai, H.; Hou, F.; Li, D. Spatio-temporal change of urban-rural vegetation phenology and its response to land surface temperature in Northeast China. *Acta Ecol. Sin.* **2020**, *40*, 4137–4145. [CrossRef]
36. Li, Y.; Guan, K.; Peng, B.; Franz, T.E.; Wardlow, B.; Pan, M. Quantifying irrigation cooling benefits to maize yield in the US Midwest. *Glob. Chang. Biol.* **2019**, *26*, 3065–3078. [CrossRef] [PubMed]
37. Li, X.; Gong, P.; Zhou, Y.; Wang, J.; Bai, Y.; Chen, B.; Hu, T.; Xiao, Y.; Xu, B.; Yang, J.; et al. Mapping global urban boundaries from the global artificial impervious area (GAIA) data. *Environ. Res. Lett.* **2020**, *15*, 094044. [CrossRef]
38. Zhou, D.; Zhao, S.; Liu, S.; Zhang, L.; Zhu, C. Surface urban heat island in China's 32 major cities: Spatial patterns and drivers. *Remote Sens. Environ.* **2014**, *152*, 51–61. [CrossRef]
39. Peng, S.; Piao, S.; Ciais, P.; Friedlingstein, P.; Ottle, C.; Bréon, F.M.; Nan, H.; Zhou, L.; Myneni, R.B. Surface urban heat island across 419 global big cities. *Environ. Sci. Technol.* **2012**, *46*, 696–703. [CrossRef]
40. Ganguly, S.; Friedl, M.A.; Tan, B.; Zhang, X.; Verma, M. Land surface phenology from MODIS: Characterization of the Collection 5 global land cover dynamics product. *Remote Sens. Environ.* **2010**, *114*, 1805–1816. [CrossRef]
41. Zhang, X.; Friedl, M.A.; Schaaf, C.B. Global vegetation phenology from Moderate Resolution Imaging Spectroradiometer (MODIS): Evaluation of global patterns and comparison with in situ measurements. *J. Geophys. Res. Biogeosci.* **2006**, *111*, G04017. [CrossRef]
42. Zhang, X.; Friedl, M.A.; Schaaf, C.B.; Strahler, A.H.; Schneider, A. The footprint of urban climates on vegetation phenology. *Geophys. Res. Lett.* **2004**, *31*, L12209. [CrossRef]
43. Huete, A.; Didan, K.; Miura, T.; Rodriguez, E.P.; Gao, X.; Ferreira, L.G. Overview of the radiometric and biophysical performance of the MODIS vegetation indices. *Remote Sens. Environ.* **2002**, *83*, 195–213. [CrossRef]
44. Dallimer, M.; Tang, Z.; Bibby, P.R.; Brindley, P.; Gaston, K.J.; Davies, Z.G. Temporal changes in greenspace in a highly urbanized region. *Biol. Lett.* **2011**, *7*, 763–766. [CrossRef]

45. Zhou, D.; Zhao, S.; Liu, S.; Zhang, L. Spatiotemporal trends of terrestrial vegetation activity along the urban development intensity gradient in China's 32 major cities. *Sci. Total Environ.* **2014**, *488*, 136–145. [CrossRef] [PubMed]
46. Brown, M.E.; de Beurs, K.M.; Vrieling, A. The response of African land surface phenology to large scale climate oscillations. *Remote Sens. Environ.* **2010**, *114*, 2286–2296. [CrossRef]
47. Buyantuyev, A.; Wu, J. Urbanization diversifies land surface phenology in arid environments: Interactions among vegetation, climatic variation, and land use pattern in the Phoenix metropolitan region, USA. *Landsc. Urban Plan.* **2012**, *105*, 149–159. [CrossRef]
48. Zhang, J.; Tong, X.; Zhang, J.; Meng, P.; Li, J.; Liu, P. Dynamics of phenology and its response to climatic variables in a warm-temperate mixed plantation. *For. Ecol. Manag.* **2021**, *483*, 118785. [CrossRef]
49. Herrmann, D.L.; Schifman, L.A.; Shuster, W.D. Widespread loss of intermediate soil horizons in urban landscapes. *Proc. Natl. Acad. Sci. USA* **2018**, *115*, 6751–6755. [CrossRef] [PubMed]
50. Grange, S.K.; Lewis, A.C.; Moller, S.J.; Carslaw, D.C. Lower vehicular primary emissions of NO_2 in Europe than assumed in policy projections. *Nat. Geosci.* **2017**, *10*, 914–918. [CrossRef]
51. Jochner, S.; Markevych, I.; Beck, I.; Traidl-Hoffmann, C.; Heinrich, J.; Menzel, A. The effects of short- and long-term air pollutants on plant phenology and leaf characteristics. *Environ. Pollut.* **2015**, *206*, 382–389. [CrossRef]
52. Honour, S.L.; Bell, J.N.; Ashenden, T.W.; Cape, J.N.; Power, S.A. Responses of herbaceous plants to urban air pollution: Effects on growth, phenology and leaf surface characteristics. *Environ. Pollut.* **2009**, *157*, 1279–1286. [CrossRef]
53. Sanz, J.; Bermejo, V.; Muntifering, R.; Gonzalez-Fernandez, I.; Gimeno, B.S.; Elvira, S.; Alonso, R. Plant phenology, growth and nutritive quality of Briza maxima: Responses induced by enhanced ozone atmospheric levels and nitrogen enrichment. *Environ. Pollut.* **2011**, *159*, 423–430. [CrossRef]
54. Skvareninova, J.; Tuharska, M.; Skvarenina, J.; Babalova, D.; Slobodnikova, L.; Slobodnik, B.; Stredova, H.; Mindas, J. Effects of light pollution on tree phenology in the urban environment. *Morav. Geogr. Rep.* **2017**, *25*, 282–290. [CrossRef]
55. Massetti, L. Assessing the impact of street lighting on Platanus x acerifolia phenology. *Urban For. Urban Green.* **2018**, *34*, 71–77. [CrossRef]
56. Matsumoto, K.; Ohta, T.; Irasawa, M.; Nakamura, T. Climate change and extension of the Ginkgo biloba L. growing season in Japan. *Glob. Chang. Biol.* **2003**, *9*, 1634–1642. [CrossRef]
57. Dai, J.; Wang, H.; Ge, Q. Multiple phenological responses to climate change among 42 plant species in Xi'an, China. *Int. J. Biometeorol.* **2013**, *57*, 749–758. [CrossRef] [PubMed]
58. Polgar, C.; Gallinat, A.; Primack, R.B. Drivers of leaf-out phenology and their implications for species invasions: Insights from Thoreau's Concord. *New Phytol.* **2014**, *202*, 106–115. [CrossRef] [PubMed]
59. Piao, S.; Tan, J.; Chen, A.; Pennuelas, J.; Ciais, P.; Liu, Q.; Janssens, I.A.; Vicca, S.; Zeng, Z.; Jeong, S.J.; et al. Leaf onset in the northern hemisphere triggered by daytime temperature. *Nat. Commun.* **2015**, *6*, 6911. [CrossRef] [PubMed]
60. Huang, Y.; Jiang, N.; Shen, M.; Guo, L. Effect of preseason diurnal temperature range on the start of vegetation growing season in the Northern Hemisphere. *Ecol. Indic.* **2020**, *112*, 106161. [CrossRef]

remote sensing

MDPI

Article

Specific Drivers and Responses to Land Surface Phenology of Different Vegetation Types in the Qinling Mountains, Central China

Jiaqi Guo [1], Xiaohong Liu [1,*], Wensen Ge [1], Xiaofeng Ni [2], Wenyuan Ma [3], Qiangqiang Lu [1] and Xiaoyu Xing [4]

[1] School of Geography and Tourism, Shaanxi Normal University, Xi'an 710119, China; guojiaqi@snnu.edu.cn (J.G.); wsge@snnu.edu.cn (W.G.); luqiang@snnu.edu.cn (Q.L.)
[2] Key Laboratory for Earth Surface Processes of the Ministry of Education, Institute of Ecology, College of Urban and Environmental Sciences, Peking University, Beijing 100871, China; nixiaofeng@pku.edu.cn
[3] School of Science, University of New South Wales, Canberra, ACT 2600, Australia; wenyuan.ma@adfa.edu.au
[4] Qinling National Botanical Garden, Xi'an 710061, China; Xingxiaoyu668@163.com
* Correspondence: xhliu@snnu.edu.cn; Tel.: +86-138-9337-8787

Citation: Guo, J.; Liu, X.; Ge, W.; Ni, X.; Ma, W.; Lu, Q.; Xing, X. Specific Drivers and Responses to Land Surface Phenology of Different Vegetation Types in the Qinling Mountains, Central China. *Remote Sens.* **2021**, *13*, 4538. https://doi.org/10.3390/rs13224538

Academic Editors: Xuanlong Ma, Jiaxin Jin, Xiaolin Zhu, Yuke Zhou and Qiaoyun Xie

Received: 9 October 2021
Accepted: 8 November 2021
Published: 11 November 2021

Abstract: Land surface phenology (LSP), as a precise bio-indicator that responds to climate change, has received much attention in fields concerned with climate change and ecology. Yet, the dynamics of LSP changes in the Qinling Mountains (QMs)—A transition zone between warm-temperate and north subtropical climates with complex vegetation structure—under significant climatic environmental evolution are unclear. Here, we analyzed the spatiotemporal dynamics of LSP for different vegetation types in the QMs from 2001 to 2019 and quantified the degree of influence of meteorological factors (temperature, precipitation, and shortwave radiation), and soil (temperature and moisture), and biological factors (maximum of NDVI and middle date during the growing season) on LSP changes using random forest models. The results show that there is an advanced trend (0.15 days/year) for the start of the growing season (SOS), a delayed trend (0.24 days/year) for the end of the growing season (EOS), and an overall extended trend (0.39 days/year) for the length of the growing season (LOS) in the QMs over the past two decades. Advanced SOS and delayed EOS were the dominant patterns leading to a lengthened vegetation growing season, followed by a joint delay of SOS and EOS, and the latter was particularly common in shrub and evergreen broadleaved forests. The growth season length increased significantly in western QMs. Furthermore, we confirmed that meteorological factors are the main factors affecting the interannual variations in SOS and EOS, especially the meteorological factor of preseason mean shortwave radiation (SWP). The grass and crop are most influenced by SWP. The soil condition has, overall, a minor influence the regional LSP. This study highlighted the specificity of different vegetation growth in the QMs under warming, which should be considered in the accurate prediction of vegetation growth in the future.

Keywords: land surface phenology; NDVI; spatiotemporal dynamics; different drivers; random forest model

1. Introduction

Vegetation phenology is the seasonal timing of lifecycle events, such as leaf emergence, flowering, leaf coloration and fall, and it has become an important topic in the field of climate and ecology as a sensitive and precise indicator that is responsive to climate warming [1,2]. Shifts in spring and autumn vegetation phenology caused by climate warming can differentially alter the length of the growing season, which affects carbon, water, and energy exchange between terrestrial ecosystems and the atmosphere [3–5]. Recent studies have reported that in addition to climatic factors, soil and biological factors also influence shifts in vegetation phenology by affecting plant growth processes in the context of ongoing global climate change [6,7], due to the poor interpretation of phenology shifts among different vegetation types [8,9]. Hence, it is essential to study the dynamics and drivers of phenology among different vegetation types to improve phenology models and enrich our understanding of the carbon cycle of terrestrial ecosystem.

With the application of remote sensing in monitoring vegetation phenology, we traditionally use the term land surface phenology (LSP) to denote the dynamic variations in vegetation land surface as observed from satellite imagery [10]. Satellite-derived LSP metrics are usually focused on the start (SOS) and end (EOS) of growing seasons [11]. Satellite-based studies have shown that SOS was advanced by 10.6 days (i.e., 5.4 days per decade) throughout Europe and by 14 days (i.e., 7.9 days per decade) in temperate China before 2000 [12,13]. However, this trend of SOS advancement may have slowed or even reversed since the 2000s. For instance, in the entire northern hemisphere, it was advanced by only 0.2 days during 2000 to 2008, but a delayed SOS was revealed in the Tibetan Plateau [14,15]. Regarding satellite-derived EOS, the published results have not always been consistent. Across the entire Northern Hemisphere, EOS was delayed at a rate of 2.2 days per decade during 2000–2008 [14]. In the Yellow River Basin, EOS was delayed by 5.6 and 3.4 days in 1982–1999 and 2000–2015, respectively [16]. In the Qinghai-Tibet Plateau, however, Wang et al. [17] reported the opposite phenology change trends, in both the east and west zones. Overall, these varied results might be due to different study areas, periods, and methods of extracting phenology metrics. However, few studies have focused on the diversity of phenology across different vegetation types. In particular, the dynamic phenology characteristics of herbaceous or shrubs and evergreen forests in the Qinling Mountains have been little studied.

To date, the processes and drivers governing LSP remain poorly understood. Several studies reported that temperature is a major driver of early spring leaf development and delayed autumn leaf fall in plants and has less control over autumn phenology than spring phenology [18,19]. The impact of precipitation on LSP processes is mainly directed at plants in arid and semiarid regions, where water deficits limit the use of light and heat conditions by plants in arid and semiarid areas [20]. Some studies further considered solar radiation (i.e., shortwave radiation) and found that increasing photosynthetic active radiation can promote earlier leaf germination and delaying leaf senescence [21]. Besides meteorological factors, soil factors and biological factors have also been shown to be drivers of LSP change processes [22–24]. Soil temperature and moisture information, due to the prevalent freezing—Thawing process of soil in alpine and arctic regions, could more directly control vegetation growth; for example, soil wetting will, to some extent, reduce the effect of soil warming on LSP changes [22,25]. Slight fluctuations in the time interval between the middle date (MD) of the growing season and the autumn phenology have a strong effect on regulating EOS [7]. Peak growth in summer (i.e., maximum NDVI during the growing season, MN) can have an impact on vegetation greening and senescence, and its unique vegetation growth patterns may result in different allocations of green or carbon across the growing season [24]. To date, how these meteorological, soil, and biological factors affect regional-scale LSP variations has not been clearly and consistently studied, which seriously affects our ability to predict LSP periods.

The responses of vegetation growth processes to climate change are inherently nonlinear [26]. Random forest (RF), as a nonparametric multivariate method, can explain nonlinear processes to a large extent [27]. The advantage of the RF model is that it can consider many predictor variables and nonlinearly determine the relative importance of each predictor variable [28]. Due to its high efficiency in handling the potentially complex relationships between LSP periods and meteorological, soil, and biological factors, RF has been widely and successful applied in recent ecology studies [6,29].

The Qinling Mountains (QMs), rich in vegetation types, represent a demarcation line of climate in China and are also an area characterized by sensitivity or response to climate change, with a significant upward temperature trend in the past half century [30]. Here, we used the satellite derived normalized difference vegetation index (NDVI) records (2001–2019) from MOD13A2 to extract the LSP dates of QMs. The objectives were (a) to explore the temporal and spatial trends of LSP in the QMs, (b) to quantify the relative contribution of SOS and EOS of different vegetation types to the length of growing season (LOS) and to determine the dominant growth pattern during the growing season, and (c) to simulate the LSP dates and assess the relative importance of meteorological, soil, and biological factors on the interannual variations in LSP. This study focuses on the specificity of different vegetation growths in the QMs, and the results are helpful for future accurate prediction of vegetation growth and to develop scientific management strategies.

2. Materials and Methods

2.1. Study Area

Qinling is the highest mountain range in the central region of China and also a geographical boundary between the north and the south of China, with an elevation range of 51 to 5120 m and a spatial range from 30.8° to 35.5°N to 102.5° to 114.6°E (Figure 1a). Its climate differs significantly between north and south, with a humid northern subtropical climate in the south and a warm temperate semi-humid and semiarid climate in the north (Figure 1a). It has also been classified as one of the critical terrestrial biodiversity areas of world significance [31], with mixed coniferous and deciduous broadleaved forests widely distributed on its northern slopes, while the southern slopes are dominated by mixed evergreen and deciduous broadleaved forests (Figure 1b).

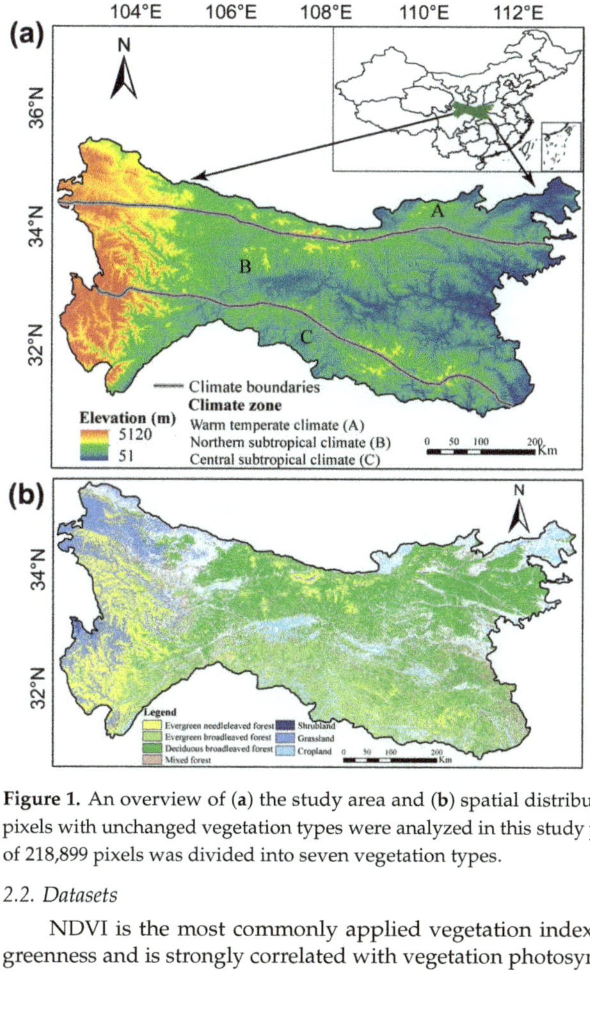

Figure 1. An overview of (**a**) the study area and (**b**) spatial distribution of vegetation types. Only pixels with unchanged vegetation types were analyzed in this study period. The dataset with a total of 218,899 pixels was divided into seven vegetation types.

2.2. Datasets

NDVI is the most commonly applied vegetation index to characterize vegetation greenness and is strongly correlated with vegetation photosynthetic activity [32]. Climate change is the main factor affecting the change of vegetation greenness, and this change can be reflected by the spectral information of NDVI images. In this study, we used the NDVI datasets generated from NOAA/AVHRR series satellite images by the NASA MODIS13A2 group (Table 1). We used this dataset to extract the LSP dates for QMs from 2001 to 2019. We also excluded areas of bare soil/sparse vegetation with an annual average NDVI of less than 0.1 [33].

Table 1. Datasets and sources in the study area.

Dataset	Spatial Resolution	Temporal Resolution	Time Span	Source
MODIS13A2 NDVI	1 km	16 days	2001–2019	The Level-1 and Atmosphere Archive and Distribution System Distributed Active Archive Center (LAADS DAAC) (https://search.earthdata.nasa.gov/search/, accessed on 15 October 2020).
Land cover (CCI-LC)	300 m	Yearly	2001–2019	http://maps.elie.ucl.ac.be/CCI/viewer/index.php, accessed on 20 October 2020
Temperature	0.1°	hourly	2001–2019	The Reanalysis (ERA5) climatic datasets (https://cds.climate.copernicus.eu, accessed on 12 November 2020)
Precipitation	0.1°	hourly	2001–2019	
Shortwave radiation	0.1°	hourly	2001–2019	
Soil temperature	0.1°	hourly	2001–2019	
Soil moisture	0.1°	hourly	2001–2019	

Meteorological data, including daily mean air temperature, daily total precipitation, and daily mean shortwave radiation and soil data, including daily mean soil temperature and moisture in 0–100 cm soil layer, from 2001 to 2019 were used in this study. These gridded data were derived from the ERA5-Land hourly data (Table 1). Moreover, we transformed the hourly climate data (24 hourly data were averaged for temperature, solar radiation, soil temperature and humidity, 24 hourly data were summed for precipitation) to daily-scale temporal resolution and resampled meteorological data to the same resolution as MODIS13A2 data. A time lag of 30 days before SOS and 60 days before EOS is defined as preseason.

The 300 m spatial resolution Climate Change Initiative Land Cover (CCI-LC) maps from 2001 to 2019 were available from the European Space Agency (ESA) (Table 1). CCI-LC discriminates 22 classes of land cover. In this study, we resampled these maps to 1 km and analyzed only pixels of unchanged vegetation types containing evergreen needle leaved forest (ENF), evergreen broadleaved forest (EBF), deciduous broadleaved forest (DBF), mixed forest (MF), shrubland (SL), grassland (GL), and cropland (CL).

2.3. Retrieval of Phenology Metrics from NDVI Time Series Data

The premise of quantitatively analyzing the phenology changes is to derive several key phenology metrics: SOS, EOS, LOS, MN, and MD (Figure 2). In this study, we firstly stacked the NDVI images from 2001 to 2019 in chronological order and smoothed the NDVI time series with a Savitzky-Golay (SG) filter for each pixel per year. The SG filter was chosen because it can best preserve the temporal vegetation dynamics and minimize atmospheric contamination and has also been integrated into the processing of the MODIS phenology product [34]. The smoothed data was used further for extracting phenology metrics of different vegetation types by detecting the inflection point (i.e., date) when the NDVI time series begins to ascend or descend for the specific year. This is the derivative method which the phenology metrics were extracted for each pixel per year, whereby the maximum value of $NDVI_{ratio}$ corresponds to the greatest change of the smoothed NDVI time series [35]. Equation (1) is given as

$$NDVI_{ratio(t)} = \frac{NDVI_{t+1} - NDVI_t}{NDVI_t} \quad (1)$$

where $NDVI_{ratio(t)}$ is the calculated relative changing rate of NDVI at time t and $NDVI_t$ is the NDVI value at time t. Occurrence dates were obtained using these smoothed NDVI time series. SOS and EOS dates were determined as the day with the maximum and minimum $NDVI_{ratio}$. LOS was determined to be the difference between EOS and SOS. MN was defined as the peak of vegetation growth, i.e., the NDVI value corresponding to $NDVI_{ratio}$ closest to zero. MD date was the middle date between the EOS date and the SOS date. The description of phenology metrics correlations is shown in Figure 2.

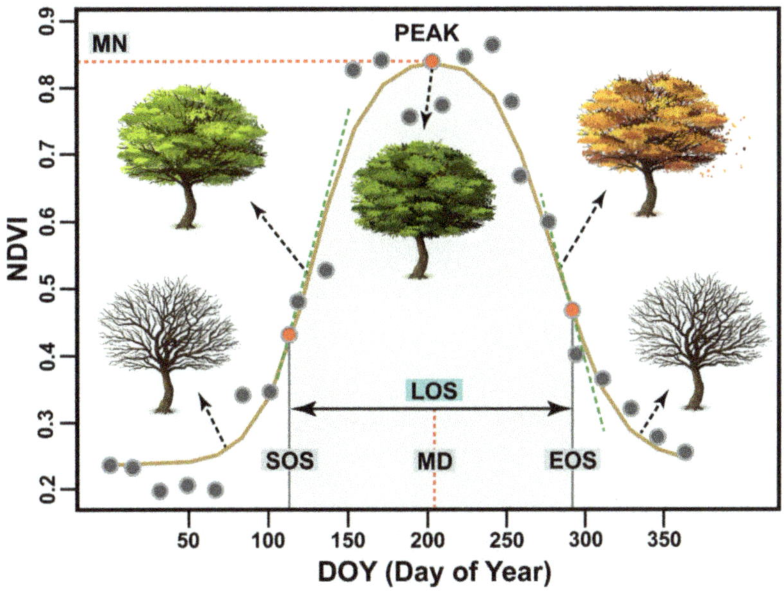

Figure 2. The description of phenology metrics correlations extracted using the NDVI time series datasets.

2.4. Method and Statistical Analysis

2.4.1. Trend Analysis

The method used in this study is shown in Figure 3. The spatiotemporal trends of LSP during 2001–2019 were estimated using *Sen's slope* method, also known as the Theil-Sen median method. The method is a robust nonparametric statistical method for trend calculation that is insensitive to measurement errors and is far more accurate than nonrobust simple linear regression [36]. *Sen's slope* was calculated using Equation (2):

$$Sen's\ slope = Median\left(\frac{x_j - x_i}{j - i}\right) \quad (2)$$

where the median is the mean value of all the slopes, and x_i and x_j represent the LSP dates of years i and j. A negative *Sen's slope* indicates an advancing trend, whereas a positive *Sen's slope* indicates a delaying trend.

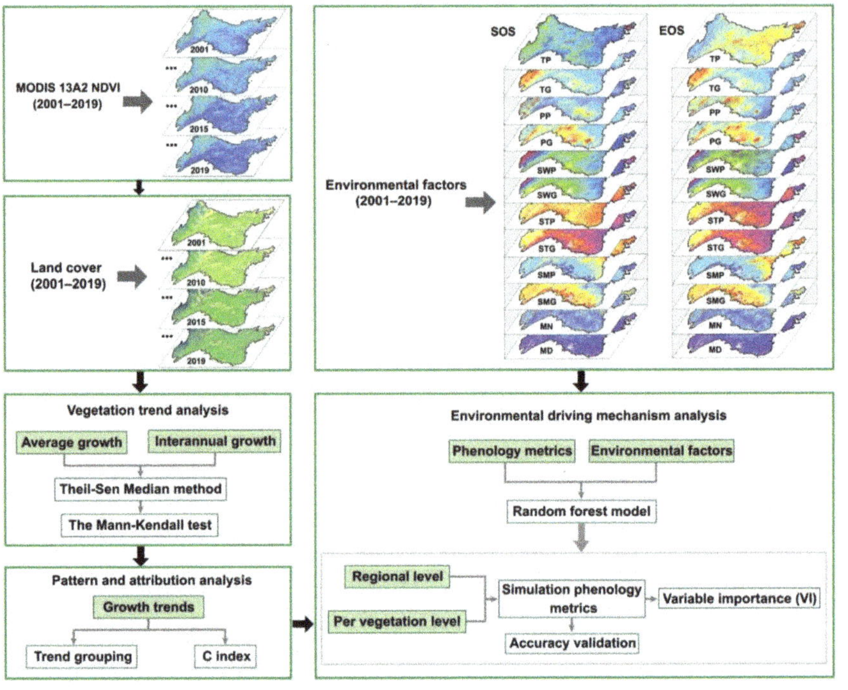

Figure 3. Flowchart of the method used in this study. MODIS 13A2 NDVI (2001–2019), in which the spatiotemporal distributions of NDVI for 2001, 2010, 2015 and 2019 are presented as examples of NDVI time series used as extract phenology metrics. Land cover (2001–2019), in which the spatiotemporal distribution of land cover for 2001, 2010, 2015 and 2019, used as an example of the changes in different vegetation types. Environmental factors, the first column indicates the drivers for SOS and the second column indicates the drivers for EOS, and shows the spatial distribution of each driver separately, which is used to example the predictors used for RF models.

Then, we used the Mann-Kendall (MK) method to test the significance of time series trends, which is a nonparametric statistical test and is robust to outliers [37]. We used the normal cumulative distribution function to determine the p-value of the MK test statistic with a significant confidence level of $p < 0.1$. In this study, we used *Sen's slope* and MK test to trend analysis and significance test the spatial distribution (average growth) and interannual variation (interannual growth) of LSP for different vegetation types from 2001 to 2019, all using MATLAB 2017a were completed.

2.4.2. Change Pattern and Relative Attribution Analysis

To further understand the seasonal changes of vegetation growth in the QMs in the past two decades, we divided the trend changes of LSP (SOS, EOS, and LOS) into six combinations, where each combination represents a pattern of vegetation growth (Table 2). We used spatial analysis to count the proportion and significance of different vegetation for each pattern and to derive the dominant pattern of seasonal changes in the growth of each vegetation. All analysis was accomplished using ArcGIS 10.4.

Table 2. Six change patterns of plant growing seasons, i.e., six combination types of SOS, EOS, and LOS change trends. The plus and minus signs represent the trend direction corresponding to SOS, EOS, or LOS, respectively.

Change Pattern	Trend of SOS	Trend of EOS	Trend of LOS
I	Advanced (−)	Delayed (+)	Lengthened (+)
II	Advanced (−)	Advanced (−)	Lengthened (+)
III	Advanced (−)	Advanced (−)	Shortened (−)
IV	Delayed (+)	Advanced (−)	Shortened (−)
V	Delayed (+)	Delayed (+)	Lengthened (+)
VI	Delayed (+)	Delayed (+)	Shortened (−)

To evaluate the symmetry of SOS and EOS for LOS changes, we used the C-index proposed by Garonna et al. [38] to calculate the relative contribution of trends in SOS and EOS for LOS changes. It was calculated as follows:

$$C = -1 + \frac{2 \cdot \left|SOS_{slope}\right|}{\left|SOS_{slope}\right| + \left|EOS_{slope}\right|} \quad (3)$$

where SOS_{slope} and EOS_{slope} are the *Sen's slope* of SOS and EOS, respectively. A positive C value indicates that the trend in LOS is mainly attributable to changes in EOS, and a negative C value indicates that the trend in LOS is mainly attributable to changes in SOS. The variation of C value is from −1 to 1.

2.4.3. Analysis of the Relative Importance of Different Drivers

Based on previous studies on the drivers of interannual variation in vegetation phenology [18–24], we selected drivers such as Table 3 to simulate SOS and EOS. These drivers are divided into three main categories: meteorological, soil and biological factors, which are further divided into preseason cumulative values and cumulative values throughout the growing season.

Table 3. Predictive variables used in the modeling of the LSP dates. The 12 predictive variables were classified into three categories: meteorological factors, soil factors, and biological factors. Meteorological factors include TP, TG, PP, PG, SWP, and SWG (6 in total). Soil factors include STP, STG, SMP, and SMG (4 in total). Biological factors include MN and MD (2 in total). Growing season is defined as the days between SOS and EOS.

Variables	SOS Drivers	EOS Drivers
Meteorological factors	Preseason average temperature * (TP)	Preseason average temperature ** (TP)
	Growing season average temperature (TG)	Growing season average temperature (TG)
	Preseason total precipitation * (PP)	Preseason total precipitation ** (PP)
	Growing season total precipitation (PG)	Growing season total precipitation (PG)
	Preseason mean shortwave radiation * (SWP)	Preseason mean shortwave radiation ** (SWP)
	Growing season mean shortwave radiation (SWG)	Growing season mean shortwave radiation (SWG)
Soil factors	Preseason soil temperature * (STP)	Preseason soil temperature ** (STP)
	Growing season soil temperature (STG)	Growing season soil temperature (STG)
	Preseason soil moisture * (SMP)	Preseason soil moisture ** (SMP)
	Growing season soil moisture (SMG)	Growing season soil moisture (SMG)
Biological factors	Maximum NDVI during growing season (MN)	Maximum NDVI during growing season (MN)
	Middle season date (MD)	Middle season date (MD)

* Predicted over a period of 30 days. ** Predicted over a period of 60 days.

In this study, the RF model was used to assess the relative importance of the drivers affecting interannual variations in LSP. First, for each pixel, we calculated the partial correlation coefficients between environmental factors (TP, TG, PP, PG, SWP, SWG, STP, STG, SMP, SMG, MN, and MD) and SOS and EOS during 0, 1, 2, … n months before SOS and EOS, and we separately derived the highest correlation with SOS and EOS for the time range of 30 days before SOS and 60 days before EOS for environmental data. Moreover, a subset of variables highly correlated with SOS and EOS was selected, and the values of the variables at selected years and locations (spatiotemporal models) were combined into a set of input feature vectors that are used as inputs for the RF algorithm. Then, these feature vectors were divided equally into two subsets, with 2/3 of the dataset used for model training (in bag) and the remaining 1/3 of the dataset used as an additional test of the RF internal computation (out of bag, i.e., OOB) to estimate the importance of each variable. Variable importance can also be measured by OOB, which compares the increases in OOB error with that variable randomly permuted and all others unchanged [39]. The importance score of a variable is as follows:

$$VI\left(X^j\right) = \frac{1}{ntree}\sum_t\left(err'OOB_t^j - errOOB_t^j\right) \tag{4}$$

where X^j is the jth variable, $ntree$ is the number of trees, $errOOB_t^j$ is the OOB error of each tree t, and $err'OOB_t^j$ is the OOB error when X^j is permuted, while all other variables remain unchanged among OOB data. For regression, the OOB error is the mean square error.

Finally, to optimize the model, the hyperparameter search was used to select the best tested hyperparameter set, and the optimal model was trained on the whole training set for accurate prediction of the LSP dates. Besides modeling multi-year-scale LSP variation for the entire region, subregional variation according to different vegetation types was also modeled to quantify the relative importance of different drivers. These models were constructed using the RF package in R statistical software.

To evaluate the predictiveness of the model and to further test the applicability of the model, we used a randomly selected subset (1/3 of the dataset) for model validation. Both the proportion of explained variance (R^2) and the root mean square error (RMSE) were used to assess the performance of the model on the complete datasets. The stability of the model fit is explained by the mean absolute error (MAE) [40]. These statistics were calculated as follows:

$$R^2 = 1 - \left[\frac{\sum_{i=1}^{n}\left(y_i - \widehat{y}_i\right)^2}{\sum_{i=1}^{n}\left(y_i - \overline{y}_i\right)^2}\right] \tag{5}$$

$$RMSE = \sqrt{\frac{\sum_{i=1}^{n}\left(y_i - \widehat{y}_i\right)^2}{n-1}} \tag{6}$$

$$MAE = \sum_{i=1}^{n}\left|\frac{y_i - \widehat{y}_i}{n}\right| \tag{7}$$

where y_i is the observed satellite-based SOS or EOS values, is $\widehat{y}_i$ the predicted RF-based SOS or EOS values, and $\overline{y}_i$ is the mean satellite-observed SOS or EOS values of all selected test pixels for 2001–2019. n is the sum of all selected test pixels.

3. Results

3.1. Spatiotemporal Variations of Phenology Metrics in the QMs

Figure 4 shows the spatial variation of the annual mean LSP and their corresponding standard deviations (Std) over the study period 2001–2019. Earlier (<90 days) sites of SOS (14.4%) were located at low elevations in the central QMs, and later (>130 days) sites (6.8%) were located at high elevations in the western QMs (Figure 4a). The earliest occurrence of SOS was for SL, with a mean SOS of 97 ± 14 days, and the latest occurrence was for GL, with a mean SOS of 109 ± 12 days (Figure 4a). The Std of SOS has significant spatial variation, with larger areas located in the southwestern QMs (57.2%) having higher Std

(>9 days) (Figure 4b). The overall spatial variation in multi-year average EOS was not significant, and in the southeastern QMs, EOS was mainly concentrated in 290–300 days, accounting for 43.0% of the entire study area (Figure 4c). The earliest EOS occurred in GL, with a mean EOS of 287 $\pm$ 11 days, and the latest occurred in EBF, with a mean EOS of 295 $\pm$ 12 days (Figure 4c). There was higher Std (>13 days) in the southern QMs (25.2%) compared with the northern QMs (Figure 4d). LOS had clear spatial differences, with the central QMs (15.5%) having the longest LOS (>210 days) and the western high-altitude areas (8.5%) had the shortest LOS (<150 days). For SL in the southern QMs (27.9%), its mean LOS was 193 $\pm$ 20 days and the std (>18 days) was also the largest (Figure 4e,f).

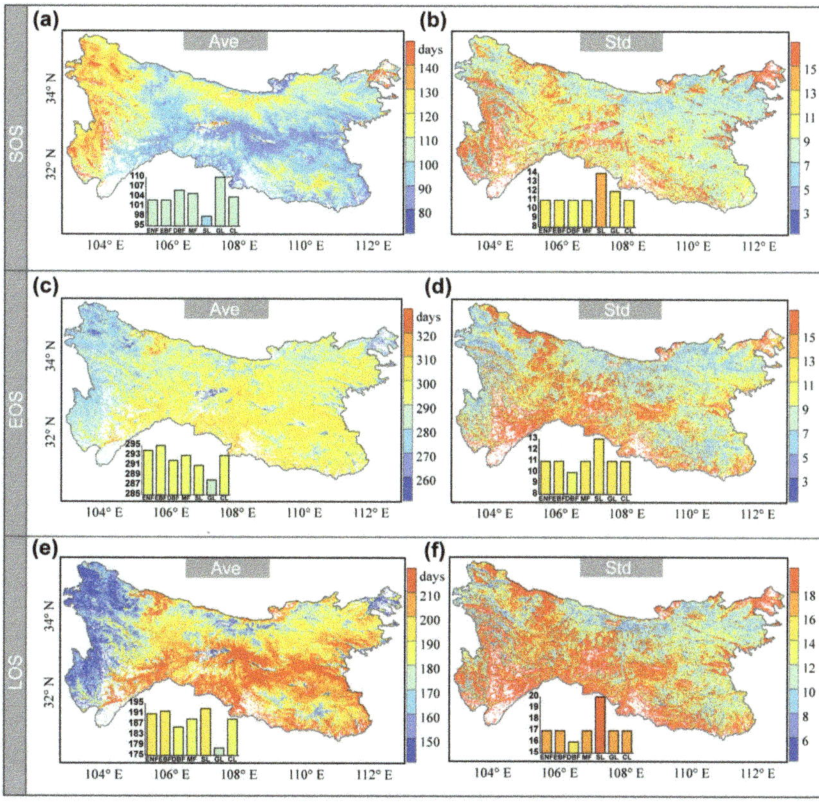

Figure 4. (**a**,**c**,**e**) Spatial distribution of the average phenology metrics from 2001 to 2019 and (**b**,**d**,**f**) standard deviation (Std) of the phenology metrics. Insets at bottom left show the histogram of the average pixel values for different vegetation types.

We also characterized the spatial distribution of LSP trends for different vegetation types from 2001 to 2019 (Figure 5). For the whole QMs, SOS was advanced in 67.8%, the average rate of advance was 1.5 days/decade, and 27.5% of the area (mostly located in the northern QMs) was significant (Figure 5a,b). DBF advanced at a rate of 1.9 days/decade and was the fastest compared to other vegetation types (Figure 5a). EOS was delayed in 72.1% of the region and significant for 42.1% of the region (mostly located in the southern QMs), with an average delay rate of 2.4 days/decade across the region (Figure 5c,d). EBF had the fastest delay rate of 3.3 days/decade (Figure 5c). The average rate of LOS lengthening across the study area was 3.9 days/decade, and 74.6% of the areas (mostly in the southwestern QMs) were lengthened (Figure 5e). The rate of LOS lengthening was 4.7 days/decade for EBF, fastest among the seven vegetation types. Of these areas that were lengthened, the change was found to be significant in 40.3% (mostly in the western QMs) of cases (Figure 5f).

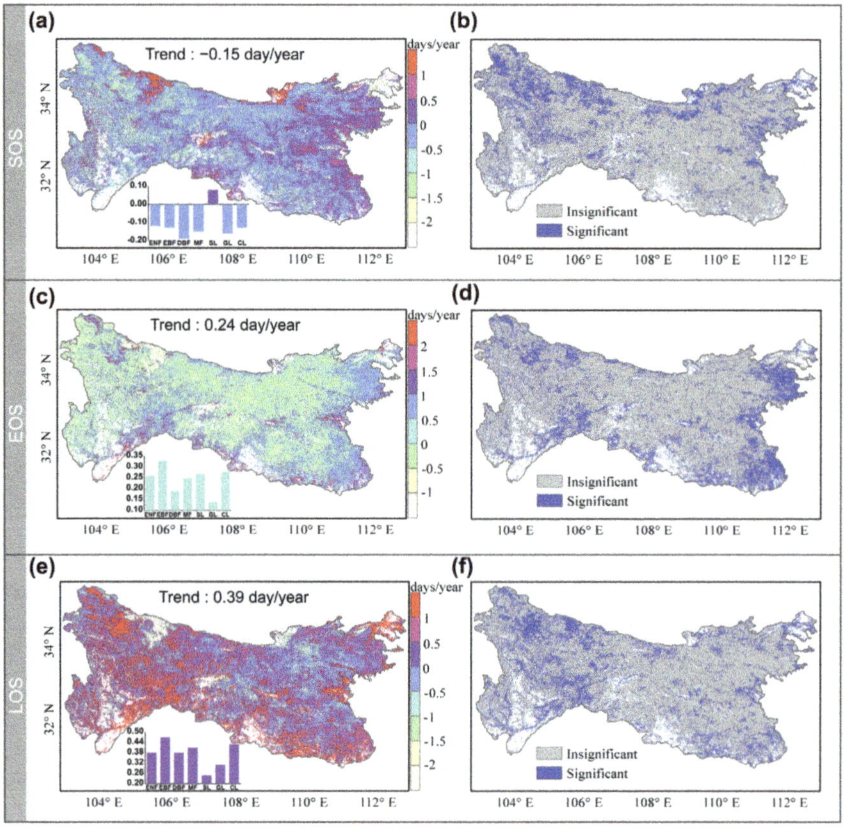

Figure 5. (**a**,**c**,**e**) Spatial distribution of phenology metrics trends from 2001 to 2019 and (**b**,**d**,**f**) significant ($p < 0.1$) changes in phenology metrics trends for the study periods. Insets at bottom left show the average pixel values of the trends of the phenology metrics for the different vegetation types.

Data of interannual variation trends and the significance of LSP for different vegetation types are shown in Figure 6. Overall, the *Sen's slope* of SOS is −0.09 days/year from 2001 to 2019, but this advance is insignificant. There was a trend of significant SOS advancement for DBF and GL, at 0.16 and 0.13 days/year, respectively. EOS shows a significant delay trend over the entire region of 0.29 days/year. The trend of EOS delay was more significant for both EBF and CL compared to other vegetation ($p < 0.05$), and the rate of EBF delay was the fastest (0.37 days/year). LOS is significantly lengthened at a rate of 0.48 days/year. ENF, EBF, DBF, MF, and CL show a more significant trend for lengthened LOS ($p < 0.05$), with the fastest being for EBF (0.68 days/year).

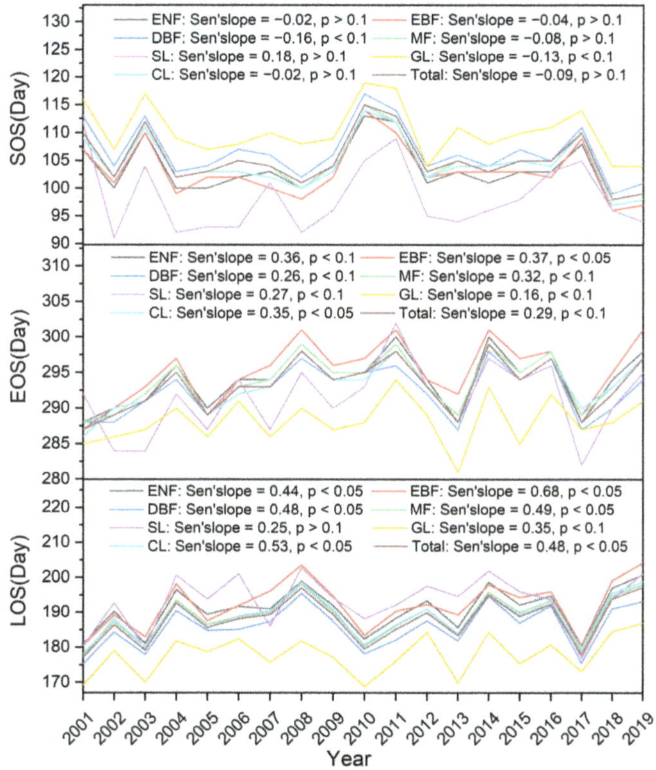

Figure 6. Interannual variations and significance of the mean phenology metrics for the entire study area and the areas covered by different vegetation types. The unit of *Sen's slope* is days/year.

3.2. Change Pattern of LSP and Relative Attribution Analysis

For all vegetation types, the dominant change pattern of the growing season was Type I, with a proportion of 48.4%, which implies that most vegetation in the QMs had an advanced SOS, delayed EOS, and lengthened LOS (Table 4). Type V showed the second largest proportion (15.2%) which meant that there were also many plant species on the QMs having delayed SOS, delayed EOS, and lengthened LOS. Types III, IV, and VI had the smallest proportions (all lower than 10.0%), which indicated that the probabilities of shortened LOS were very low for all plants on the QMs. For five of these vegetation types (ENF, EBF, MF, SL, and CL), the dominant change pattern of the growing season was Type I, followed by Type V. This indicates that these types of plants on the QMs had delayed EOS and lengthened LOS. The main change pattern for DBF and GL was also Type I, with Type II being the second most prevalent. This implies that there is some DBF and GL showing advanced SOS, advanced EOS, and lengthened LOS.

Figure 7 shows the significance of each pattern. For all vegetation types, lengthened LOS, advanced SOS, and delayed EOS were significant in terms of Type I, II, and V change patterns, respectively. Types I, II, and V were the top three patterns in terms of percentage, as seen in Table 4, which were also the three patterns of LOS lengthening. In the Type I, lengthened LOS is significant for most vegetation types (EBF, MF, SL, GL, and CL). In terms of Type II change, the SOS of ENF, DBF, MF, GL, and CL were all significantly advanced, and advanced SOS resulted in lengthened LOS. Fewer changes in LSP trends were significant in terms of Type III, IV, and VI changes, only SL and GL showed significant Type IV changes, and SL (15.5%) and GL (12.2%) accounted for a large proportion of Type IV changes. In terms of Type V, the EOS of all six vegetation types (ENF, EBF, DBF, MF, GL, and CL) was significantly delayed. Delayed EOS resulted in lengthened LOS, a pattern which also happens to be the second largest in terms of proportion, and this pattern is also the one we should be concerned about.

Table 4. The percentage of phenology metrics datasets consisting of each pixel showing different change patterns in the growing seasons for all the vegetation types. Change patterns refer to the trend groupings in Table 2.

Change Patterns	All the Vegetation Types	ENF	EBF	DBF	MF	SL	GL	CL
I	48.4%	46.6%	50.1%	49.3%	47.2%	43.2%	44.8%	48.7%
II	12.0%	10.9%	9.5%	13.7%	11.8%	6.9%	15.2%	10.5%
III	7.3%	6.8%	6.2%	8.6%	7.3%	3.1%	7.6%	5.9%
IV	8.7%	9.0%	7.7%	7.9%	8.8%	15.5%	12.2%	9.3%
V	15.2%	16.8%	18.0%	12.4%	16.5%	23.5%	12.7%	17.2%
VI	8.5%	9.9%	8.5%	8.1%	8.4%	7.8%	7.5%	8.6%
Total	100.0%	100.0%	100.0%	100.0%	100.0%	100.0%	100.0%	100.0%

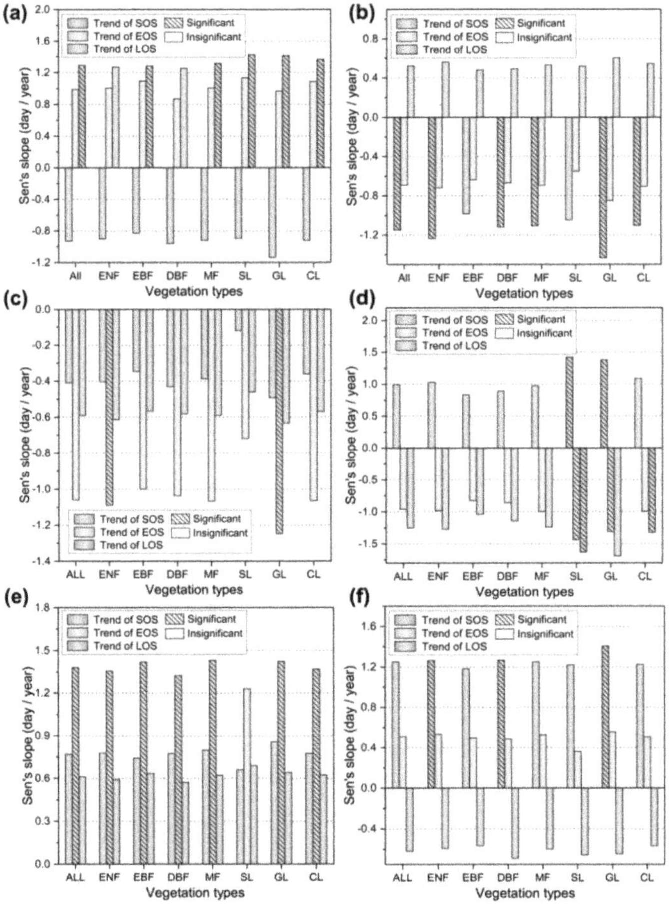

Figure 7. Trends and significance of each pattern during the growing season for different vegetation types. (**a**) growing season pattern I. (**b**) growing season pattern II. (**c**) growing season pattern III. (**d**) growing season pattern IV. (**e**) growing season pattern V. (**f**) growing season pattern VI. Note: "All" refers to all vegetation in the QMs.

For the entire study area, the calculated C-index values were negative for 106,385 pixels and positive for 112,514 pixels (Table 5 and Figure S1). Pixels with C-index values less than 0 are mainly distributed in the easternmost and southernmost parts of the QMs, which indicates that their LOS variations are mainly controlled by SOS shifts. All other regions have pixels with values over 0, which indicates that they are controlled by EOS shifts (Figure S1, in Supplementary Materials). The percentage of LOS changes controlled by SOS and EOS is 48.6% and 51.4%, respectively (Table 5). This also shows that LOS trends, except for DBF and GL, were mainly controlled by the shift in EOS for each vegetation type. The percentage of LOS changes controlled by SOS shifts was 53.4% and 52.0% for DBF and GL, respectively. The largest percentage of changes in LOS of all vegetation types controlled by EOS was 58.3% (SL), and the smallest was 46.6% (DBF).

Table 5. The percentage of datasets in which LOS change was primarily attributable to the shift in SOS or EOS. For most vegetation types, the percentages show that the trend in LOS was mainly controlled by the shift in EOS.

All the Vegetation Types	SOS Controlled	EOS Controlled	Total
ENF	46.9%	53.1%	100%
EBF	43.4%	56.6%	100%
DBF	53.4%	46.6%	100%
MF	47.2%	52.8%	100%
SL	41.7%	58.3%	100%
GL	52.0%	48.0%	100%
CL	45.9%	54.1%	100%
The whole area	48.6%	51.4%	100%

3.3. Drivers of Interannual Variations in LSP

For the QMs, different drivers affect the interannual variability in SOS and EOS (Figure 8 and Table 6). The SWP, MD, and STP are the three most important factors influencing the interannual SOS variation, and the relative importance accounts for 54.4% of total (Figure 8a and Table 6). The total percentage of TG, PP, TP, STG, and PG was 39.2%, and the effect of TG and PP on the interannual SOS variation was almost the same. The remaining four variables (SWG, SMG, SMP, and MN) have a very small effect on interannual SOS variation, and their combined percentage was only 6.4%. Figure 8b and Table 6 also show that SWP, PP, and MD are the three most important factors influencing the interannual EOS variation, with a total relative importance of 54.0%, and the influence of SWP is much stronger than that of PP and MD. The effects of TP, SWG, STP, PG, TG, and STG on the interannual EOS variation totaled 41.9%, and the effects of TP and SWG on the interannual EOS variation were not very different, with a relative importance of 10.2% and 9.9%, respectively. There was also little difference in the relative importance of PG, TG, and STG. The effect of the single variable of STP on the interannual EOS variation (7.2%) is much larger than the sum of SMP, SMG, and MN (4.1%).

Table 6. The top three dominant drivers affecting interannual variations in LSP. These three dominant factors are derived from the ranking of the importance scores of the variables (VI). Different vegetation types have different dominant drivers.

LSP	All the Vegetation Types	First Dominant Driver	Second Dominant Driver	Third Dominant Driver
SOS	ENF	SWP	MD	STP
	EBF	MD	SWP	PP
	DBF	MD	SWP	STP
	MF	SWP	MD	STP
	SL	STP	TP	SWP
	GL	SWP	MD	STP
	CL	SWP	MD	TG
	the whole area	SWP	MD	STP

Table 6. *Cont.*

LSP	All the Vegetation Types	First Dominant Driver	Second Dominant Driver	Third Dominant Driver
EOS	ENF	SWP	TP	PP
	EBF	SWP	MD	PP
	DBF	SWP	PP	TP
	MF	SWP	PP	MD
	SL	SWP	PP	MD
	GL	SWP	SWG	TP
	CL	SWP	TP	PP
	the whole area	SWP	PP	MD

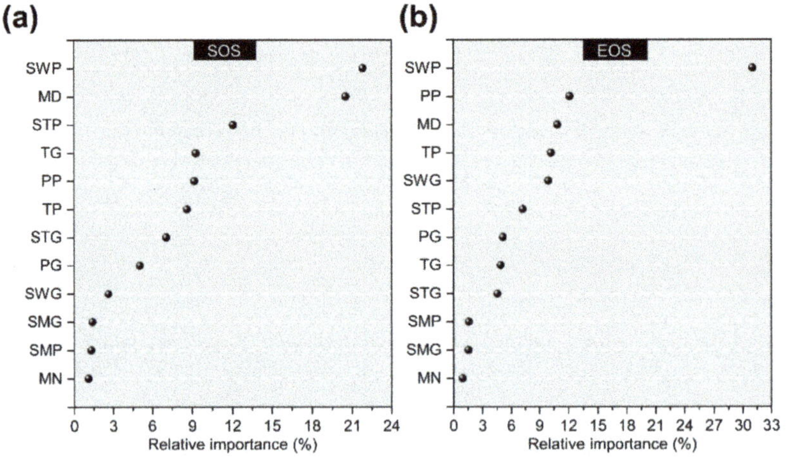

Figure 8. Relative importance of drivers of affecting interannual variations in LSP for the entire study area, in decreasing order. (**a**) The ranking of the drivers of affecting interannual variations in SOS. (**b**) The ranking of the drivers of affecting interannual variations in EOS. The specific values for the relative importance of each driver are in Table S1. Note: The abbreviated variable names are the same as in Table 3.

The drivers influencing interannual variations in LSP of different vegetation types were assessed (Figure 9 and Table 6). Figure 9a and Table 6 show that the main drivers affecting the interannual SOS variation of ENF, MF, and GL were SWP, MD, and STP, and the relative importance of SWP in these three vegetation types was ranked as GL (28.6%) > MF (22.1%) > ENF (21.2%). The interannual SOS variation of EBF, DBF, and CL was mainly influenced by MD and SWP. The main drivers influencing the interannual SOS variation of SL were STP, TP, and SWP, and STP (29.2%) was the most important factor influencing the interannual SOS variation of SL. As shown in Figure 9b and Table 6, in terms of the interannual EOS variation, SWP had the strongest effect on GL (36.4%) and the slightest effect on SL (20.5%). The main drivers of interannual EOS variation of ENF, DBF, and CL are SWP, TP, and PP. SWP, PP and MD are the main drivers of interannual variations in EOS for EBF, MF and SL. Besides SWP, which is the most important driver, PP is the second factor affecting MF and SL, and MD is the second factor affecting EBF with a relative importance of 14.5%. Moreover, the main factors affecting the interannual EOS variation of GL are SWP, SWG, and TP, and the relative importance of SWG and TP is 11.5% and 11.2%, respectively.

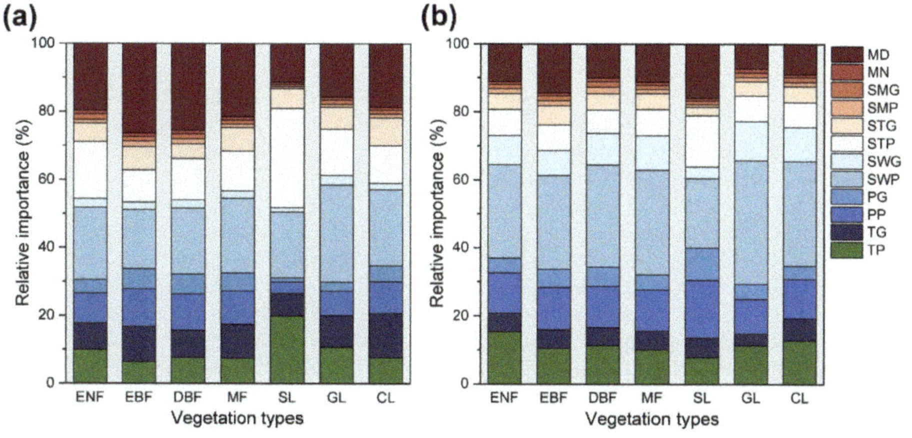

Figure 9. The relative importance of drivers that affect interannual variations in LSP for different vegetation types. Relative importance was derived from the importance scores of variables (VI) based on RF models established for different vegetation types: (**a**) SOS. (**b**) EOS. Different colors indicate different factors. Note: SOS or EOS as response variable in the RF model. The specific values for the relative importance of each driver are in Table S1.

We used 19 years of data to construct an RF model for seven vegetation types (Table S2), and we randomly sampled 1/3 of the pixel dataset to assess their linear relationships (Figure S2). The actual and predicted SOS displayed good linear relationships, with their correlation coefficient R2 values ranging from 0.900 to 0.938, RMSE values ranging from 4.90 to 6.52, and MAE values ranging from 3.93 to 5.15 (Figure S2a). Both the actual and predicted EOS also show good linear relationships, with their correlation coefficients R2 values ranging from 0.911 to 0.942, RMSE values ranging from 4.23 to 5.63, and MAE values ranging from 3.26 to 3.74 (Figure S2b). These results indicate that it is appropriate to use RF models to analyze interannual variations in LSP in the QMs.

4. Discussion

4.1. Dynamics Changes in LSP in the QMs

Understanding the interannual variations in vegetation phenology and its trends is important for recognizing the patterns of vegetation growth dynamics as a response to climate warming. Our study showed that there is an advanced trend (1.5 days/decade) for SOS, a delayed trend (2.4 days/decade) for EOS, and an overall extended trend (3.9 days/decade) for LOS in the QMs during 2001–2019. In comparison with previous studies on phenology changes, there were different degrees of advanced SOS, delayed EOS, and lengthened LOS in different study areas and periods [32,41]. For example, during 1982–2006, the SOS was advanced by 0.56 days/decade, while the EOS delayed trend rate was 5.5 days/decade, and the growing season was significantly longer by 6.06 days/decade in North America [42]. In the Tibetan Plateau region, SOS was advanced at a rate of 0.17 days/decade, EOS was delayed at a rate of 5.29 days/decade, and LOS was lengthened at a rate of 5.46 days/decade for the period 1981–2017 [41]. These results are not entirely consistent with those of studies conducted for the QMs, and the differences may be mainly due to their different target periods and the different methods of phenology extraction. However, other investigations reported that, compared to 1982–1999, the phenology trend slowed down in 2000–2008 and the changes were not highly significant [14,43]. Meanwhile, our results also show a slower change in phenology trends over the last 20 years in the QMs, and the magnitude of SOS advance is also smaller than that of EOS delay. This observation is similar to the results of Wang et al. and Xia et al. [33,44], which indicates that the satellite-observed phenology change rates slowed down during a global warming hiatus between 1998 and 2012.

We also found that the trend in phenology changes of different vegetation is also highly variable, which is related to the microclimate of different regions or the geographical variation of plant origin [45]. Our results show a significantly greater trend of SOS changes in SL than other vegetation types, indicating that the earlier the SOS, the more significant the trend in SOS variations, and that this difference may be related to plant pollination type, life type, phylogenetic and wood type, etc. [46]. However, the trend of SOS changes in CL was weaker than in other vegetation types, and this trend was insignificant because farmers controlled the sowing time in each year, resulting in significantly smaller variability in crop phenology than in field observed plants [47]. The results show that the trend of delayed EOS is significantly stronger for EBF than other vegetation types, mainly because EBF is mainly located in the region south of the QMs, with a humid northern subtropical climate. Some researchers have shown that in subtropical mountainous and hilly areas, broadleaved forests can grow longer under the same similar climatic conditions compared to coniferous forests [48]. The study also found a significantly stronger trend in the lengthening of growing season for trees than for shrubs and herbs, which is the same as the findings of Zhu et al. [49] but in contrast to those of Ge et al. [50], who reported that the interannual variation trend for trees in China from the 1960s to the 2000s was significantly weaker than for herbaceous plants, and this difference in trend was due to differences in the study area and the species of the plants themselves.

4.2. Asymmetry in Contributions of SOS and EOS Trends to LOS

We found the asymmetry in contributions of the SOS and EOS trends to LOS variations by counting the percentage of pixels with positive and negative C-index values. The results show that SOS trends control 48.4% of LOS variations and EOS trends control 51.4% of LOS variations, which shows a stronger association between EOS trends and LOS variations compared to SOS (Figure S1). Previous studies illustrated that the lengthened growing season was mainly driven by delayed autumn phenology, which is consistent with our results [14,38,49]. However, other researchers found that it is the changes in SOS, and not EOS, that dominate the changes in growing season length [51,52]. It can be seen that there are differences in previous studies regarding the attribution of LOS variations. To investigate the reasons for such differences, we divided the trends of SOS, EOS, and LOS into six change patterns (Table 2 and Figure 7). Our results show that in addition to the main growth pattern of Type I (SOS advanced, EOS delayed, and LOS lengthened), 15.2% of the regions had Type V (SOS delayed, EOS delayed, and LOS lengthened), and the delayed EOS was significant in this pattern. Another 12% of the regions showed Type II (SOS advanced, EOS advanced, and LOS lengthened) growth pattern, and advanced SOS was significant. However, since the percentage of the region of the growth pattern Type II is smaller than that of Types I and V, it is still the trend of delayed EOS that dominates the variation in LOS for the whole study area, leading to asymmetry of the relative contribution of SOS and EOS to LOS.

In addition, we found that ENF, EBF, MF, SL, and CL were all controlled by EOS trends, while the variations in LOS for two vegetation types, DBF and GL, were controlled by SOS trends (Table 5). As Figure 4e shows, the growing season lengths of DBF and GL are short, and there are previous studies demonstrating that the effect of EOS shifts on vegetation with short growing season cycles is insignificant [53]. The percentage of growth pattern type II is higher than other vegetation types in DBF and GL, and the advance in SOS is also significant, resulting in SOS dominating the variation in LOS (Table 4). Therefore, we suggest that the asymmetry in SOS and EOS trends contributing to LOS is related to vegetation types, and that future studies should focus on vegetation types to accurately model and predict vegetation phenology periods.

4.3. Analysis of the Drivers of Interannual Variations in LSP

Previous studies showed that the interaction of meteorological, soil, and biological factors influenced the interannual variability of LSP [6,54]. Our results suggest that SWP is the most important driver of interannual variations in SOS and EOS across the QMs (Figure 8). This is mainly because shortwave radiation compensates for the lack of chilling demand during plant physiological dormancy through day length, i.e., longer daylight hours, and has an critical effect on SOS by delaying the accumulation of abscisic acid and slowing down the rate of leaf senescence, and also on EOS [21]. We also found that SWP and MD contributed a total of 42.3% to the interannual variations in SOS, and besides SWP, MD was also an important driver of the interannual SOS variation (Tables 6 and S1). This is due to the fluctuation of the time interval between SOS and MD, which depends on the different developmental stages of the phenology events to a large extent and on the specific differences in the life history of the plant, and needs to be explained by the phenotypic plasticity of the individual and the adaptation to the environment [55]. Therefore, we suggest that the effect of MD on interannual SOS variation varies considerably among vegetation individuals. STP also contributed 12.0% importance in explaining the interannual SOS variation (Table S1). This is mainly due to the increased soil temperature, which accelerated the rate of leaf tip emergence and whole leaf expansion, thus promoting SOS [56]. Moreover, SWP and PP together explain the importance of 43.2% of the interannual EOS variation (Table S1). This is mainly because preseason shortwave radiation and precipitation control the availability of sunlight and water in vegetation, respectively, and reduced precipitation affects water transport capacity, which limits the photosynthetic rate of leaf, leading to lower utilization of light and water conditions by plants and affecting the interannual variation in EOS [20,57]. The combined contribution of MD, SWP, and PP to interannual EOS variation was also found to be as high as 54.0% (Table S1), suggesting that the lifecycle of vegetation is strongly regulated by its own rhythms under improved hydrothermal conditions, and that biological rhythms play a critical role in interannual EOS variation [7].

Furthermore, our study shows that the effect of each driver on interannual variations in LSP was varied for different vegetation types (Figure 9 and Table 6). For example, SWP is the most important driver for ENF, MF, GL, and CL; MD is the most important driver for EBF and DBF; and STP is the most important driver for SL (Table 6). This difference is mainly due to the diversity of plant physiological structures and the different adaptive strategies of plants to environmental changes [58]. SWP has the greatest effect on the interannual SOS variation in GL, which is mainly distributed in higher parts of the QMs and receives abundant solar radiation. The strong solar radiation promotes root activity and advances SOS [59]. The greatest contribution of MD to interannual SOS variation in EBF is related to the fact that EBF grows mainly in the southern part of the QMs, where its deeply rooted system and water conservation adaptations combine to reduce water stress under the influence of a humid northern subtropical climate. This adaptation to environmental changes is strongly regulated by its own rhythms, such that EBF is most affected by MD [60]. The effect of STP in SL is mainly due to the preseason accumulation of soil temperatures susceptible to specific thresholds that accelerate soil thaw and vegetation wake, triggering SOS [61]. SWP was the main driver of interannual EOS variation for all vegetation types, with GL being most influenced by SWP (Table 6). This is because abundant solar radiation increases surface evaporation and reduces water availability in grasslands, which subsequently inhibits vegetation growth, resulting in the EOS of GL being most influenced by SWP [62]. PP and MD have the strongest effect on interannual EOS variation in SL (Table 6). To our best knowledge, SL is mainly distributed in semi-humid and semiarid areas, and the control of plant metabolism by water stress affects its transpiration and photosynthesis, resulting in impaired ATPase synthesis and accelerated chlorophyll degradation. Meanwhile the adaptation of vegetation to such adversity changes also affects the interannual variations in EOS [63]. The relationship between regional climate and vegetation phenology growth will be further explored in future studies.

4.4. Evaluation of RF Model

We validated the accuracy of our RF model (Figure S2). The R^2 values of the linear regression formed by the predicted values and the observations inversions are both greater than 0.900 and 0.911, respectively. When predicting the dates of SOS or EOS, both RMSE and MAE are relatively small, showing that our RF model has good predictive performance. Machine learning, as a nonparametric multivariate approach, can integrate complex relationships between multiple spatial and temporal LSP dates and climate into a single model for predicting SOS and EOS. Then, the main drivers of SOS and EOS can be identified by estimating the importance of each variable [6]. However, it should be noted that although we have tried our best to adjust the hyperparameters of the algorithm to prevent overfitting in the model, some errors still appear in the test datasets (Table S2). Therefore, to obtain a better fit, it is necessary to further refine the study area and tree species in the future. Further study should compare different algorithms to better simulate the phenology period.

5. Conclusions

This study used the phenology metrics of vegetation in the QMs extracted from satellite NDVI data to analyze the spatiotemporal trends of LSP during 2001–2019, and to identify the dominant growth patterns of different vegetation types during the growing season. Furthermore, driving factors influencing interannual variations in LSP were emphatically investigated using the RF model. The main conclusions were as follows:

(1) The average advance of SOS across QMs was 1.5 days/decade, with a significant advance in SOS observed for 27.5% of pixels. EOS was delayed by 2.4 days/decade, with a significant delay in EOS observed for 42.1% of pixels. LOS was lengthened by 3.9 days/decade, with a significant LOS lengthening observed for 40.3% of pixels.

(2) The dominant pattern of change in the growing season for different vegetation types was advanced SOS, delayed EOS, and lengthened LOS, and this pattern had the highest percentage in evergreen broadleaved forests. The percentage of area shows that the patterns of delayed SOS and EOS and lengthened LOS were the highest percentage in shrubs.

(3) For the whole QMs, LOS changes were mainly controlled by EOS, and the percentage was 51.4%. For deciduous broadleaved forests and grasses, LOS changes were attributed to SOS, while for other vegetation types, they were attributed to EOS.

(4) SWP was found to be the most important factor influencing SOS and EOS, and grass and crop most influenced by SWP. Interannual variations in SOS were more influenced by biological factors (MD) than in EOS. The interannual variability of EOS is more influenced by preseason precipitation (PP) than SOS.

Supplementary Materials: The following are available online at https://www.mdpi.com/2072-4292/13/22/4538/s1, Figure S1: The primary factor (SOS or EOS) controlling change in LOS across the entire study area calculated by the C-index, Figure S2: Red fitting curves between predicted LSP dates from RF model and observed LSP dates from MODIS13A2-NDVI, Table S1: The relative importance of each driver affecting interannual variation in LSP, Table S2: Correlation results of random forest models built for different vegetation type cover areas.

Author Contributions: J.G.: Formal analysis, Methodology, Writing—original draft. X.L.: Conceptualization, Funding acquisition, Methodology, Supervision, Writing—original draft. W.G.: Data curation, Visualization. X.N.: Data curation, Visualization. W.M.: Writing—review & editing. Q.L.: Writing—review & editing. X.X.: Resources. All authors have read and agreed to the published version of the manuscript.

Funding: This research was supported by the National Natural Science Foundation of China (Grants 41971104) and partially supported by the Fundamental Research Funds for the Central Universities (Projects GK202107009 and 2021TS015).

Institutional Review Board Statement: Not applicable.

Informed Consent Statement: Not applicable.

Data Availability Statement: The datasets generated and/or analyzed during the study are available from the corresponding author upon reasonable request.

Acknowledgments: I am very grateful to the reviewers and editors for their suggestions on the revision of the manuscript.

Conflicts of Interest: The authors declare no conflict of interest.

Abbreviations

Abbreviations

LSP	Land surface phenology
QMs	Qinling Mountains
NDVI	Normalized difference vegetation index
SOS	The start of the growing season
EOS	The end of the growing season
LOS	The length of the growing season
ENF	Evergreen needleleaved forest
EBF	Evergreen broadleaved forest
DBF	Deciduous broadleaved forest
MF	Mixed forest
SL	Shrubland
GL	Grassland
CL	Cropland
TP	Preseason average temperature
TG	Growing season average temperature
PP	Preseason total precipitation
PG	Growing season total precipitation
SWP	Preseason mean shortwave radiation
SWG	Growing season mean shortwave radiation
STP	Preseason soil temperature
STG	Growing season soil temperature
SMP	Preseason soil moisture
SMG	Growing season soil moisture
MD	The middle date of the growing season
MN	Maximum NDVI during growing season
RF	Random forest
OOB	Out of bag

References

1. Güsewell, S.; Furrer, R.; Gehrig, R.; Pietragalla, B. Changes in temperature sensitivity of spring phenology with recent climate warming in Switzerland are related to shifts of the preseason. *Glob. Chang. Biol.* **2017**, *23*, 5189–5202. [CrossRef]
2. Fu, Y.H.; Piao, S.; Vitasse, Y.; Zhao, H.; De Boeck, H.J.; Liu, Q.; Yang, H.; Weber, U.; Hänninen, H.; Janssens, I.A. Increased heat requirement for leaf flushing in temperate woody species over 1980–2012: Effects of chilling, precipitation and insolation. *Glob. Chang. Biol.* **2015**, *21*, 2687–2697. [CrossRef]
3. Lang, W.; Chen, X.; Qian, S.; Liu, G.; Piao, S. A new process-based model for predicting autumn phenology: How is leaf senescence controlled by photoperiod and temperature coupling? *Agric. For. Meteorol.* **2019**, *268*, 124–135. [CrossRef]
4. Thackeray, S.J.; Henrys, P.A.; Hemming, D.; Bell, J.R.; Botham, M.S.; Burthe, S.; Helaouet, P.; Johns, D.G.; Jones, I.D.; Leech, D.I.; et al. Phenological sensitivity to climate across taxa and trophic levels. *Nature* **2016**, *535*, 241–245. [CrossRef] [PubMed]
5. Zeng, Z.; Piao, S.; Li, L.; Zhou, L.; Ciais, P.; Wang, T.; Li, Y.; Lian, X.; Wood, E.; Mao, J. Climate mitigation from vegetation biophysical feedbacks during the past three decades. *Nat. Clim. Chang.* **2017**, *7*, 432–436. [CrossRef]
6. Rodriguez-Galiano, V.F.; Sanchez-Castillo, M.; Dash, J.; Atkinson, P.M.; Ojeda-Zujar, J. Modelling interannual variation in the spring and autumn land surface phenology of the European forest. *Biogeosciences* **2016**, *13*, 3305–3317. [CrossRef]
7. Zu, J.; Zhang, Y.; Huang, K.; Liu, Y.; Chen, N.; Cong, N. Biological and climate factors co-regulated spatial-temporal dynamics of vegetation autumn phenology on the Tibetan Plateau. *Int. J. Appl. Earth Obs. Geoinf.* **2018**, *69*, 198–205. [CrossRef]

8. Keenan, T.F.; Gray, J.; Friedl, M.A.; Toomey, M.; Bohrer, G.; Hollinger, D.Y.; Munger, J.W.; Keefe, J.O.; Schmid, H.P.; Wing, I.S. Net carbon uptake has increased through warming-induced changes in temperate forest phenology. *Nat. Clim. Chang.* **2014**, *4*, 598–604. [CrossRef]
9. Wu, C.; Hou, X.; Peng, D.; Gonsamo, A.; Xu, S. Land surface phenology of China's temperate ecosystems over 1999–2013: Spatial–temporal patterns, interaction effects, covariation with climate and implications for productivity. *Agric. For. Meteorol.* **2016**, *216*, 177–187. [CrossRef]
10. Beurs, K.M.D.; Henebry, G.M. Land surface phenology and temperature variation in the International Geosphere-Biosphere Program high-latitude transects. *Glob. Chang. Biol.* **2010**, *11*, 779–790. [CrossRef]
11. Cleland, E.E.; Chuine, I.; Menzel, A.; Mooney, H.A.; Schwartz, M.D. Shifting plant phenology in response to global change. *Trends Ecol. Evol.* **2007**, *22*, 357–365. [CrossRef]
12. Stoeckli, R.; Vidale, P.L. European plant phenology and climate as seen in a 20-year AVHRR land-surface parameter dataset. *Int. J. Remote Sens.* **2004**, *25*, 3303–3330. [CrossRef]
13. Piao, S.; Fang, J.; Zhou, L.; Ciais, P.; Zhu, B. Variations in satellite-derived phenology in China's temperate vegetation. *Glob. Chang. Biol.* **2010**, *12*, 672–685. [CrossRef]
14. Jeong, S.J.; Chang-Hoi, H.O.; Gim, H.J.; Brown, M.E. Phenology shifts at start vs. end of growing season in temperate vegetation over the Northern Hemisphere for the period 1982–2008. *Glob. Chang. Biol.* **2011**, *17*, 2385–2399. [CrossRef]
15. Shen, M.; Tang, Y.; Chen, J.; Zhu, X.; Zheng, Y. Influences of temperature and precipitation before the growing season on spring phenology in grasslands of the central and eastern Qinghai-Tibetan Plateau. *Agric. For. Meteorol.* **2011**, *151*, 1711–1722. [CrossRef]
16. Yuan, M.; Zhao, L.; Lin, A.; Li, Q.; Qu, S. How do climatic and non-climatic factors contribute to the dynamics of vegetation autumn phenology in the Yellow River Basin, China? *Ecol. Indic.* **2020**, *112*, 106112. [CrossRef]
17. Wang, C.; Guo, H.; Zhang, L.; Liu, S.; Qiu, Y.; Sun, Z. Assessing phenological change and climatic control of alpine grasslands in the Tibetan Plateau with MODIS time series. *Int. J. Biometeorol.* **2015**, *59*, 11–23. [CrossRef]
18. Flynn, D.; Wolkovich, E.M. Temperature and photoperiod drive spring phenology across all species in a temperate forest community. *New Phytol.* **2018**, *219*, 1353–1362. [CrossRef]
19. Liu, Q.; Piao, S.; Fu, Y.H.; Gao, M.; Peñuelas, J.; Janssens, I.A. Climatic Warming Increases Spatial Synchrony in Spring Vegetation Phenology Across the Northern Hemisphere. *Geophys. Res. Lett.* **2019**, *46*, 1641–1650. [CrossRef]
20. Prevéy, J.S.; Seastedt, T.R. Seasonality of precipitation interacts with exotic species to alter composition and phenology of a semi-arid grassland. *J. Ecol.* **2014**, *102*, 1549–1561. [CrossRef]
21. Way, D.A.; Montgomery, R.A. Photoperiod constraints on tree phenology, performance and migration in a warming world. *Plant Cell Environ.* **2015**, *38*, 1725–1736. [CrossRef] [PubMed]
22. Ztürk, M.; Bolat, L.; Ergün, A. Influence of air-soil temperature on leaf expansion and LAI of Carpinus betulus trees in a temperate urban forest patch. *Agric. For. Meteorol.* **2015**, *200*, 185–191. [CrossRef]
23. Jin, Z.; Zhuang, Q.; He, J.S.; Luo, T.; Shi, Y. Phenology shift from 1989 to 2008 on the Tibetan Plateau: An analysis with a process-based soil physical model and remote sensing data. *Clim. Chang.* **2013**, *119*, 435–449. [CrossRef]
24. Zhou, Y. Asymmetric Behavior of Vegetation Seasonal Growth and the Climatic Cause: Evidence from Long-Term NDVI Dataset in Northeast China. *Remote Sens.* **2019**, *11*, 2107. [CrossRef]
25. Yang, M.; Nelson, F.E.; Shiklomanov, N.I.; Guo, D.; Wan, G. Permafrost degradation and its environmental effects on the Tibetan Plateau: A review of recent research. *Earth-Sci. Rev.* **2010**, *103*, 31–44. [CrossRef]
26. Park, H.; Jeong, S.J.; Ho, C.H.; Kim, J.; Brown, M.E.; Schaepman, M.E. Nonlinear response of vegetation green-up to local temperature variations in temperate and boreal forests in the Northern Hemisphere. *Remote Sens. Environ.* **2015**, *165*, 100–108. [CrossRef]
27. Breiman, L. Random forests. *Mach. Learn.* **2001**, *45*, 5–32. [CrossRef]
28. Li, Z.; Liang, M.; Li, Z.; Pierre, M.; Tong, X.; Zhang, J.; Dong, L.; Zheng, Y.; Ma, W.; Zhao, L. Plant functional groups mediate effects of climate and soil factors on species richness and community biomass in grasslands of Mongolian Plateau. *J. Plant Ecol.* **2021**, *14*, 679–691. [CrossRef]
29. Zhao, M.; Peng, C.; Xiang, W.; Deng, X.; Tian, D.; Zhou, X.; Yu, G.; He, H.; Zhao, Z. Plant phenological modeling and its application in global climate change research: Overview and future challenges. *Environ. Rev.* **2013**, *21*, 1–14. [CrossRef]
30. Deng, C.; Bai, H.; Ma, X.; Zhao, T.; Huang, X. Spatiotemporal differences in the climatic growing season in the Qinling Mountains of China under the influence of global warming from 1964 to 2015. *Theor. Appl. Climatol.* **2019**, *138*, 1899–1911. [CrossRef]
31. Bai, W.G.; Zhang, L. Phytocoenological characteristics and community classification of Abies fargesii forests in Qinling Mountains. *J. Beijing For. Univ.* **2007**, *S2*, 222–226.
32. Myneni, R.B.; Keeling, C.D.; Tucker, C.J.; Asrar, G.; Nemani, R.R. Increased plant growth in the northern high latitudes from 1981 to 1991. *Nature* **1997**, *386*, 698–702. [CrossRef]
33. Wang, X.; Xiao, J.; Li, X.; Cheng, G.; Ma, M.; Zhu, G.; Altaf Arain, M.; Andrew Black, T.; Jassal, R.S. No trends in spring and autumn phenology during the global warming hiatus. *Nat. Commun.* **2019**, *10*, 2389. [CrossRef] [PubMed]
34. Cao, R.; Yang, C.; Shen, M.; Chen, J.; Ji, Z.; Cong, W.; Wei, Y. A simple method to improve the quality of NDVI time-series data by integrating spatiotemporal information with the Savitzky-Golay filter. *Remote Sens. Environ.* **2018**, *217*, 244–257. [CrossRef]
35. Kaur, R.; Kiran, G.S.; Shah, M.N.; Mistry, N.V.; Mohan, S. Applicability of Smoothing Techniques in Generation of Phenological Metrics of Tectona grandis L. Using NDVI Time Series Data. *Remote Sens.* **2021**, *13*, 3343.

36. Gocic, M.; Trajkovic, S. Analysis of changes in meteorological variables using Mann-Kendall and Sen's slope estimator statistical tests in Serbia. *Glob. Planet. Chang.* **2013**, *100*, 172–182. [CrossRef]
37. Mann, H.B. Nonparametric test against trend. *Econometrica* **1945**, *13*, 245–259. [CrossRef]
38. Garonna, I.; Jong, R.D.; Wit, A.D.; Mücher, C.A.; Schmid, B.; Schaepman, M.E. Strong contribution of autumn phenology to changes in satellite-derived growing season length estimates across Europe (1982–2011). *Glob. Chang. Biol.* **2015**, *20*, 3457–3470. [CrossRef]
39. Abdel-Rahman, E.M.; Ahmed, F.B.; Ismail, R. Random forest regression and spectral band selection for estimating sugarcane leaf nitrogen concentration using EO-1 Hyperion hyperspectral data. *Int. J. Remote Sens.* **2013**, *34*, 712–728. [CrossRef]
40. Tian, F.; Cai, Z.; Jin, H.; Hufkens, K.; Scheifinger, H.; Tagesson, T.; Smets, B.; Van Hoolst, R.; Bonte, K.; Ivits, E.; et al. Calibrating vegetation phenology from Sentinel-2 using eddy covariance, PhenoCam, and PEP725 networks across Europe. *Remote Sens. Environ.* **2021**, *260*, 112456. [CrossRef]
41. Sun, Q.; Li, B.; Zhou, G.; Jiang, Y.; Yuan, Y. Delayed autumn leaf senescence date prolongs the growing season length of herbaceous plants on the Qinghai-Tibetan Plateau. *Agric. For. Meteorol.* **2020**, *284*, 107896. [CrossRef]
42. Wang, X.; Piao, S.; Xu, X.; Ciais, P.; Macbean, N.; Myneni, R.B.; Li, L. Has the advancing onset of spring vegetation green-up slowed down or changed abruptly over the last three decades? *Glob. Ecol. Biogeogr.* **2015**, *24*, 621–631. [CrossRef]
43. Zeng, H.; Jia, G.; Epstein, H. Recent changes in phenology over the northern high latitudes detected from multi-satellite data. *Environ. Res. Lett.* **2011**, *6*, 45508–45518. [CrossRef]
44. Xia, H.; Qin, Y.; Feng, G.; Meng, Q.; Liu, G. Forest Phenology Dynamics to Climate Change and Topography in a Geographic and Climate Transition Zone: The Qinling Mountains in Central China. *Forests* **2019**, *10*, 1007. [CrossRef]
45. Wang, H.; Ge, Q.; Dai, J.; Tao, Z. Geographical pattern in first bloom variability and its relation to temperature sensitivity in the USA and China. *Int. J. Biometeorol.* **2015**, *59*, 961–969. [CrossRef]
46. Panchen, Z.A.; Primack, R.B.; Nordt, B.; Ellwood, E.R.; Stevens, A.D.; Renner, S.S.; Willis, C.G.; Fahey, R.; Whittemore, A.; Du, Y. Leaf out times of temperate woody plants are related to phylogeny, deciduousness, growth habit and wood anatomy. *New Phytol.* **2014**, *203*, 1208–1219. [CrossRef]
47. Kurukulasuriya, P.; Mendelsohn, R. Impact And Adaptation Of South-East Asian Farmers To Climate Change: Conclusions And Policy Recommendations. *Clim. Chang. Econ.* **2017**, *8*, 1740007. [CrossRef]
48. Qiu, B.W.; Zhong, M. Spatiotemporal variability of vegetation phenology with reference to altitude and climate in the subtropical mountain and hill region, China. *Chin. Sci. Bull.* **2013**, *58*, 2883–2892. [CrossRef]
49. Zheng, Z.; Zhu, W.; Chen, G.; Jiang, N.; Fan, D.; Zhang, D. Continuous but diverse advancement of spring-summer phenology in response to climate warming across the Qinghai-Tibetan Plateau. *Agric. For. Meteorol.* **2016**, *223*, 194–202. [CrossRef]
50. Ge, Q.; Wang, H.; Rutishauser, T.; Dai, J. Phenological response to climate change in China: A meta-analysis. *Glob. Chang. Biol.* **2015**, *21*, 265–274. [CrossRef]
51. Park, T.; Ganguly, S.; T Mmervik, H.; Euskirchen, E.S.; H Gda, K.A.; Karlsen, S.R.; Brovkin, V.; Nemani, R.R.; Myneni, R.B. Changes in growing season duration and productivity of northern vegetation inferred from long-term remote sensing data. *Environ. Res. Lett.* **2016**, *11*, 84001. [CrossRef]
52. Shen, M.; Piao, S.; Dorji, T.; Liu, Q.; Cong, N.; Chen, X.; An, S.; Wang, S.; Wang, T.; Zhang, G. Plant phenological responses to climate change on the Tibetan Plateau: Research status and challenges. *Natl. Sci. Rev.* **2015**, *2*, 454–467. [CrossRef]
53. Cong, N.; Huang, K.; Zhang, Y. Unsynchronized Driving Mechanisms of Spring and Autumn Phenology Over Northern Hemisphere Grasslands. *Front. For. Glob. Chang.* **2021**, *3*, 610162. [CrossRef]
54. Wang, J.; Sun, H.; Xiong, J.; He, D.; Cheng, W.; Ye, C.; Yong, Z.; Huang, X. Dynamics and Drivers of Vegetation Phenology in Three-River Headwaters Region Based on the Google Earth Engine. *Remote Sens.* **2021**, *13*, 2528. [CrossRef]
55. Mcwatters, H.G.; Devlin, P.F. Timing in plants—A rhythmic arrangement. *Febs. Lett.* **2011**, *585*, 1474–1484. [CrossRef]
56. Stone, P.J.; Sorensen, I.B.; Jamieson, P.D. Effect of soil temperature on phenology, canopy development, biomass and yield of maize in a cool-temperate climate. *Field Crop. Res.* **1999**, *63*, 169–178. [CrossRef]
57. Chen, X.; Ciais, P.; Maignan, F.; Zhang, Y.; Zhang, H. Vapor Pressure Deficit and Sunlight Explain Seasonality of Leaf Phenology and Photosynthesis Across Amazonian Evergreen Broadleaved Forest. *Glob. Biogeochem. Cycles* **2021**, *35*, e2020GB006893. [CrossRef]
58. Diez, J.M.; Ibáñez, I.; Miller-Rushing, A.J.; Mazer, S.J.; Crimmins, T.M.; Crimmins, M.A.; Bertelsen, C.D.; Inouye, D.W. Forecasting phenology: From species variability to community patterns. *Ecol. Lett.* **2012**, *15*, 545–553. [CrossRef]
59. Li, K.; Wang, C.; Sun, Q.; Rong, G.; Tong, Z.; Liu, X.; Zhang, J. Spring Phenological Sensitivity to Climate Change in the Northern Hemisphere: Comprehensive Evaluation and Driving Force Analysis. *Remote Sens.* **2021**, *10*, 1972. [CrossRef]
60. Zhao, A.; Zhang, A.; Cao, S.; Liu, X.; Liu, J.; Cheng, D. Responses of vegetation productivity to multi-scale drought in Loess Plateau, China. *Catena* **2017**, *163*, 165–171. [CrossRef]
61. Chuine, I. A unified model for budburst of trees. *J. Theor. Biol.* **2000**, *207*, 337–347. [CrossRef] [PubMed]
62. Chen, B.Z.; Zhang, H.F.; Yan, J.W.; Wang, G.Y.; Zhou, T.M.; Xu, G.; Innes, J.L.; Che, M.L.; Dou, X.M. Spatial and temporal variations in the end date of the vegetation growing season throughout the Qinghai-Tibetan Plateau from 1982 to 2011. *Agric. For. Meteorol.* **2014**, *189–190*, 81–90.
63. Gerten, D.; Schaphoff, S.; Lucht, W. Potential future changes in water limitations of the terrestrial biosphere. *Clim. Chang.* **2007**, *80*, 277–299. [CrossRef]

remote sensing

MDPI

Article

Exploring the Applicability and Scaling Effects of Satellite-Observed Spring and Autumn Phenology in Complex Terrain Regions Using Four Different Spatial Resolution Products

Fangxin Chen [1], Zhengjia Liu [1,2,3,*], Huimin Zhong [1,2] and Sisi Wang [4]

1 Key Laboratory of Regional Sustainable Development Modeling, Institute of Geographic Sciences and Natural Resources Research, Chinese Academy of Sciences, Beijing 100101, China; N185527002@fzu.edu.cn (F.C.); zhonghm.20s@igsnrr.ac.cn (H.Z.)
2 College of Resources and Environment, University of Chinese Academy of Sciences, Beijing 100049, China
3 State Key Laboratory of Remote Sensing Science, Aerospace Information Research Institute, Chinese Academy of Sciences, Beijing 100101, China
4 National Remote Sensing Center of China, Beijing 100036, China; wangsisi@nrscc.gov.cn
* Correspondence: liuzj@igsnrr.ac.cn

Abstract: The information on land surface phenology (LSP) was extracted from remote sensing data in many studies. However, few studies have evaluated the impacts of satellite products with different spatial resolutions on LSP extraction over regions with a heterogeneous topography. To bridge this knowledge gap, this study took the Loess Plateau as an example region and employed four types of satellite data with different spatial resolutions (250, 500, and 1000 m MODIS NDVI during the period 2001–2020 and ~10 km GIMMS3g during the period 1982–2015) to investigate the LSP changes that took place. We used the correlation coefficient (r) and root mean square error (RMSE) to evaluate the performances of various satellite products and further analyzed the applicability of the four satellite products. Our results showed that the MODIS-based start of the growing season (SOS) and end of the growing season (EOS) were highly correlated with the ground-observed data with r values of 0.82 and 0.79, respectively ($p < 0.01$), while the GIMMS3g-based phenology signal performed badly ($r < 0.50$ and $p > 0.05$). Spatially, the LSP that was derived from the MODIS products produced more reasonable spatial distributions. The inter-annual averaged MODIS SOS and EOS presented overall advanced and delayed trends during the period 2001–2020, respectively. More than two-thirds of the SOS advances and EOS delays occurred in grasslands, which determined the overall phenological changes across the entire Loess Plateau. However, both inter-annual trends of SOS and EOS derived from the GIMMS3g data were opposite to those seen in the MODIS results. There were no significant differences among the three MODIS datasets (250, 500, and 1000 m) with regard to a bias lower than 2 days, RMSE lower than 1 day, and correlation coefficient greater than 0.95 ($p < 0.01$). Furthermore, it was found that the phenology that was derived from the data with a 1000 m spatial resolution in the heterogeneous topography regions was feasible. Yet, in forest ecosystems and areas with an accumulated temperature ≥ 10 °C, the differences in phenological phase between the MODIS products could be amplified.

Keywords: land surface phenology; data suitability; satellite data; spatial scaling effects; the Loess Plateau

Citation: Chen, F.; Liu, Z.; Zhong, H.; Wang, S. Exploring the Applicability and Scaling Effects of Satellite-Observed Spring and Autumn Phenology in Complex Terrain Regions Using Four Different Spatial Resolution Products. *Remote Sens.* **2021**, *13*, 4582. https://doi.org/10.3390/rs13224582

Academic Editors: Xuanlong Ma, Xiaolin Zhu, Jiaxin Jin, Yuke Zhou and Qiaoyun Xie

Received: 24 September 2021
Accepted: 11 November 2021
Published: 15 November 2021

1. Introduction

Land surface phenology (LSP) has been recognized as one of the most effective indicators of climate change [1–4] and is closely related to animal migration, gross primary production, and crop productivity [5–7]. Methods for measuring phenology include ground observations (i.e., PhenoCam network and phenology network) and satellite observations [8–10]. Ground observations usually only reflect the phenological information of local ecological communities [11–13]. However, satellite remote sensing has the potential to continuously observe the variation in vegetation phenology at multiple scales [14–17].

Phenology has been widely monitored in different types of remote sensing data in attempts to understand the interactions between vegetation and climate change during the past few decades [18–22]. A variety of vegetation indexes were used to monitor vegetation phenology in previous studies [23–25]. The two most used vegetation indices are the enhanced vegetation index (EVI) and the normalized difference vegetation index (NDVI) [26–29]. Meanwhile, several sets of freely accessible remote sensing products with different resolutions were released, such as the third generation GIMMS (GIMMS3g) NDVI with a spatial resolution of 1/12 degree that is derived from the Advanced Very-High-Resolution Radiometer (AVHRR) series satellites [30,31]; the Systeme Probatoire d'Observation de la Tarre (SPOT) NDVI with a 1 km spatial resolution [32]; and the moderate-resolution imaging spectroradiometer (MODIS) NDVI with spatial resolutions of 250, 500, or 1000 m [33,34]. Among them, GIMMS3g NDVI is the latest and longest-used product, and MODIS NDVI has several different spatial resolutions. These products have been widely used for studies involving phenology extraction [15]. Recent studies showed that the temporal and spatial variation trends of vegetation phenology that were observed in some areas by GIMMS3g NDVI and MODIS NDVI were consistent, but the conclusions that were reached in other regions were opposite [35,36]. The topography has a significant impact on the phenology of different product identifications [37]. Furthermore, the spatial phenological heterogeneity of data with a different resolution increases with the increase in landscape fragmentation [5]. However, the impacts of satellite products with different spatial resolutions on LSP extraction over regions with a heterogeneous topography have not been well clarified.

In general, data with a finer spatial resolution possess more information about the seasonality and phenology properties of vegetation [38–40]. Meanwhile, data with a fine spatial resolution have the problem of providing a larger amount of data, having a slow computation speed, and being time-consuming [41]. The coarse spatial resolution data is more suitable for monitoring phenology at a landscape scale [42]. Moreover, there will be some differences in vegetation phenology that are estimated from the coarse spatial resolution data to the fine spatial resolution data [5]. Thus, it is important to select remote sensing products with an appropriate spatial resolution in order to investigate vegetation phenology changes.

In this study, we explored the applicability of 250, 500, and 1000 m MODIS NDVI and GIMMS3g NDVI across the entire Loess Plateau, which is a typically ecological fragile region with a heterogeneous topography. The aims of this study were to (1) investigate the spatial and temporal patterns of vegetation phenology in the Loess Plateau, (2) analyze the applicability of different types of satellite data in complex terrain regions, and (3) explore the factors that influence the differences in the phenology of multiple datasets.

2. Materials and Methods

2.1. Study Area

The Loess Plateau lies in the north of China, covering an area of 62.4×10^4 km^2 (Figure 1). The region is dominated by a continental monsoon climate. The annual accumulated temperature $\geq$10 °C (annual AT10) increases from ~50 °C·d in the high-elevation western part to ~5800 °C·d in the southern part. Additionally, the annual mean precipitation varies from 50 mm in the northwest to 700 mm in the southeast. The terrain of the Loess Plateau varies significantly, and the altitude ranges from 80 m in the southeast to 5200 m in the west. Land use data with a spatial resolution of 500 m from the MODIS land cover product (MCD12Q1) show that there are seven land cover types in the entire study area; these are grasslands (68.5%), croplands (21.2%), forests (5.1%), barren regions (3.3%), urban and built-up areas (1.8%), and water bodies (0.1%).

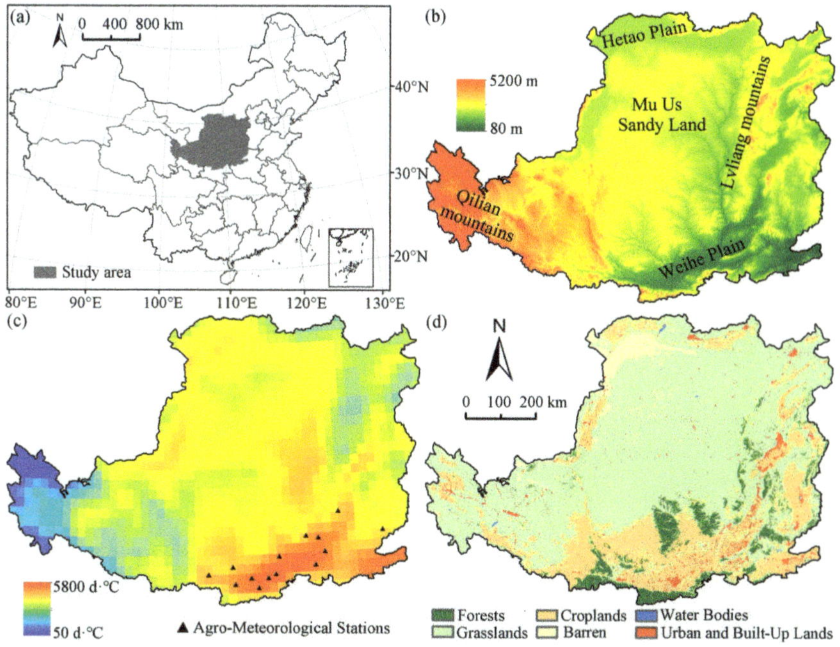

Figure 1. (**a**) Location of the Loess Plateau; (**b**) digital elevation model (DEM) showing the location of the Weihe Plain, Lvliang Mountains, Mu Us Sandy Land, Hetao Plain, and Qilian Mountains; (**c**) the pattern of accumulated temperature $\geq$ 10 °C (AT10); and (**d**) land cover types in 2001.

2.2. Data Resources and Preprocessing

In this study, we used four types of satellite data: MODIS NDVI with spatial resolutions of 250, 500, and 1000 m and GIMMS3g NDVI with a spatial resolution of ~10 km. These data have been widely applied for vegetation phenology extraction at regional to global scales [43–45]. MODIS data for the period 2001–2020 were acquired from NASA (ftp://ladsweb.nascom.nasa.gov/allData/6/, accessed on 15 April 2021). MODIS NDVI with a temporal resolution of 16 days is a gridded level 3 product. GIMMS3g NDVI was obtained from the NASA Earth Exchange platform for the period 1982–2015 (https://nex.nasa.gov/nex/, accessed on 15 April 2021). The spatial and temporal resolution of the GIMMS3g NDVI was 15 days and 1/12 degree (~10 km), respectively. Both the MODIS and GIMMS3g time-series datasets were used to identify the start of the growing season (SOS) and the end of the growing season (EOS).

The daily mean air temperature with a spatial resolution of 0.25 degrees for the period 1982–2020 was obtained from the Copernicus Climate Change Service Climate Data Store (CDS) (https://cds.climate.copernicus.eu/#!/home, accessed on 19 June 2021). The daily air temperature was used to obtain AT10 in order to analyze the interaction between phenology and temperature.

The annual MODIS land cover type product at a 500 m spatial resolution was used to examine the influence of vegetation type on the LSP. The classification scheme that is used for the product is the International Geosphere–Biosphere Program (IGBP). Additionally, the year used for the latest MODIS land cover type data was 2019. The land cover type product for the period 2001–2019 was obtained from NASA's Land Processes Distributed Active Archive Center (LP DAAC) (https://lpdaac.usgs.gov/products/mcd12q1v006/, accessed on 19 June 2021). The spatial resolution of the product was resampled to 250 m, 1000 m, and 1/12 degree using nearest-neighbor interpolation. The Shuttle Radar Topography Mission (SRTM) digital elevation model (DEM) dataset was achieved at the

CGIAR (https://srtm.csi.cgiar.org/, accessed on 11 July 2021). The DEM dataset was used to understand the distribution of the topography in the Loess Plateau. The terrain of the central and western parts of the Loess Plateau is heterogeneous, while the terrain of the Weihe Plain is flat [46].

The 271 detailed ground-observed records in the Loess Plateau were obtained from 14 agro-meteorological stations from 2001 to 2013 (a small set of data with low quality was rejected). The distributions and information of the ground-observed sites can be found in Figure 1a and Table 1. The observed records were used to validate the vegetation phenology that was derived from MODIS data and more information could be used as prior knowledge of the research area [47].

Table 1. The description of the ground-observed sites.

Site Name	Longitude (°E)	Latitude (°N)	Altitude (m)	Data Range
Fengxiang	107.38	34.51	779	2001–2013
Yongshou	108.15	34.70	1006	2001–2013
Wugong	108.22	34.25	429	2001–2013
Xianyang	108.71	34.40	473	2001–2013
Changan	108.92	34.15	435	2001–2013
Lintong	109.23	34.40	418	2001–2013
Weinan	109.46	34.50	357	2001–2013
Baishui	109.58	34.95	482	2001–2013
Hancheng	110.45	35.46	446	2001–2013
Ruicheng	110.71	34.70	503	2001–2013
Wanrong	110.83	35.40	609	2001–2013
Yuncheng	111.02	35.03	380	2001–2013
Linfen	111.50	36.06	450	2001–2013
Jincheng	112.83	35.51	726	2001–2013

2.3. Methods

The quality of the NDVI time series was first examined based on the QA information. The LSP was not produced if three serial periods of NDVI data were contaminated by clouds. Second, to reduce the impacts of noise from cloud contamination or other poor atmospheric conditions, the MODIS NDVI time series were smoothed using the modified Savitsky–Golay algorithm (mSG) with the help of a specific MODIS data layer named "composite day of the year" [19,48]. Similarly, the GIMMS3g NDVI data were also smoothed. However, the GIMMS3g data lacked the layer of "composite day of the year"; we thus regarded the 1st and 16th days of each month as the "day of year (DOY)" for each image. The mSG algorithm is a simple but robust method that is based on the Savitsky–Golay algorithm [48,49]. Finally, the smoothed NDVI growth curve was used to estimate the SOS and EOS with the following logistic model [34]:

$$y(t) = \frac{c}{1 + e^{a+bt}} + d, \tag{1}$$

where t is the DOY, $y(t)$ represents the NDVI value at time t, a and b are the fitting parameters, c is the difference between the maximum and minimum NDVI values, and d is the initial background vegetation index value. Next, the SOS and EOS were produced from the rate of change in curvature:

$$K = -\frac{b^2cz(1-z)(1+z)^3}{\left[(1+z)^4 + (bcz)^2\right]^{3/2}} \tag{2}$$

$$K' = b^3cz\left\{\frac{3z(1-z)(1+z)^3\left[2(1+z)^3 + b^2c^2z\right]}{\left[(1+z)^4 + (bcz)^2\right]^{5/2}} - \frac{(1+z)^2\left(1+2z-5z^2\right)}{\left[(1+z)^4 + (bcz)^2\right]^{\frac{3}{2}}}\right\} \tag{3}$$

where K represents the curvature, $z = e^{a+bt}$, and K' is the rate of change of K.

In order to acquire the deviation and the correlation characteristics between different products, the root mean square error (RMSE) and correlation coefficient (r) were calculated using Equations (4) and (5), respectively:

$$RMSE = \sqrt{\frac{\sum_{i=1}^{n}\left(X_i - \overline{X}\right)^2}{n}} \quad (4)$$

where $\overline{X}$ is the mean value of X and $\left(X_i - \overline{X}\right)$ represents the deviation value, that is, the bias;

$$\mathrm{r} = \frac{{}_{i=1}^{n}\left(X_i - \overline{X}\right)\left(Y_i - \overline{Y}\right)}{\sqrt{\sum_{i=1}^{n}\left(X_i - \overline{X}\right)^2 \sum_{i=1}^{n}\left(Y_i - \overline{Y}\right)^2}} \quad (5)$$

where X_i and Y_i correspond to two different datasets.

Based on the geographical location information of agro-meteorological stations, we first extracted the SOS and EOS pixels (e.g., 3 × 3 homogeneous pixels) around the center of each station, and then averaged these values to obtain the mean SOS and EOS of each station [50]. Finally, we employed RMSE and r to investigate the correlation between the satellite-based and ground-observed phenology.

3. Results

3.1. The Performances of Satellite-Based SOS and EOS

This study first verified the performances of the SOS and EOS that were produced from the 250 m MODIS NDVI and GIMMS3g NDVI against the ground-observed data from 2001 to 2013. The results showed that a good agreement was observed between the MODIS-derived phenology and ground-observed data, where the r values of the MODIS SOS and EOS were 0.83 and 0.79, respectively ($p < 0.01$, Figure 2). However, the phenology that was estimated from GIMMS3g NDVI showed poor consistency. The r values of the GIMMS3g SOS and EOS were only 0.49 and 0.21, and both the p-values were more than 0.05.

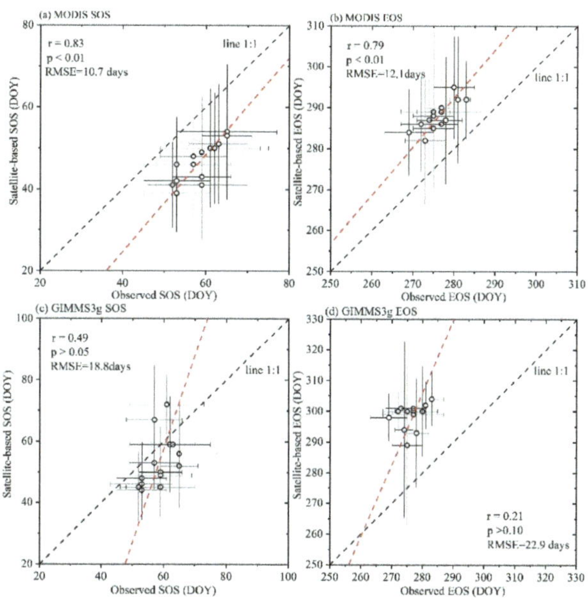

Figure 2. Validations of the (**a**) SOS and (**b**) EOS that were derived from the 250 m MODIS NDVI and the (**c**) SOS and (**d**) EOS that were derived from the GIMMS3g NDVI.

3.2. Spatial Patterns of Vegetation Phenology

Figure 3 presents the SOS and EOS of the Loess Plateau in 2001. The spatial patterns of the vegetation phenology that were produced from the GIMMS3g data were quite different from those of the MODIS data. However, three sets of MODIS phenology results with different resolutions gave a similar spatial distribution (Figure A1). The spring phenology of croplands in the Weihe Plain was the earliest, with an average SOS of DOY 46. Most SOSs in the grassland region of the central Loess Plateau were later than DOY 170. In particular, the SOS that was derived from MODIS NDVI in the Mu Us Sandy Land region was earlier than that in the surrounding areas. The early EOS in the Loess Plateau was mainly concentrated in the southern region, while the late EOS was mainly concentrated in the western region and the Lvliang mountains. Among the land cover types, the EOS of almost 80% of the croplands was between DOY 260 and 290, the EOS date of the forests was the earliest, and that of the grasslands was the latest.

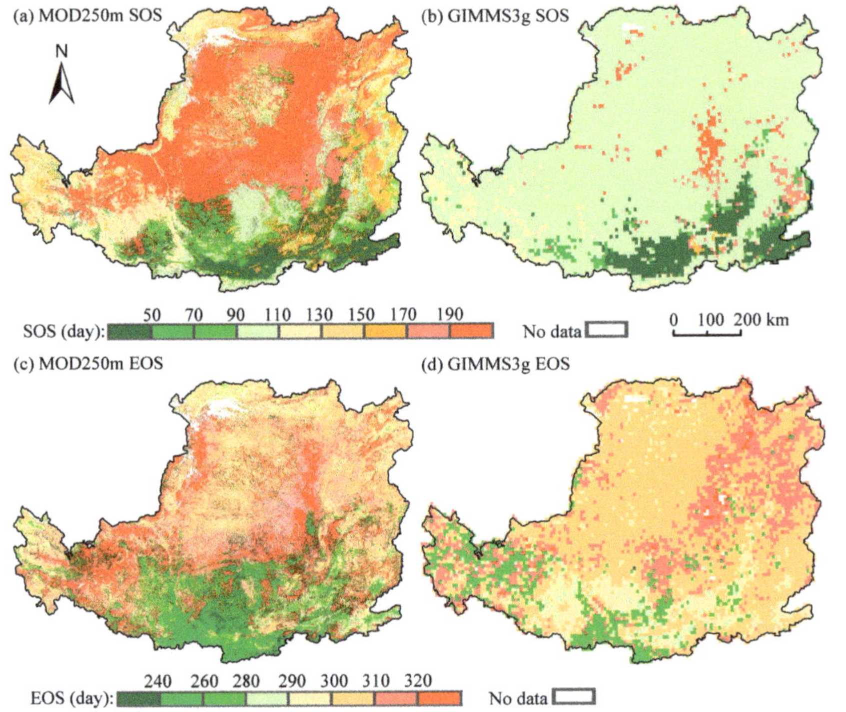

Figure 3. Spatial patterns of the SOSs were estimated from (**a**) 250 m MODIS and (**b**) GIMMS3g data from 2001; spatial patterns of the EOSs were estimated from (**c**) 250 m MODIS and (**d**) GIMMS3g data from 2001.

Compared with the MODIS-derived phenology, the spatial distribution of the phenology period that was identified by the GIMMS3g data was more concentrated. Moreover, the SOS values that were derived from the GIMMS3g product were largely consistent with the phenology that was derived from the MODIS data, which was mainly distributed in the Weihe Plain. The SOS that was produced by the GIMMS3g NDVI data in the northern part of the Loess Plateau was concentrated in DOY 90–110, and the spatial details were greatly neglected. Additionally, there was a significant difference between the GIMMS3g EOS and the MODIS EOS in the south–central region of the Loess Plateau. The difference ranged from 20 days to more than 60 days.

Figure 4a,b presents the raw and smoothed NDVI in heterogeneous areas, where the land use types are grassland and forest, respectively. Figure 4c shows that the time series of NDVI in the Weihe Plain, where the land use type was cropland. Based on the time-series data, we found that no matter what year it was, the inflection point of the time series of the original GIMMS3g data was always concentrated in the seventh or eighth period's data in heterogeneous areas. However, this phenomenon did not occur in flat areas. This indicates that the problem of GIMMS3g data is one of the important reasons for the spatially aggregated distribution of GIMMS3g phenology.

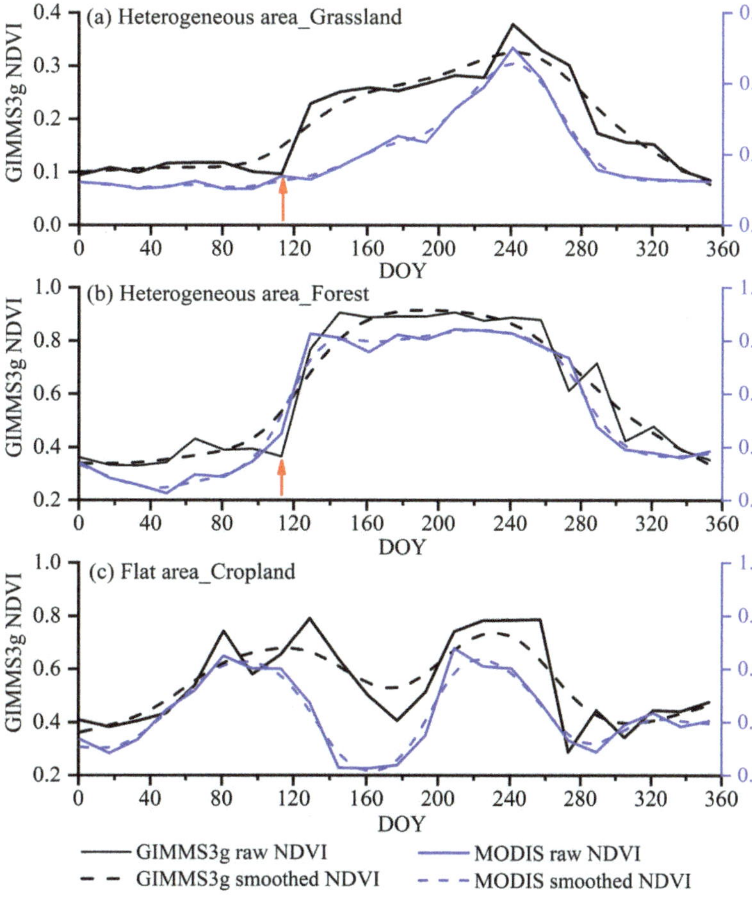

Figure 4. Time series of (**a**) grassland, (**b**) forest, and (**c**) cropland NDVIs from the GIMMS3g and MODIS products.

In order to compare the differences between the GIMMS3g-derived and MODIS-derived phenology in flat areas, we calculated the differences of the two datasets in the Weihe Plain during the period 2001–2015 (Figure 5). The findings showed that the differences in SOS (GIMMS3g SOS–250 m MODIS SOS) were mainly less than 5 days and 10 days in 43.49% and 70.66%. The GIMMS3g data performed well at monitoring SOS over the flat areas. In addition, we found that the frequency with which the differences in SOS were greater than 20 days decreased significantly and became close to 0. In the results showing the SOS differences greater than 25 days, the value of the GIMMS3g SOS was always greater than that of the 250 m MODIS SOS. This shows that GIMMS3g SOS tended to be later than MODIS SOS. However, differences in the EOS (GIMMS3g EOS–250 m MODIS EOS) that were less than 5 days or 10 days were only 13.85% and 29.88%. In addition, for the entire Loess Plateau, the proportion of differences in EOS between the GIMMS3g and 250 m MODIS data that were less than 5 days was still less than 20%. This indicated that the consistency of the GIMMS3g EOS and the MODIS EOS was poor, even in flat areas.

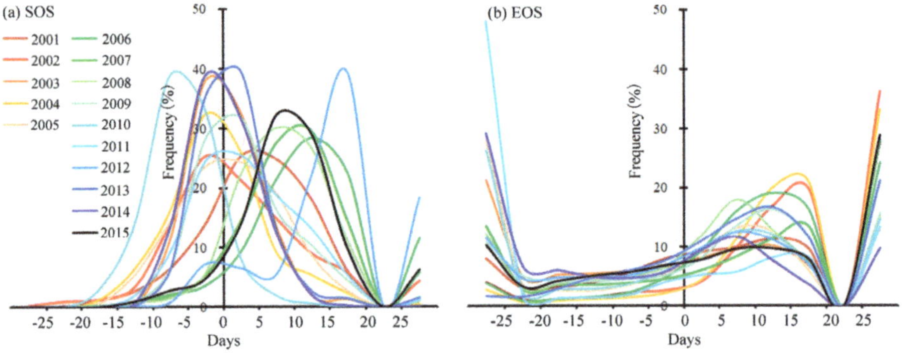

Figure 5. The distribution of the (**a**) SOS and (**b**) EOS differences between the 250 m MODIS and GIMMS3g phenology during the period 2001–2015.

3.3. Temporal Variation in Vegetation Phenology

The inter-annual trends of the SOS and EOS during the period 1982–2020 are presented in Figure 6. The results showed that the inter-annual trends of the SOS and EOS from GIMMS3g were the reverse of those of the MODIS data. The SOS showed a trend of postponing and the EOS presented a trend of advancing from 1982 to 2015 based on the GIMMS3g data. In addition, the SOS (EOS) of the GIMMS3g data delay (advance) trend during the period 2001–2015 was more significant, and the trend lines K of the SOS and EOS were 0.34 and −0.14, respectively. In contrast, the SOS (EOS) showed an advanced (delayed) trend based on the MODIS data, and the trend line K of the MODIS SOS and EOS was −0.63 and 0.19, respectively, during the period 2001–2015.

In the comparison of MODIS products, our findings showed that the phenological periods that were derived from MODIS products with various spatial resolutions gave only small differences. The average difference between the 500 m MODIS SOS (EOS) and 250 m MODIS SOS (EOS) was only 1.2 (0.3) days. The correlation coefficient and RMSE between the 500 m MODIS and 250 m MODIS results were greater than 0.99 and less than 0.60, respectively (Table 2). Moreover, the average difference between the 1000 m MODIS SOS (EOS) and 250 m MODIS slightly increased to 1.7 (1.4) days. The correlation coefficient and RMSE between the 1000 m MODIS and 250 m MODIS results were greater than 0.95 and approximately equal to 1.0, respectively. This demonstrated that there was little difference between the 1000 m MODIS NDVI and the 250 m MODIS NDVI. Therefore, the 1000 m MODIS NDVI could be used to monitor the LSP of the Loess Plateau. However, the GIMMS3g product may not be able to accurately monitor heterogeneous areas, such as the Loess Plateau.

Figure 7 shows the spatial trend of the phenology over the Loess Plateau that was calculated from the MODIS data during the period 2001–2020, which passed the significance test of $\alpha = 0.05$. The area with a significantly advanced SOS was about 16.7×10^4 km^2, which was about a quarter of the area of the Loess Plateau. Areas with a significantly advanced SOS were mainly in the central and northeastern regions of the Loess Plateau. The area of the delayed SOS was only one-third that of the advanced SOS. Meanwhile, the delayed SOS was mainly distributed across the croplands of the Weihe and Hetao Plains. Moreover, the area with a significantly delayed EOS was about 9.3×10^4 km^2, which was more than three times that of the area with an advanced EOS. More than two-thirds of the advanced SOS and delayed EOS occurred in grasslands, which determined the overall phenological changes in the Loess Plateau. Although the land cover types in the Loess Plateau changed dramatically, the phenological changes were still dominated by the areas where the land use did not change. Only about 20% of the SOS or EOS significant trends occurred in areas with changes in the land cover type.

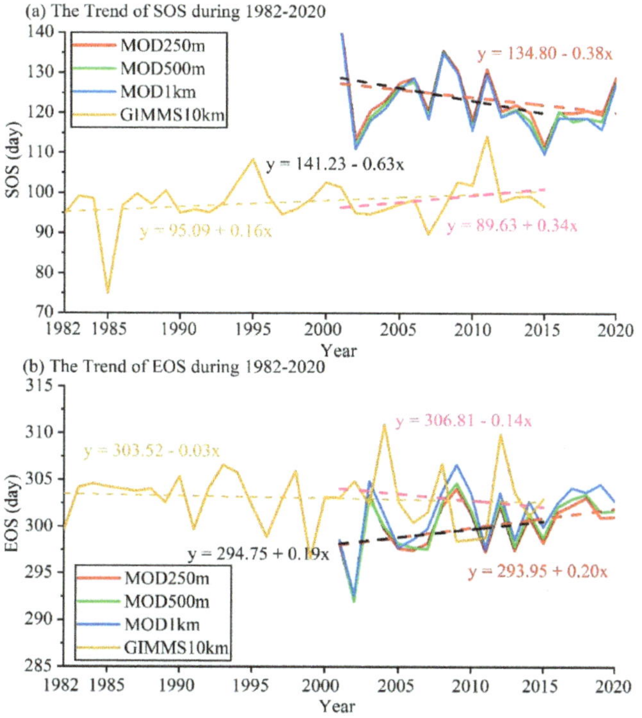

Figure 6. Inter-annual trends of the (**a**) SOS and (**b**) EOS that were estimated from the GIMMS3g and MODIS products during the period 1982–2020. The yellow, red, green, and blue solid lines represent the SOS and EOS values that were inferred from the GIMMS3g NDVI (1982–2015) and MODIS NDVI (2001–2020). The yellow dashed line shows the GIMMS3g-based SOS and EOS trends during the period 1982–2015. The pink dashed line shows the GIMMS3g-based SOS and EOS trends during the period 2001–2015. The black and red dashed line shows the 250 m MODIS-based SOS and EOS trends during the periods 2001–2015 and 2001–2020, respectively.

Table 2. The correlation coefficient and RMSE that were used for the comparison of the vegetation phenology between different MODIS products.

	500 m MODIS–250 m MODIS			1000 m MODIS–250 m MODIS		
	Bias	Correlation Coefficient	RMSE	Bias	Correlation Coefficient	RMSE
SOS	−1.2	0.9977 **	0.5665	−1.7	0.9957 **	0.9101
EOS	0.3	0.9952 **	0.3176	1.4	0.9775 **	0.8241

The symbol ** indicates significance at the 0.01 level.

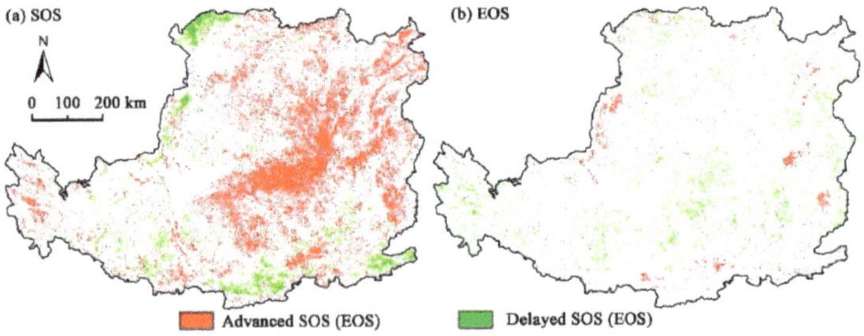

Figure 7. Spatial trends of the (**a**) SOS and (**b**) EOS from the 250 m MODIS time-series data during the period 2001–2020.

3.4. *Impact Factors on MODIS Products*

3.4.1. Influences of Vegetation on MODIS Products

To examine the possible causes of the difference in LSP from the MODIS data, we investigated the differences in the SOSs and EOSs between MODIS products that were used on different land cover types (Figure 8). Our findings showed that the largest difference between the 250 m MODIS SOS and the 1000 m MODIS SOS (1000 m MODIS SOS–250 m MODIS SOS) in forests was 3.5 days, which was larger than the difference found in grasslands (1.9 days) and croplands (0.6 days). Additionally, the standard deviation of the inter-annual difference between the 250 m and 1000 m MODIS SOS in forests was 1.1 days, which was the largest value that was obtained among all vegetation types. The differences that were obtained between the 250 m MODIS EOS and the 1000 m MODIS EOS were 0.9 days (forests), 1.2 days (grasslands), and 1.1 days (croplands). The standard deviation of the inter-annual difference was the largest in forests. In addition, we found that the differences in the SOS between the 250 m MODIS and 500 m MODIS (500 m MODIS SOS–250 m MODIS SOS) were both less than one day.

3.4.2. Influences of AT10 on the Phenology

Figure 9 presents the variations in the differences that were obtained between multiple sets of the SOS and EOS with an AT10 from 1 January to 30 April and from 1 September to 31 October, respectively. As the AT10 from January to April increased, the difference between each MODIS SOS gradually increased. In particular, the difference between the 1000 m MODIS SOS and the 250 m MODIS SOS (1000 m MODIS SOS–250 m MODIS SOS) was greater than the difference between the 500 m MODIS SOS and the 250 m MODIS SOS (500 m MODIS SOS–250 m MODIS SOS). The standard deviation of the 1000 m MODIS SOS was greater than that of the 500 m MODIS SOS. The relationship between the EOS and AT10 was found to be opposite to that of the SOS and AT10. Our results showed that with the increase in AT10 from September to October, the difference between each MODIS EOS gradually decreased. No matter whether the SOS or EOS were used, the difference in vegetation phenology between the 250 m and 1000 m products was less than 3 days. Additionally, we found that there was almost no difference between the 250 m and 500 m vegetation phenology.

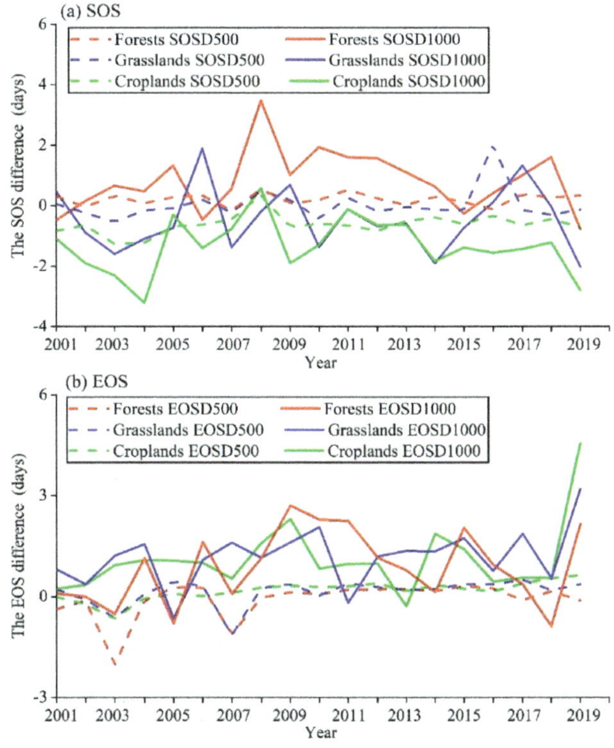

Figure 8. Inter-annual trends of the differences in the (**a**) SOS and (**b**) EOS between different land cover types over the Loess Plateau. The dashed line represents the difference in the SOS between the 250 m SOS and 500 m SOS for different land cover types during the period 2001–2019. The solid line represents the difference in SOS between the 250 m SOS and 1000 m SOS in different land cover types during the period 2001–2019.

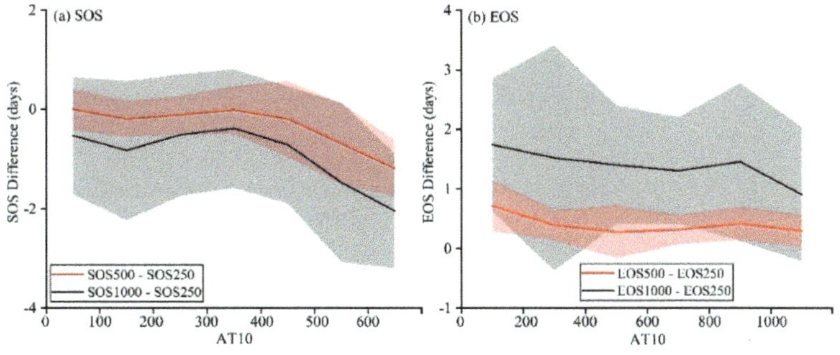

Figure 9. The variation in the differences between multiple sets of the (**a**) SOS and (**b**) EOS with an AT10 from January to April and from September to October, respectively. The red line shows the difference between the 500 m MODIS SOS (EOS) and the 250 m MODIS SOS (EOS). The black line shows the difference between the 1000 m MODIS SOS (EOS) and the 250 m MODIS SOS (EOS). The red and gray shadows indicate the standard deviations of the red and black lines, respectively.

4. Discussion

4.1. Difference between Satellite-Based LSP and Observations

Based on the MODIS NDVI and the layer named "composite day of the year," we calculated three sets of MODIS phenology with different spatial resolutions from 2001 to 2020. The GIMMS3g data lacked a "composite day of the year" layer, therefore we regarded the 1st and 16th days of each month as the DOY and calculated the GIMMS3g phenology from 1982 to 2015. We used the ground-observed data during the period 2001–2013 to verify the performances of the phenology that was derived from the MODIS and GIMMS3g data. Our results showed that the MODIS-derived phenology and ground-observed data had a good agreement, but that the correlation between the GIMMS3g-derived phenology and ground-observed data was very bad. This indicates that the SOS and EOS that were identified by the MODIS NDVI were more robust, but that the GIMMS3g phenology performed badly.

Moreover, the MODIS SOS that was identified by the logistics model was earlier than the SOS that was observed on the ground, while the MODIS EOS was later than the ground-observed data. Similarly, previous studies also found that in the north of China, the SOS that was identified by the logistics model based on SPOT satellite data was earlier than the SOS observed on the ground [49]. We suggest that this phenomenon was caused by the phenological recognition algorithm and the time resolution of the data used. The finer the temporal resolution of the image is, the more accurate the identified phenological phase will be [51–54]. However, among the existing vegetation index products, product data with a high time resolution are still limited. Therefore, an alternative suggestion is that using a proper method may make up for the lack of advanced or delayed phenological phases, such as the cumulative NDVI [49,55]. However, it is also worth noting that different methods will cause different problems.

The agreement between the EOS and the ground-observed data was not as good as for the SOS, especially GIMMS3g EOS. A previous study found that the NDVI at harvest time will be increased due to the noise-reduction algorithm [56]. The land use type of the Weihe Plain is mainly cropland, and the locations of the agro-meteorological stations are distributed around the Weihe Plain. This may lead to worse accuracy for the EOS in cultivated land than for the SOS. In addition, due to the limitation of data from observation stations, the phenological results that were gained in other areas cannot be verified. Compared with other studies, we found that the MODIS phenology was similar to those of other studies in terms of their spatial patterns [49,57].

4.2. Comparisons of Different Product Data

Effectiveness and using the smallest possible amount of data are matters that must be considered first in experiments. In this study, we compared four sets of remote sensing product data with different resolutions for the extraction of phenology in the Loess Plateau. Notably, the three sets of MODIS results with different resolutions showed good consistency. From the correlation coefficient and RMSE, we found that there was little difference between the 1000 m MODIS and 250 m MODIS results. This indicated that lower-resolution data could achieve the same effect as relatively higher-resolution data. Moreover, computers could process the 1000 m MODIS data much faster. Therefore, the comprehensive performance of the 1000 m MODIS data was better. If errors within the range of 1–3 days are allowed, we can use the 1000 m MODIS NDVI rather than the 250 m MODIS NDVI in future phenological studies.

In contrast, the GIMMS3g products were less effective in the Loess Plateau. From the time-series phenological results of the GIMMS3g products, we observed that the phenology was delayed in spring and advanced in autumn. However, the phenological results based on the MODIS products showed the opposite situation. As we know, many studies have confirmed that global warming advances spring phenology and delays autumn phenology [58–60]. This indicates that the phenological results obtained using the GIMMS3g data may contain errors in complex-terrain regions, such as the Loess Plateau. Additionally, we found that the strong spatial homogeneity of the GIMMS3g NDVI was the main reason for these phenological differences. Except for croplands, from the time series of the original data, no matter what year was used, the inflection point of the GIMMS3g data was always concentrated in the seventh or eighth period, and this led to the maximum curvature of the smoothed timing signal focus in this period. Therefore, the SOS that was identified by the GIMMS3g NDVI data in the northern part of the Loess Plateau was concentrated in DOY 90–110 (Figure A2).

However, previous studies showed that coarse-resolution SOS was comparable with finer-resolution SOS in homogeneous areas [5,42]. In this study, we calculated the difference between the MODIS and GIMMS3g results in the Weihe Plain during the period 2001–2015. Similarly, our findings showed that the GIMMS3g data performed well in monitoring the SOS over flat areas. Furthermore, the cropping intensity in a large area of the Weihe Plain during 1982–2013 changed from cropping twice a year to a single cropping taking place each year [61]. Due to the effectiveness of the GIMMS3g image in the Weihe Plain, the GIMMS3g SOS also showed a delayed trend that was the same as the MODIS phenological trend. Except for the flat area, the large heterogeneous areas in the Loess Plateau were affected by the original GIMMS3g data. Therefore, the original GIMMS3g data from the heterogeneous area incorrectly identified the phenological trend in the whole Loess Plateau.

Moreover, although the inflection point did not occur at the end of the vegetation growth of the raw GIMMS3g data, the proportion of the differences in EOSs between the GIMMS3g and MODIS data that were less than 5 days was still less than 20% for the entire Loess Plateau. This means that the GIMMS3g data had a weak ability to predict the SOS and EOS in areas with complex terrain, while they could better monitor the change in the SOS in relatively flat areas. If the GIMMS3g product cannot be replaced in an experiment on phenology production, other phenology estimation methods may be used, such as the dynamic threshold method [62–64]. Based on the maximum and minimum NDVI each year, the dynamic threshold method was used to determine the SOS and EOS in spring and autumn with the threshold ratio. This method could reduce the deviation in phenological estimation that is caused by the mutation of the temporal signal to a certain extent [65,66].

4.3. Factors for the Differences from MODIS Products

The differences in vegetation phenology that were determined from MODIS products with different spatial resolutions were mainly due to the land-cover types and temperatures involved. This finding can also be seen in other regions and ecosystems [67–70]. It is usually the case that when the same time resolution of the image is used, the finer the spatial resolution is, the more accurate the phenology properties will be. For example, in forest ecosystems, the vegetation phenology that is identified by MODIS images with a 1000 m resolution is more variable than that identified using images with a 500 m resolution. The main reason for this may be that the structures and functions of forest ecosystems are more complex than those of other natural ecosystems, such as cropland ecosystems [71,72]. Croplands are homogeneous and mainly affected by crop management. In contrast, forests are usually controlled by multiple environmental factors. Previous studies also suggested that the performance of coarse-resolution images over homogeneous areas is better than that over other regions [42,73,74].

Temperature is the main factor that leads to the advancement of spring phenology and the delay of autumn phenology [75–77]. Further, we found that the higher the AT10 during the early growth season was, the greater the differences and variability between the SOSs were. Meanwhile, the lower the AT10 during the late growth season was, the greater the differences and volatility of the EOSs were. Areas with a high AT10 from January to April were mainly distributed south of the Weihe Plain. We suggested that the main reason for this phenomenon was that there are more types of ecosystems in the southeast of the Loess Plateau, while the northwest of the plateau has a relatively homogenous ecosystem. Due to differences in the sensitivity of various types of vegetation to AT10, the SOS in areas with a relatively high AT10 showed greater differences. In addition, areas with a low AT10 from September to October were mainly concentrated in the southwest of the Loess Plateau. The huge elevation fluctuation in this area may be the reason for the phenological differences that were seen in products with different spatial resolutions. Although the AT10 and vegetation type or terrain had an impact on the data, the maximum averaged differences of the SOS and EOS between the 250 m MODIS products and the 1000 m MODIS products were less than three days. Additionally, the phenological difference remained within an acceptable range.

5. Conclusions

Based on the 250, 500, and 1000 m MODIS data during the period 2001–2020 and the ~10 km GIMMS3g data during the period 1982–2015, as well as the rate of change in the curvature of the logistic models, this study investigated the applicability and spatial scaling effects of various remote sensing products with different spatial resolutions on phenology extraction in a complex-terrain region. Our study showed that the MODIS products performed better in phenology analysis across the Loess Plateau, and the phenology results that were derived from the different MODIS products showed only small differences. However, the GIMMS3g-based SOSs that were derived from logistic models had a good performance in the flat region (i.e., the Weihe Plain) but a poor performance in regions with a more heterogeneous topography. Additionally, the performances of the GIMMS3g-based EOSs across the whole Loess Plateau were poor. The phenology results that were derived from the MODIS data presented advanced SOS trends and delayed EOS trends during the period 2001–2020 for the entire Loess Plateau. However, both the SOS and EOS trends that were identified by the GIMMS3g products were the opposite. Our finding emphasized that the 1000 m MODIS product can be used to extract phenology from areas with a complex terrain, such as the Loess Plateau, and almost no difference was found in the phenology extraction between the 500 m MODIS product and the 250 m MODIS products. Furthermore, we also investigated the effects of vegetation and AT10 on the spatiotemporal variability of vegetation phenology, which could help us to understand the driving factors of such phenological changes in the future.

Author Contributions: Conceptualization and methodology, F.C., Z.L. and S.W.; software, F.C.; validation, F.C.; formal analysis, F.C. and H.Z.; investigation, F.C. and H.Z.; resources, F.C., Z.L. and S.W.; data curation, F.C.; writing—original draft preparation, F.C.; writing—review and editing, Z.L. and S.W.; visualization, Z.L. and F.C.; supervision, Z.L.; project administration, Z.L.; funding acquisition, Z.L. All authors have read and agreed to the published version of the manuscript.

Funding: This work was funded by the National Key Research and Development Program of China (grant no. 2017YFC0504701), National Natural Science Foundation of China (grant nos. 41971218 and 41601582), and the Strategic Priority Research Program of the Chinese Academy of Sciences (grant nos. XDA23070302 and XDA28130400).

Data Availability Statement: The data that we used in this study can be requested by contacting the corresponding author.

Acknowledgments: We appreciate three anonymous reviewers and their valuable suggestions and comments.

Conflicts of Interest: The authors declare no conflict of interest.

Appendix A

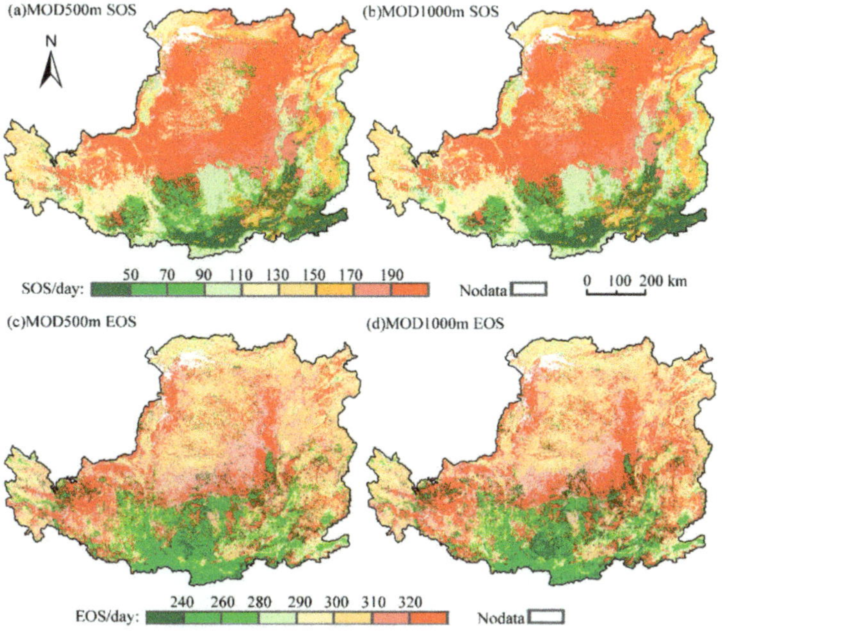

Figure A1. Spatial patterns of the SOS were estimated from (**a**) 500 m MODIS and (**b**) 1000 m MODIS from 2001; spatial patterns of the EOS were estimated from (**c**) 500 m MODIS and (**d**) 1000 m MODIS from 2001.

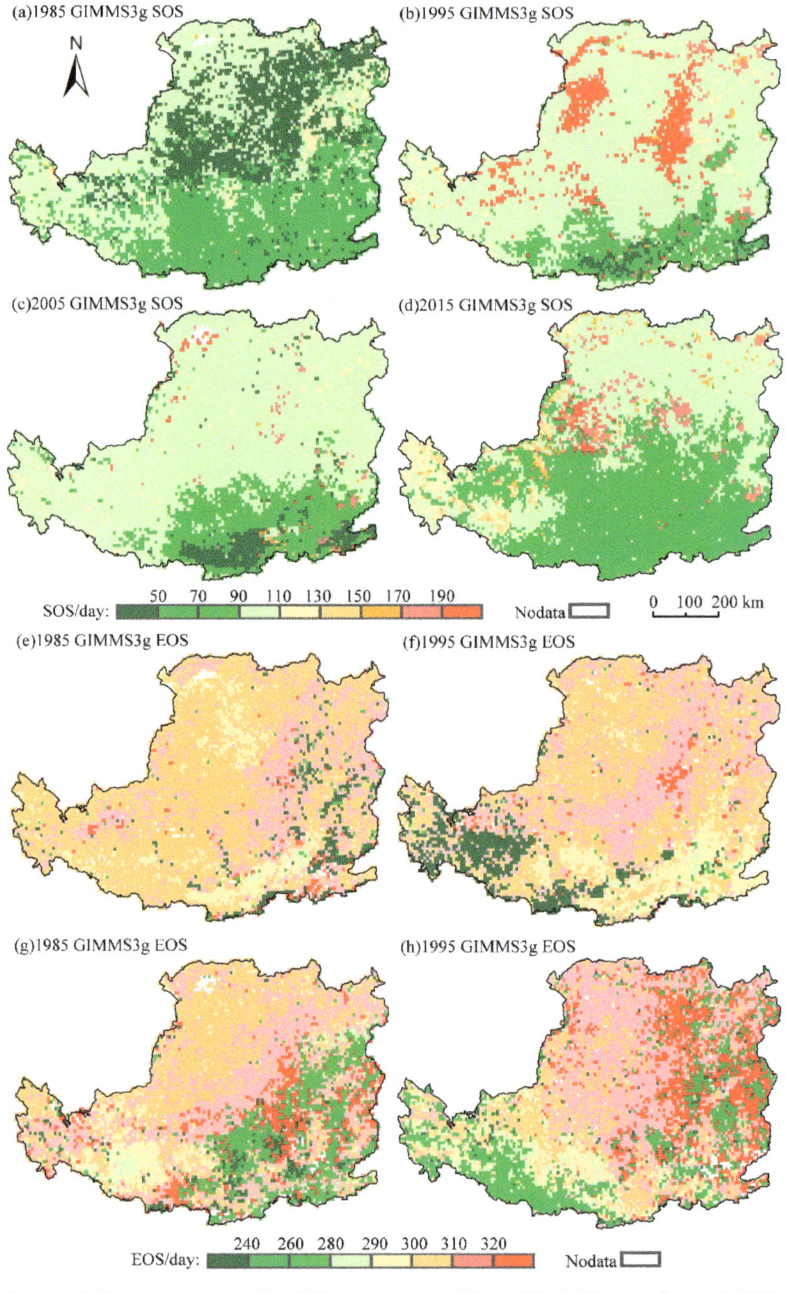

Figure A2. Spatial patterns of the SOS were estimated from GIMMS3g data from (**a**) 1985, (**b**) 1995, (**c**) 2005, and (**d**) 2015; spatial patterns of the EOS were estimated from GIMMS3g data from (**e**) 1985, (**f**) 1995, (**g**) 2005, and (**h**) 2015.

References

1. Wang, X.; Xiao, J.; Li, X.; Cheng, G.; Ma, M.; Zhu, G.; Arain, M.A.; Black, T.A.; Jassal, R.S. No trends in spring and autumn phenology during the global warming hiatus. *Nat. Commun.* **2019**, *10*, 2389. [CrossRef] [PubMed]
2. Shen, M. Spring phenology was not consistently related to winter warming on the Tibetan Plateau. *Proc. Natl. Acad. Sci. USA* **2011**, *108*, E91–E92. [CrossRef] [PubMed]
3. Zhang, G.; Zhang, Y.; Dong, J.; Xiao, X. Green-up dates in the Tibetan Plateau have continuously advanced from 1982 to 2011. *Proc. Natl. Acad. Sci. USA* **2013**, *110*, 4309–4314. [CrossRef]
4. Maignan, F.; Bréon, F.-M.; Bacour, C.; Demarty, J.; Poirson, A. Interannual vegetation phenology estimates from global AVHRR measurements: Comparison with in situ data and applications. *Remote Sens. Environ.* **2008**, *112*, 496–505. [CrossRef]
5. Peng, D.; Zhang, X.; Zhang, B.; Liu, L.; Liu, X.; Huete, A.R.; Huang, W.; Wang, S.; Luo, S.; Zhang, X.; et al. Scaling effects on spring phenology detections from MODIS data at multiple spatial resolutions over the contiguous United States. *ISPRS J. Photogramm. Remote Sens.* **2017**, *132*, 185–198. [CrossRef]
6. Smith, P.C.; de Noblet-Ducoudré, N.; Ciais, P.; Peylin, P.; Viovy, N.; Meurdesoif, Y.; Bondeau, A. European-wide simulations of croplands using an improved terrestrial biosphere model: Phenology and productivity. *J. Geophys. Res.* **2010**, *115*, 640. [CrossRef]
7. Emmenegger, T.; Hahn, S.; Bauer, S. Individual migration timing of common nightingales is tuned with vegetation and prey phenology at breeding sites. *BMC Ecol.* **2014**, *14*, 9. [CrossRef]
8. Richardson, A.D.; Hufkens, K.; Milliman, T.; Aubrecht, D.M.; Chen, M.; Gray, J.; Johnston, M.R.; Keenan, T.; Klosterman, S.T.; Kosmala, M.; et al. Tracking vegetation phenology across diverse North American biomes using PhenoCam imagery. *Sci. Data* **2018**, *5*, 180028. [CrossRef]
9. Schwartz, M.; Betancourt, J.L.; Weltzin, J.F. From Caprio's lilacs to the USA National Phenology Network. *Front. Ecol. Environ.* **2012**, *10*, 324–327. [CrossRef]
10. Norris, J.R.; Walker, J.J. Solar and sensor geometry, not vegetation response, drive satellite NDVI phenology in widespread ecosystems of the western United States. *Remote Sens. Environ.* **2020**, *249*, 112013. [CrossRef]
11. Peng, D.; Zhang, X.; Wu, C.; Huang, W.; Gonsamo, A.; Huete, A.R.; Didan, K.; Tan, B.; Liu, X.; Zhang, B. Intercomparison and evaluation of spring phenology products using National Phenology Network and AmeriFlux observations in the contiguous United States. *Agric. For. Meteorol.* **2017**, *242*, 33–46. [CrossRef]
12. Cao, M.; Sun, Y.; Jiang, X.; Li, Z.; Xin, Q. Identifying leaf phenology of deciduous broadleaf forests from PhenoCam images using a convolutional neural network regression method. *Remote Sens.* **2021**, *13*, 2331. [CrossRef]
13. Bórnez, K.; Richardson, A.D.; Verger, A.; Descals, A.; Peñuelas, J. Evaluation of VEGETATION and PROBA-V phenology using PhenoCam and eddy covariance data. *Remote Sens.* **2020**, *12*, 3077. [CrossRef]
14. Wang, Y.; Luo, Y.; Shafeeque, M. Interpretation of vegetation phenology changes using daytime and night-time temperatures across the Yellow River Basin, China. *Sci. Total Environ.* **2019**, *693*, 133553. [CrossRef] [PubMed]
15. Zhang, X.; Friedl, M.A.; Schaaf, C.B. Global vegetation phenology from moderate resolution imaging spectroradiometer (MODIS): Evaluation of global patterns and comparison with in situ measurements. *J. Geophys. Res.* **2006**, *111*, 367–375. [CrossRef]
16. Shen, X.; Liu, B.; Xue, Z.; Jiang, M.; Lu, X.; Zhang, Q. Spatiotemporal variation in vegetation spring phenology and its response to climate change in freshwater marshes of Northeast China. *Sci. Total Environ.* **2019**, *666*, 1169–1177. [CrossRef]
17. Moulin, S.; Kergoat, L.; Viovy, N.; Dedieu, G. Global-scale assessment of vegetation phenology using NOAA/AVHRR satellite measurements. *J. Clim.* **1997**, *10*, 1154–1170. [CrossRef]
18. Wang, H.; Wu, C.; Ciais, P.; Peñuelas, J.; Dai, J.; Fu, Y.; Ge, Q. Overestimation of the effect of climatic warming on spring phenology due to misrepresentation of chilling. *Nat. Commun.* **2020**, *11*, 4945. [CrossRef]
19. Atkinson, P.M.; Jeganathan, C.; Dash, J.; Atzberger, C. Inter-comparison of four models for smoothing satellite sensor time-series data to estimate vegetation phenology. *Remote Sens. Environ.* **2012**, *123*, 400–417. [CrossRef]
20. Shilong, P.; Fang, J.Y.; Zhou, L.M.; Ciais, P.; Zhu, B. Variations in satellite-derived phenology in China's temperate vegetation. *Glob. Chang. Biol.* **2006**, *12*, 672–685.
21. Cong, N.; Wang, T.; Nan, H.; Ma, Y.; Wang, X.; Myneni, R.B.; Piao, S. Changes in satellite-derived spring vegetation green-up date and its linkage to climate in China from 1982 to 2010: A multimethod analysis. *Glob. Chang. Biol.* **2013**, *19*, 881–891. [CrossRef]
22. Cleland, E.E.; Chuine, I.; Menzel, A.; Mooney, H.A.; Schwartz, M.D. Shifting plant phenology in response to global change. *Trends Ecol. Evol.* **2007**, *22*, 357–365. [CrossRef]
23. Pan, Y.; Li, L.; Zhang, J.; Liang, S.; Zhu, X.; Sulla-Menashe, D. Winter wheat area estimation from MODIS-EVI time series data using the Crop Proportion Phenology Index. *Remote Sens. Environ.* **2012**, *119*, 232–242. [CrossRef]
24. Wang, C.; Chen, J.; Wu, J.; Tang, Y.; Shi, P.; Black, T.A.; Zhu, K. A snow-free vegetation index for improved monitoring of vegetation spring green-up date in deciduous ecosystems. *Remote Sens. Environ.* **2017**, *196*, 1–12. [CrossRef]
25. Yang, W.; Kobayashi, H.; Wang, C.; Shen, M.; Chen, J.; Matsushita, B.; Tang, Y.; Kim, Y.; Bret-Harte, M.S.; Zona, D.; et al. A semi-analytical snow-free vegetation index for improving estimation of plant phenology in tundra and grassland ecosystems. *Remote Sens. Environ.* **2019**, *228*, 31–44. [CrossRef]
26. Wong, C.Y.; D'Odorico, P.; Bhathena, Y.; Arain, M.A.; Ensminger, I. Carotenoid based vegetation indices for accurate monitoring of the phenology of photosynthesis at the leaf-scale in deciduous and evergreen trees. *Remote Sens. Environ.* **2019**, *233*, 111407. [CrossRef]

27. Hou, X.; Niu, Z.; Gao, S.; Huang, N. Monitoring vegetation phenology in farming-pastoral zone using SPOT-VGT NDVI data. *Trans. Chin. Soc. Agric. Eng.* **2013**, *29*, 142–150.
28. Fabio, F.; Christian, R.; Tobias, J.; Gabriel, A.; Stefan, W. Alpine grassland phenology as seen in AVHRR, VEGETATION, and MODIS NDVI time series—A comparison with in situ measurements. *Sensors* **2008**, *8*, 2833–2853.
29. Cao, R.; Chen, J.; Shen, M.; Tang, Y. An improved logistic method for detecting spring vegetation phenology in grasslands from MODIS EVI time-series data. *Agric. For. Meteorol.* **2015**, *200*, 9–20. [CrossRef]
30. Tong, X.; Tian, F.; Brandt, M.; Liu, Y.; Zhang, W.; Fensholt, R. Trends of land surface phenology derived from passive microwave and optical remote sensing systems and associated drivers across the dry tropics 1992–2012. *Remote Sens. Environ.* **2019**, *232*, 111307. [CrossRef]
31. Markon, C.J.; Fleming, M.D.; Binnian, E.F. Characteristics of vegetation phenology over the Alaskan landscape using AVHRR time-series data. *Polar Rec.* **1995**, *31*, 179–190. [CrossRef]
32. Guyon, D.; Guillot, M.; Vitasse, Y.; Cardot, H.; Hagolle, O.; Delzon, S.; Wigneron, J.-P. Monitoring elevation variations in leaf phenology of deciduous broadleaf forests from SPOT/VEGETATION time-series. *Remote Sens. Environ.* **2011**, *115*, 615–627. [CrossRef]
33. Tan, B.; Morisette, J.T.; Wolfe, R.E.; Gao, F.; Ederer, G.A.; Nightingale, J.; Pedelty, J.A. An enhanced TIMESAT algorithm for estimating vegetation phenology metrics from MODIS data. *IEEE J. Sel. Top. Appl. Earth Obs. Remote Sens.* **2011**, *4*, 361–371. [CrossRef]
34. Zhang, X.; Friedl, M.A.; Schaaf, C.B.; Strahler, A.H.; Hodges, J.C.F.; Gao, F.; Reed, B.C.; Huete, A. Monitoring vegetation phenology using MODIS. *Remote Sens. Environ.* **2003**, *84*, 471–475. [CrossRef]
35. Ye, W.; van Dijk, A.I.J.M.; Huete, A.; Yebra, M. Global trends in vegetation seasonality in the GIMMS NDVI3g and their robustness. *Int. J. Appl. Earth Obs. Geoinf.* **2021**, *94*, 102238. [CrossRef]
36. Zhang, J.; Zhao, J.; Wang, Y.; Zhang, H.; Zhang, Z.; Guo, X. Comparison of land surface phenology in the Northern Hemisphere based on AVHRR GIMMS3g and MODIS datasets. *ISPRS J. Photogramm. Remote Sens.* **2020**, *169*, 1–16. [CrossRef]
37. Xia, H.; Qin, Y.; Feng, G.; Meng, Q.; Cui, Y.; Song, H.; Ouyang, Y.; Liu, G. Forest phenology dynamics to climate change and topography in a geographic and climate transition zone: The Qinling mountains in central China. *Forests* **2019**, *10*, 1007. [CrossRef]
38. Shen, M.; Zhang, G.; Cong, N.; Wang, S.; Kong, W.; Piao, S. Increasing altitudinal gradient of spring vegetation phenology during the last decade on the Qinghai–Tibetan Plateau. *Agric. For. Meteorol.* **2014**, *189–190*, 71–80. [CrossRef]
39. Richardson, A.D.; Keenan, T.F.; Migliavacca, M.; Ryu, Y.; Sonnentag, O.; Toomey, M. Climate change, phenology, and phenological control of vegetation feedbacks to the climate system. *Agric. For. Meteorol.* **2013**, *169*, 156–173. [CrossRef]
40. Liu, L.; Cao, R.; Shen, M.; Chen, J.; Zhang, X. How does scale effect influence spring vegetation phenology estimated from satellite-derived vegetation indexes? *Remote Sens.* **2019**, *11*, 2137. [CrossRef]
41. Ives, A.R.; Zhu, L.; Wang, F.; Zhu, J.; Morrow, C.J.; Radeloff, V.C. Statistical inference for trends in spatiotemporal data. *Remote Sens. Environ.* **2021**, *266*, 112678. [CrossRef]
42. Zhang, X.; Wang, J.; Gao, F.; Liu, Y.; Schaaf, C.; Friedl, M.; Yu, Y.; Jayavelu, S.; Gray, J.; Liu, L.; et al. Exploration of scaling effects on coarse resolution land surface phenology. *Remote Sens. Environ.* **2017**, *190*, 318–330. [CrossRef]
43. Potter, C.; Alexander, O. Changes in vegetation phenology and productivity in alaska over the past two decades. *Remote Sens.* **2020**, *12*, 1546. [CrossRef]
44. Ma, X.; Huete, A.; Tran, N.N. Interaction of seasonal sun-angle and savanna phenology observed and modelled using MODIS. *Remote Sens.* **2019**, *11*, 1398. [CrossRef]
45. Deng, G.; Zhang, H.; Yang, L.; Zhao, J.; Guo, X.; Ying, H.; Rihan, W.; Guo, D. estimating frost during growing season and its impact on the velocity of vegetation greenup and withering in northeast China. *Remote Sens.* **2020**, *12*, 1355. [CrossRef]
46. Fu, B.; Chen, L.; Ma, K.; Zhou, H.; Wang, J. The relationships between land use and soil conditions in the hilly area of the loess plateau in northern Shaanxi, China. *CATENA* **2000**, *39*, 69–78. [CrossRef]
47. Xiao, D.; Tao, F.; Liu, Y.; Shi, W.; Wang, M.; Liu, F.; Zhang, S.; Zhu, Z. Observed changes in winter wheat phenology in the North China Plain for 1981–2009. *Int. J. Biometeorol.* **2013**, *57*, 275–285. [CrossRef]
48. Chen, J.; Jönsson, P.; Tamura, M.; Gu, Z.; Matsushita, B.; Eklundh, L. A simple method for reconstructing a high-quality NDVI time-series data set based on the Savitzky–Golay filter. *Remote Sens. Environ.* **2004**, *91*, 332–344. [CrossRef]
49. Wu, C.; Hou, X.; Peng, D.; Gonsamo, A.; Xu, S. Land surface phenology of China's temperate ecosystems over 1999–2013: Spatial–temporal patterns, interaction effects, covariation with climate and implications for productivity. *Agric. For. Meteorol.* **2016**, *216*, 177–187. [CrossRef]
50. Liu, Z.; Wang, L.; Wang, S. Comparison of different GPP models in China using MODIS image and ChinaFLUX Data. *Remote Sens.* **2014**, *6*, 10215–10231. [CrossRef]
51. Wang, J.; Dong, J.; Liu, J.; Huang, M.; Li, G.; Running, S.W.; Smith, W.K.; Harris, W.; Saigusa, N.; Kondo, H.; et al. Comparison of gross primary productivity derived from GIMMS NDVI3g, GIMMS, and MODIS in southeast Asia. *Remote Sens.* **2014**, *6*, 2108–2133. [CrossRef]
52. Song, X.-P.; Huang, W.; Hansen, M.C.; Potapov, P. An evaluation of Landsat, Sentinel-2, Sentinel-1 and MODIS data for crop type mapping. *Sci. Remote Sens.* **2021**, *3*, 100018. [CrossRef]
53. Fensholt, R.; Proud, S.R. Evaluation of earth observation based global long term vegetation trends—Comparing GIMMS and MODIS global NDVI time series. *Remote Sens. Environ.* **2012**, *119*, 131–147. [CrossRef]

54. Tarnavsky, E.; Garrigues, S.; Brown, M.E. Multiscale geostatistical analysis of AVHRR, SPOT-VGT, and MODIS global NDVI products. *Remote Sens. Environ.* **2008**, *112*, 535–549. [CrossRef]
55. Hou, X.; Gao, S.; Niu, Z.; Xu, Z. Extracting grassland vegetation phenology in North China based on cumulative SPOT-VEGETATION NDVI data. *Int. J. Remote Sens.* **2014**, *35*, 3316–3330. [CrossRef]
56. Cao, R.; Chen, Y.; Shen, M.; Chen, J.; Zhou, J.; Wang, C.; Yang, W. A simple method to improve the quality of NDVI time-series data by integrating spatiotemporal information with the Savitzky-Golay filter. *Remote Sens. Environ.* **2018**, *217*, 244–257. [CrossRef]
57. Wang, H.; Liu, G.-H.; Li, Z.-S.; Ye, X.; Wang, M.; Gong, L. Driving force and changing trends of vegetation phenology in the Loess Plateau of China from 2000 to 2010. *J. Mt. Sci.* **2016**, *13*, 844–856. [CrossRef]
58. Körner, C.; Basler, D. Phenology under global warming. *Science* **2010**, *327*, 1461–1462. [CrossRef]
59. Fu, Y.H.; Zhao, H.; Piao, S.; Peaucelle, M.; Peng, S.; Zhou, G.; Ciais, P.; Huang, M.; Menzel, A.; Penuelas, J.; et al. Declining global warming effects on the phenology of spring leaf unfolding. *Nature* **2015**, *526*, 104–107. [CrossRef]
60. Roberts, A.M.; Tansey, C.; Smithers, R.J.; Phillimore, A.B. Predicting a change in the order of spring phenology in temperate forests. *Glob. Chang. Biol.* **2015**, *21*, 2603–2611. [CrossRef]
61. Qiu, B.; Lu, D.; Tang, Z.; Song, D.; Zeng, Y.; Wang, Z.; Chen, C.; Chen, N.; Huang, H.; Xu, W. Mapping cropping intensity trends in China during 1982–2013. *Appl. Geogr.* **2017**, *79*, 212–222. [CrossRef]
62. Cai, Y.; Li, X.; Zhang, M.; Lin, H. Mapping wetland using the object-based stacked generalization method based on multi-temporal optical and SAR data. *Int. J. Appl. Earth Obs. Geoinf.* **2020**, *92*, 102164. [CrossRef]
63. Tang, H.; Li, Z.; Zhu, Z.; Chen, B.; Zhang, B.; Xin, X. Variability and climate change trend in vegetation phenology of recent decades in the Greater Khingan Mountain area, Northeastern China. *Remote Sens.* **2015**, *7*, 11914–11932. [CrossRef]
64. Huang, X.; Liu, J.; Zhu, W.; Atzberger, C.; Liu, Q. The optimal threshold and vegetation index time series for retrieving crop phenology based on a modified dynamic threshold method. *Remote Sens.* **2019**, *11*, 2725. [CrossRef]
65. Descals, A.; Verger, A.; Yin, G.; Penuelas, J. A threshold method for robust and fast estimation of land-surface phenology using google earth engine. *IEEE J. Sel. Top. Appl. Earth Obs. Remote Sens.* **2021**, *14*, 601–606. [CrossRef]
66. You, X.; Meng, J.; Zhang, M.; Dong, T. Remote sensing based detection of crop phenology for agricultural zones in China using a new threshold method. *Remote Sens.* **2013**, *5*, 3190–3211. [CrossRef]
67. Zhang, Q.; Kong, D.; Shi, P.; Singh, V.P.; Sun, P. Vegetation phenology on the Qinghai-Tibetan Plateau and its response to climate change (1982–2013). *Agric. For. Meteorol.* **2018**, *248*, 408–417. [CrossRef]
68. Gao, M.; Wang, X.; Meng, F.; Liu, Q.; Li, X.; Zhang, Y.; Piao, S. Three-dimensional change in temperature sensitivity of northern vegetation phenology. *Glob. Chang. Biol.* **2020**, *26*, 5189–5201. [CrossRef]
69. Qiao, C.; Shen, S.; Cheng, C.; Wu, J.; Jia, D.; Song, C. Vegetation phenology in the Qilian mountains and its response to temperature from 1982 to 2014. *Remote Sens.* **2021**, *13*, 286. [CrossRef]
70. Ganguly, S.; Friedl, M.A.; Tan, B.; Zhang, X.; Verma, M. Land surface phenology from MODIS: Characterization of the Collection 5 global land cover dynamics product. *Remote Sens. Environ.* **2010**, *114*, 1805–1816. [CrossRef]
71. Walker, J.J.; de Beurs, K.M.; Wynne, R.H.; Gao, F. Evaluation of Landsat and MODIS data fusion products for analysis of dryland forest phenology. *Remote Sens. Environ.* **2012**, *117*, 381–393. [CrossRef]
72. Berra, E.F.; Gaulton, R. Remote sensing of temperate and boreal forest phenology: A review of progress, challenges and opportunities in the intercomparison of in-situ and satellite phenological metrics. *For. Ecol. Manag.* **2021**, *480*, 118663. [CrossRef]
73. Bajocco, S.; Dragoz, E.; Gitas, I.; Smiraglia, D.; Salvati, L.; Ricotta, C. Mapping forest fuels through vegetation phenology: The role of coarse-resolution satellite time-series. *PLoS ONE* **2015**, *10*, e0119811.
74. Tian, J.; Zhu, X.; Wu, J.; Shen, M.; Chen, J. Coarse-resolution satellite images overestimate urbanization effects on vegetation spring phenology. *Remote Sens.* **2020**, *12*, 117. [CrossRef]
75. Hänninen, H.; Kramer, K. A framework for modelling the annual cycle of trees in boreal and temperate regions. *Silva. Fenn.* **2007**, *41*, 167–205. [CrossRef]
76. Piao, S.; Tan, J.; Chen, A.; Fu, Y.H.; Ciais, P.; Liu, Q.; Janssens, I.A.; Vicca, S.; Zeng, Z.; Jeong, S.-J.; et al. Leaf onset in the northern hemisphere triggered by daytime temperature. *Nat. Commun.* **2015**, *6*, 6911. [CrossRef]
77. Estrella, N.; Sparks, T.H.; Menzel, A. Trends and temperature response in the phenology of crops in Germany. *Glob. Chang. Biol.* **2007**, *13*, 1737–1747. [CrossRef]

remote sensing

Article

Detecting the Turning Points of Grassland Autumn Phenology on the Qinghai-Tibetan Plateau: Spatial Heterogeneity and Controls

Yanzheng Yang [1], Ning Qi [2], Jun Zhao [3], Nan Meng [1,4], Zijian Lu [1,5], Xuezhi Wang [6], Le Kang [7], Boheng Wang [7], Ruonan Li [1,4], Jinfeng Ma [1] and Hua Zheng [1,4,*]

1 State Key Laboratory of Urban and Regional Ecology, Research Center for Eco-environmental Sciences, Chinese Academy of Sciences, Beijing 100085, China; yangyzh@rcees.ac.cn (Y.Y.); nanmeng_st@rcees.ac.cn (N.M.); 1000480075@smail.shnu.edu.cn (Z.L.); rnli@rcees.ac.cn (R.L.); jfma@rcees.ac.cn (J.M.)
2 School of Information Science & Technology, Beijing Forestry University, Beijing 100083, China; ningqi830@bjfu.edu.cn
3 China Aero Geophysical Survey & Remote Sensing Center for Natural Resources, Beijing 100083, China; zhaojun01@mail.cgs.gov.cn
4 University of Chinese Academy of Sciences, Beijing 100049, China
5 School of Environmental and Geographical Sciences, Shanghai Normal University, Shanghai 200234, China
6 Computer Network Information Center, Chinese Academy of Sciences, Beijing 100083, China; wxz@cnic.cn
7 East China Inventory and Planning Institute of the State Administration of Forestry and Grassland, Hangzhou 310019, China; hdybhc@163.com (L.K.); vancywang@126.com (B.W.)
* Correspondence: zhenghua@rcees.ac.cn; Tel.: +86-010-62849134

Citation: Yang, Y.; Qi, N.; Zhao, J.; Meng, N.; Lu, Z.; Wang, X.; Kang, L.; Wang, B.; Li, R.; Ma, J.; et al. Detecting the Turning Points of Grassland Autumn Phenology on the Qinghai-Tibetan Plateau: Spatial Heterogeneity and Controls. *Remote Sens.* **2021**, *13*, 4797. https://doi.org/10.3390/rs13234797

Academic Editors: Alfredo Huete, Xuanlong Ma, Jiaxin Jin, Xiaolin Zhu, Yuke Zhou and Qiaoyun Xie

Received: 12 October 2021
Accepted: 23 November 2021
Published: 26 November 2021

Abstract: Autumn phenology, commonly represented by the end of season (EOS), is considered to be the most sensitive and crucial productivity indicator of alpine and cold grassland in the Qinghai-Tibetan Plateau. Previous studies typically assumed that the rates of EOS changes remain unchanged over long time periods. However, pixel-scale analysis indicates the existence of turning points and differing EOS change rates before and after these points. The spatial heterogeneity and controls of these turning points remain unclear. In this study, the EOS turning point changes are extracted and their controls are explored by integrating long time-series remote sensing images and piecewise regression methods. The results indicate that the EOS changed over time with a delay rate of 0.08 days/year during 1982–2015. The rates of change are not consistent over different time periods, which clearly highlights the existence of turning points. The results show that temperature contributed most strongly to the EOS changes, followed by precipitation and insolation. Furthermore, the turning points of climate, human activities (e.g., grazing, economic development), and their intersections are found to jointly control the EOS turning points. This study is the first quantitative investigation into the spatial heterogeneity and controls of the EOS turning points on the Qinghai-Tibetan Plateau, and provides important insight into the growth mechanism of alpine and cold grassland.

Keywords: autumn phenology; turning point; climate changes; human activities; Qinghai-Tibetan Plateau

1. Introduction

Vegetation phenology refers to periodically recurring growth patterns [1], and sheds a unique light on how ecosystems respond to climate change [2–4]. Shifts in phenology trends can affect the carbon budget, water flux, and energy balance from a regional to global scale [5]. Regional warming in alpine regions has led to several significant phenology changes, including advancement of the start of the growing season (SOS) in spring and a delay of the end of season (EOS) in autumn, as well as extensions of the growing season [6]. Phenology changes in turn provide strong feedback to climate systems, which can affect the regional carbon and water cycles [7]. The advancement of SOS and its controls have been detected in numerous satellite data and observations [8,9]. However, emerging evidence has shown that autumn phenology may contribute more strongly to the growth season length extension than spring phenology, leading to an increase of biomass accumulation [10,11]. Autumn phenology plays a critical role in carbon and nitrogen cycling [12]; it is thus important to track the spatial dynamics of autumn phenology to obtain more accurate information regarding growth season length variations and improve the modeling of biochemical cycles at vegetation-climate intersections [13,14].

With its distinctive geographical and hydrothermal condition, the Qinghai-Tibetan Plateau is regarded as one of the planet's most vulnerable alpine and cold ecosystems because of its strong sensitivity to climate change and has thus become a hotspot of international research [15]. Some degree of consensus has been achieved in recent decades regarding EOS changes on the Qinghai-Tibetan Plateau. Previous studies have shown an overall lengthening of the growing season and extension of the EOS on the Qinghai-Tibetan Plateau due to the warming and increasingly humid climate [16,17]. Studies based on a limited number of phenological observations reported that the EOS exhibited advancement trends on a regional scale [18]. Moreover, EOS changes have significantly affected the gross primary productivity (GPP) and evapotranspiration (ET) of alpine and cold ecosystems [15]. Some evidence has demonstrated that EOS is not only controlled by climate conditions and human activities [19,20] but also depends on the previous growth stage (i.e., SOS, annual peak growth time) [21,22], which make the EOS variation controls complex and difficult to constrain. Additional studies are therefore required to more clearly reveal the mechanism of EOS changes.

The major challenge of EOS studies arises from the poorly understood control mechanism. Previous studies have recognized that warmer temperatures and inadequate autumn solar radiation enhance vegetation growth [22,23]. Daytime and nighttime temperatures have different impacts on the alpine grassland EOS. However, the effects of higher preseason precipitation or longer sunshine duration on the EOS changes remain unclear [6]. The intersection of a wide variety of climate variables complicates this interpretation. Furthermore, some evidence has shown that human activities (especially grassland grazing) can alter vegetation phenology [24,25], but the superimposed effects of ecological protection and grazing make this effect difficult to quantify.

Recent advances in time-series analysis have demonstrated that ecosystem status changes are gradual but ultimately lead to qualitative changes [26]. The concept of turning points has opened a new research direction of ecosystem status change. Land cover changes, extreme climate, and human disturbances often occur abruptly and can result in ecosystem status changes [27,28], whereas increasing human pressure or grazing may more gradually change the ecosystem. Some previous studies demonstrated that the trend rates of EOS changes tend to vary over long periods, whereas turning points (sometimes referred to as breaking points) are more distinct, with different change rates occurring before and after these points [29]. A turning point of the Qinghai-Tibetan Plateau has traditionally been defined in the year 2000 or the entire study period is taken as a whole [20], but notable variations can be detected at the pixel-scale, which have not been previously reported.

This paper investigates the Qinghai-Tibetan Plateau as a study area to (1) detect the existence of EOS turning points in different subregions, (2) quantify the determined climatic factors before and after the turning points, and (3) explore the contribution of climate change and human activities (grazing, economic development) to the EOS turning points. The detection of EOS turning points at the pixel and regional scale not only enriches the understanding of the EOS controls on alpine and cold grassland but also provides further details to reveal the EOS change mechanisms over different periods and their controls on the Qinghai-Tibetan Plateau.

2. Materials and Methods

2.1. Study Area

The Qinghai-Tibetan Plateau is situated in southwestern China and covers all of Tibet and the Qinghai provinces and is also a part of the Xinjiang, Sichuan, Gansu, and Yunnan provinces (Figure 1). The Qinghai-Tibetan Plateau is considered the third pole in the world, has an average altitude of >4000 m, and is characterized by a plateau monsoon climate with low temperatures, low precipitation, and strong insolation. More than 54% of the Qinghai-Tibetan Plateau area has a total annual precipitation below 400 mm and temperatures below 0 °C [22]. This region is known as the Asia water tower and is home to the headstreams of the Yangtze, Yellow, Lantsang, and Indus rivers. The alpine, cold, and dry climatic conditions lead to unique vegetation types on the Qinghai-Tibetan Plateau. A climate gradient exists from warm-humid in the southeast to cold-dry in the northwest, along which the vegetation types transition from forestland, meadow, steppe, and desert. The grassland, which includes meadow, steppe, and desert steppe, and covers 51.05% of the Qinghai-Tibetan Plateau area, is the most important ecosystem and sensitive to climate change. An understanding of grassland dynamics under the climate and human disturbance conditions is crucial for regional ecological security.

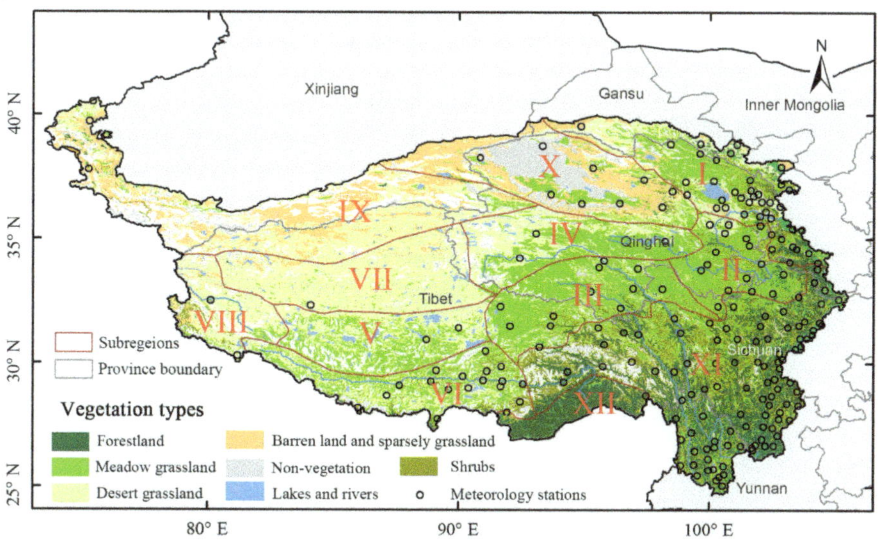

Figure 1. Study area and geographical subregions. The black circles represent the locations of 209 meteorological stations on the Qinghai-Tibetan Plateau.

We divided the entire Qinghai-Tibetan Plateau into 12 subregions (Figure 1, Table 1) based on the bio-geographical division proposed by Zheng et al. [30]. The grassland distribution was extracted according to a China vegetation map (scale = 1:100,000) [31], eliminating subregions X, XI, and XII, for which the main vegetation types are desert, forestland, and forestland, respectively. Only the remaining nine subregions (I–IX) are analyzed in this study, covering meadow, steppe, and desert grassland (Table 1). Of these nine subregions, we focused in detail on subregion I, which has the highest annual accumulated temperature above 0 °C ($AGDD_0$) and medium moisture index (MI). Subregions II and III had relatively high MI values that decreased from southeast to northwest. Each subregion exhibited unique climatic conditions and economic development levels, as well as different vegetation responses to climate and human activities.

Table 1. Description of the Qinghai-Tibetan Plateau subregions. $AGDD_0$: Annual accumulated temperature above 0 °C. MI: Moisture index calculated by the ratio of the mean annual precipitation to the annual equilibrium evapotranspiration.

ID	Subregion Names	$AGDD_0$ Means (°C)	MI Means	Main Provinces
I	Alpine temperate steppe of the Qinghai Lake Basin	1311.45	0.62	Qinghai, Gansu
II	Alpine meadow steppe on the Zoige Plateau	981.29	1.01	Qinghai, Sichuan
III	Alpine meadow steppe on the Yushu-Naqu Plateau	670.04	0.91	Qinghai, Tibet
IV	Alpine meadow steppe on the sources of the Yangtze and Yellow rivers	496.14	0.57	Qinghai
V	Alpine and cold grassland on the Southern Chang Tang Plateau	824.56	0.45	Tibet
VI	Alpine temperate grassland of the Brahmaputra River Basin	917.33	0.59	Tibet
VII	Alpine and cold grassland on the Northern Chang Tang Plateau	618.61	0.38	Tibet
VIII	Alpine and cold grassland on the Upper Indus River Basin	827.01	0.24	Tibet
IX	Alpine and cold desert grassland of the Kunlun Mountains	571.07	0.35	Tibet, Xinjiang
X	Alpine desert in the Qaidam Basin	1699.63	0.18	Qinghai
XI	Alpine forestland in the Hengduan Mountain	2043.25	1.14	Sichuan, Yunnan
XII	Subtropical forestland in the southern Tibet	3941.97	1.86	Tibet

2.2. Data Source

The GIMMS NDVI3g dataset provided by NASA was used to estimate the EOS on the Qinghai-Tibetan Plateau. The dataset was available from 1982 to 2015 with an 8-km spatial resolution and 15-day temporal resolution [32]. Some previous processes (e.g., calibration, noise removal) were performed for this version to better detect the vegetation dynamics [32]. This dataset has been widely used to detect long-term vegetation dynamics [33–35]. Due to the normalized difference vegetation index (NDVI) data might be misrepresented by snow [36]; we used the average temperature of a sequence of five days less than 0 °C to screen out the pixels that might be covered by snow. Temperature, precipitation, and insolation data from 1982–2015 were extracted from the China meteorological forcing dataset (1979–2015) downloaded from the Big Earth Data Platform for Three Poles with a spatial resolution of 0.1° and temporal resolution of 3 h (http://poles.tpdc.ac.cn/, Accessed on 15 August 2021) [37].

Human activities, including grazing density and economic development, were quantified using economic statistic data. For example, the grazing density were represented by the number of large animals (one large animal equal to five sheep unit) and sheep, and uniformly converted into sheep units. The economic development levels were quantified as the production of primary, secondary, and tertiary sectors. These data come from the statistical yearbooks of Qinghai and Tibet from 1982 to 2015.

2.3. Retrieval of EOS

Numerous methods have been used to fit the NDVI changes from seasonal vegetation cycles. After comparing the fitting results of HANTS [38], Polyfit [39], and Double logistic [40] in the nine subregions, we found that the RMES of HANTS ($1.26 \pm 0.24 \times 10^{-5}$) and Polyfit ($1.28 \pm 0.24 \times 10^{-5}$) were similar and smaller than the Double logistic results ($1.93 \pm 0.43 \times 10^{-5}$) (Figure A1). HANTS and Polyfit, were therefore selected as the two most simple and effective methods to fit the NDVI change curves. Dynamic thresholds were adopted to determine the EOS. Further details of these two fitting methods are described below.

The HANTS method involves the harmonic analysis of a time series, is adapted from the fast Fourier transform, and eliminates cloud noise using the least square method [38,40]. The HANTS methods can quickly smooth the data, remove outliers, and fill gaps of missing data. The following Equation (1) was used to fit the NDVI seasonal fluctuation curve:

$$NDVI(t) = a_0 + \sum_{i=1}^{n} a_i cos(2\pi t - \varphi_i) \tag{1}$$

where t is the Julian date, a_0 is the average of all $NDVI$ observations, and φ_i and a_i are the phase and amplitude of the curve, respectively.

The Polyfit method adopts a polynomial function to fit the *NDVI* records [39]. The following sixth order Equation (2) is used to describe the *NDVI* curve:

$$NDVI(t) = a_0 + a_1 t + a_2 t^2 + \ldots + a_6 t^6 \quad (2)$$

where a_0–a_6 are regression coefficients determined using the Levenberg–Marquardt method.

The EOS was determined by the day when the smoothed curve of the 34-year mean passed a designated threshold. We first fitted the *NDVI* changes with HANTS and Polyfit methods and then calculated the $NDVI_{ratio}$ (described in Equation (3)) for 365 days with multi-year mean *NDVI* values, next detected the time *t* with the minimum $NDVI_{ratio}$ and used the corresponding *NDVI*(*t* + 1) at time (*t* + 1) as the *NDVI* threshold for the EOS. Finally, we obtained the EOS for 34 years using the threshold:

$$NDVI_{ratio}(t) = \frac{NDVI(t+1) - NDVI(t)}{NDVI(t)} \quad (3)$$

2.4. Quantification of the EOS Trends, Turning Points, and Controls

After extracting the EOS at the pixel scale, we first quantified the tendency of the EOS changes using greenness changes methods, and then detected the turning points using the piecewise regression method. The mean EOS values and EOS trends within the subregions were calculated as the EOS and EOS trends at the subregion level. The turning points at a subregion and province level were calculated by the majority values.

The EOS trends were calculated using the greenness rate of change [41]. The EOS was considered delayed if the slope was a positive value; otherwise, the EOS advanced.

$$slope = \frac{n \times \sum_{i=1}^{n}(i \times NDVI) - \sum_{i=1}^{n} i \sum_{i=1}^{n} NDVI_i}{n \times \sum_{i=1}^{n} i^2 - (\sum_{i=1}^{n} i)^2} \quad (4)$$

where *i* is the order of the year, *n* is the number of years, $NDVI_i$ is the *NDVI* in the ith year, and the slope is the vegetation change rate. Alternatively, we can use the unary linear regression, in which the *P* values and confidence levels can be calculated.

Turning points were identified by piecewise regression [42] analysis, as defined in Equation (5), which can be used to detect sudden and sharp changes in directionality. This method has been widely applied for analyzing vegetation dynamics [19,43,44].

$$y = \begin{cases} \beta_0 + \beta_1 t + \varepsilon \ t \leq \alpha \\ \beta_0 + \beta_1 t + \beta_2 (t - \alpha) + \varepsilon \ t > \alpha \end{cases} \quad (5)$$

where *t* is the order of the year, α is the estimated turning point of the vegetation change trend determined using the least square error method, β_1 and $(\beta_1 + \beta_2)$ are the change rates before and after the turning points, respectively, and ε is the residual error. We performed *t*-tests to check the significance of the piecewise regressions.

Redundancy analysis (RDA) is a powerful analysis technique that could be applied in separating the contributions of climate, human activities, and their intersections to the EOS changes. RDA is a method to extract and summarize the variation in a set of response variables that can be explained by a set of explanatory variables [45]. In this study, RDA was performed with the vegan package in R language [46]. In RDA, climatic variables or human activity variables were chosen as predictors to maximize the extent of their correlation with the EOS changes as the response variable. RDA had been widely used in ecology-related studies [47,48]. The turning points of human activities were also calculated with Equation (5). The relationships between the turning points of the EOS and climatic variables were quantified using partial regression analysis or the correlation coefficient.

3. Results

3.1. EOS Spatial Distribution and Variation Characteristics

The obtained EOS presented high spatial heterogeneity across the grassland of the Qinghai-Tibetan Plateau during 1982–2015 (Figure 2). The EOS results extracted using the HANTS and Polyfit methods were not consistent, but their spatial distribution trends were similar (Figure 2a,b). The mean multi-year EOS began on the 291th day of the year (end of September) and spanned nearly one month from the southeast to the northwest (Figures 2 and 3a). The EOS started early (around the 277th day) in subregion IX, which has the highest elevation and lowest $AGDD_0$ values, and started late (around the 300th day) in subregions II and III, which are characterized by relatively warm-humid conditions. In the central Qinghai-Tibetan Plateau (subregion V), the EOS occurred on the 295th day. In subregion I, the EOS was early in the west and late in the east with a mean EOS on the 292th day. The spatial heterogeneity variations were significantly controlled by the MI ($EOS = 16.55 \times MI + 287.28$, $R^2_{adj} = 0.20^{**}$), with an early EOS in the drought subregions (IV, VII, VIII, and IX) and late EOS in the relatively humid subregions (II, III, and VI). The EOS spatial heterogeneity was essentially insensitive to $AGDD_0$.

The mean EOS on the Qinghai-Tibetan Plateau exhibited a slow delayed trend with an average rate of 0.08 days/year. The EOS results extracted using the HANTS and Polyfit methods presented similar patterns (Figure 2d,e). Using these two fitting methods, 60.2% of the study area presented delay trends (27.8% area is significant), while 39.8% of the study area presented advance trends (13.4% area is significant). The EOS trends differed between the nine subregions during 1982–2015 (Figure 3b), showing a delay in the northwest and an advance in the southeast. Subregions I and IX showed significantly delayed trends with more than 0.20 days/year. The EOS of subregion II, with a main land use type of wetland, was also delayed by a rate of 0.12 days/year. The EOS in subregion VIII, which is characterized by alpine, cold, and dry climatic conditions, presented a negative trend with the fastest variation rate (−0.12 days/year) compared with the other subregions. The EOS of subregion IV showed an advanced trend in the north but delayed trend in the south, with a mean EOS trend of 0.02 days/year. The EOS in subregions III and V showed slight advanced trends of -0.02 and −0.01 days/year, respectively. Subregions VI and VII both presented a slightly delayed trend with an average of approximately 0.04 days/year.

3.2. Detection of EOS Turning Points in the Subregions

The EOS changed over time and presented delayed trends during 1982–2015, but the rates of change were not fixed in each subregion over different periods, and notable turning points were observed (48.2% is significant) (Figure 4c). For example, in subregion I, the turning point occurs in the year 1994, for which the EOS was delayed before 1994 and slightly advanced after 1994 (Figure 4f). Similarly, subregion II showed a delayed EOS before 2002 and then a slightly advanced EOS after 2002. In the remaining subregions (III, IV, VI, and IX), the change trends were similar and the turning point year was 1994, where the EOS was delayed prior to 1994, suddenly advanced in 1995, and then maintained the previous change trend until 2015. The turning point trends in subregions V, VII, and VIII occurred in 1994, 1994, and 1999 respectively, but were not significant. These results demonstrate that the EOS changes clearly exhibit turning points and a wide range of EOS change trends with significant spatial heterogeneity on the Qinghai-Tibetan Plateau. The pattern of EOS turning points extracted by HANTS and Polyfit (Figure 4a,b,d,e) have a small difference in subregion I and VI.

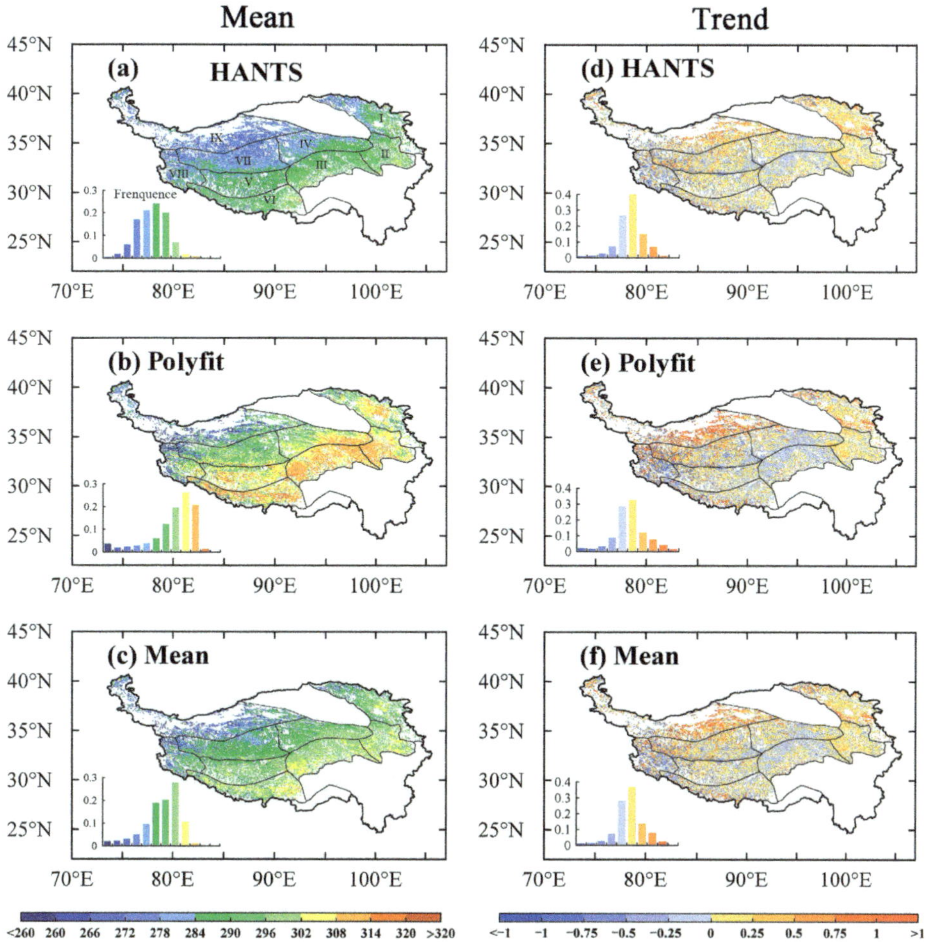

Figure 2. Distributions of the end of season (EOS) (calendar day) and their change trends (days/year). Multi-year means of the EOS obtained using the (**a**) HANTS method and (**b**) Polyfit method. (**c**) Average values of the HANTS and Polyfit methods. Trends of the EOS obtained using the (**d**) HANTS methods and (**e**) Polyfit method. (**f**) Average values of the EOS trends obtained using the HANTS and Polyfit methods.

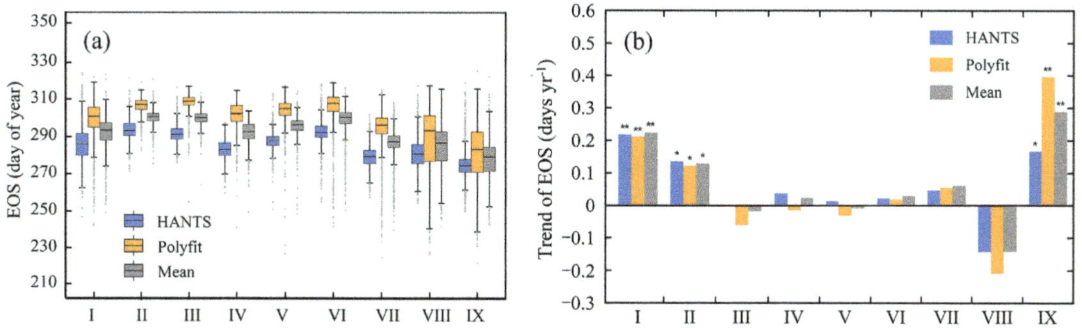

Figure 3. Bar graphs of the (**a**) mean EOS values and (**b**) their trends. "**" indicates $p < 0.01$ and "*" indicates $p < 0.05$.

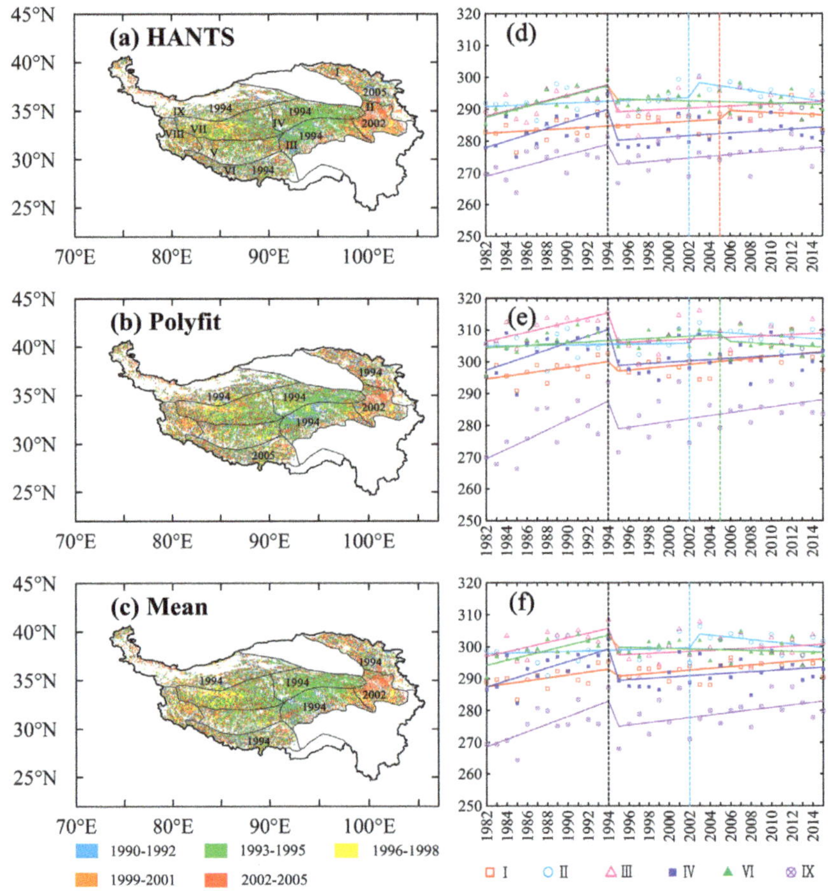

Figure 4. Turning point distributions and variations for subregions. (**a–c**) EOS turning points distributions with results of HANTS, Polyfit, and their mean values. The numbers represent the modes of the turning years in each subregion. (**d–f**) Scatter plots and results of piecewise regressions with the results of HANTS, Polyfit, and their mean values. The vertical dashed lines represent the turning points in the different subregions.

3.3. EOS Variations Controlled by Climatic Variables before and after Turning Points

The EOS changes exhibited close relationships with the climatic variables, but the dominant climatic variable differed in each subregion before and after its associated turning point (Figure 5). Temperature was the dominant control over the EOS changes in most subregions (I, II, IV, VI, VII, VIII, and IX) before and after the turning point year. In contrast, subregion III showed that the EOS was mainly controlled by the precipitation. Central subregion VII showed that the EOS was jointly controlled by the effects of temperature and precipitation. The area where the EOS changes was controlled by temperature covered the largest proportion, followed by precipitation and insolation (Figure 5d). The results indicate that the proportions controlled by each climate variable changed before and after the turning point years. For example, the EOS in subregion V was controlled by precipitation before the turning point, which then switched to temperature (Figure A2). The EOS of only approximately 40% of the area in subregion VI was significantly controlled by temperature prior to the turning point, which thereafter increased to 70%. Contribution of climates to EOS variation are similar with the fitting results of HANTs and Polyfit (Figure A3).

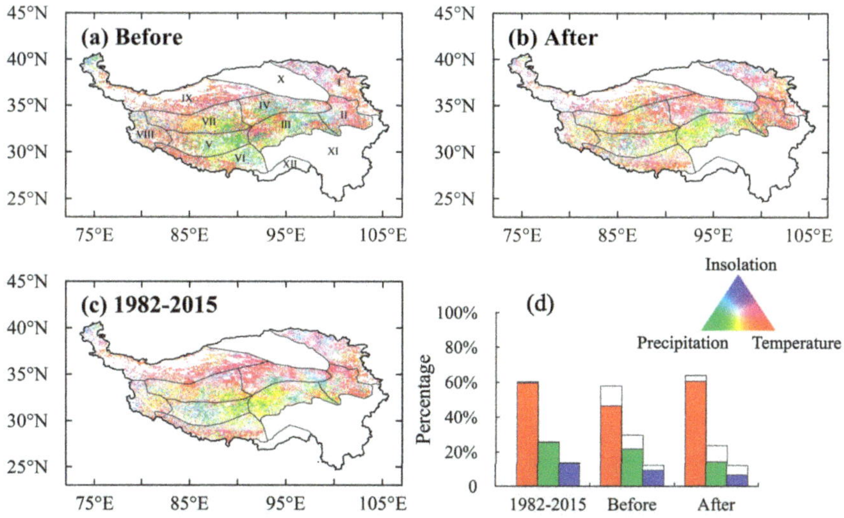

Figure 5. Relative influence of different climate variables (temperature, precipitation, and insolation) on EOS changes before (**a**) and after (**b**) the turning point, and over the entire study period (**c**) with the EOS means values of HANTS and Polyfit methods. (**d**) Area proportions controlled by the different dominant climate variables.

3.4. Controls on the EOS Turning Points

The changes of the annual EOS turning points are consistent with the turning points of the climate variables in most subregions (Table 2). In subregions I and II, the year of the EOS turning point coincides with the year of the insolation and precipitation turning points, respectively. Furthermore, the years of the EOS and temperature turning points are consistent in subregions III–IX. The major determining climatic variable for the EOS turning points is precipitation, followed by temperature and insolation. The relationship with the EOS turning point and insolation is generally weak ($R^2 < 0.05$).

Table 2. Correlation coefficients and P values between the turning points of the EOS and the turning points of climate variables.

The EOS Turning Points versus Climate Turning Points	R^2	p Value
EOS~temperature	0.331	<0.01
EOS~precipitation	0.378	<0.01
EOS~insolation	0.038	0.76

The relationship between the EOS and human activities was studied at the province level owing to limited statistical data in certain counties and subregions. The economic data show a consistent turning point with the EOS. Before the turning point year (~1996 for Qinghai and ~1995 for Tibet), Qinghai maintained a large amount of sheep, which reflected high grazing activity, and the economic development was slow with low production in the primary, secondary, and tertiary sectors. However, after the turning point year, the grazing intensity decreased and reached a stable change rate, whereas the economy developed rapidly, especially in the secondary sector. For Tibet, the grazing intensity was small before the turning point year but showed a rapid growth rate after the turning point. Similar to Qinghai, Tibet experienced fast economic growth after the turning point, especially in the tertiary sector.

At the province level, the annual EOS was closely related to climate, human activities, and their intersections (Table 3). For Qinghai, a combination of the turning points of climate and human activities can explain 78.86% of the EOS turning points changes, with climate independently accounting for 40.22% and human activities accounting for 10.45%. The intersections of climate and human activities can explain 28.19% of the EOS variation in Qinghai. In Tibet, the EOS change due to climate (66.17%) was larger than that in Qinghai and the effect of human activities (6.8%) was weaker. The climate and human activities intersections in Tibet (9.98%) were also smaller than in Qinghai.

Table 3. Independent contributions of the turning points of climate, human activities, and their intersections to the annual turning points variations of EOS at the province level.

Provinces	Climate Independent (%)	Human Activities Independent (%)	Climate-Human Activities Intersections (%)
Qinghai	40.22	10.45	28.19
Tibet	66.17	6.80	9.98

4. Discussion

4.1. Controls on the EOS and EOS Turning Points

This study is the first to demonstrate pixel-scale spatial heterogeneity of the EOS turning points and explain the turning point controls. The results indicate that the joint effects of climate variables and human activities are the main controls of the EOS turning points. The response of the EOS to environmental changes is complex. Some previous studies indicated that temperature plays a crucial role in EOS regulation [49] however, we show that the temperature control over the EOS is regulated by precipitation and insolation in the meadow and grassland ecotones. The cause of the turning points in most subregions is the abrupt change of temperature and precipitation. The results also reveal that insolation contributed considerably to the EOS changes, which is consistent with some previous reports that the EOS and its relation with GPP is mainly limited by insolation [50]. Other studies have reported that meadow shrinkage, decreased land cover, land albedo changes, and permafrost and seasonal frozen soil dynamics intersect with climate change, which alters the EOS trends [51].

Grazing is the most important human activity that affects grassland dynamics on the Qinghai-Tibetan Plateau [52]. The spatial heterogeneity of community increases, community function alteration, and biodiversity loss are considered to be some of the key disturbances that result in grassland degradation [53,54]. The pika population could also increase the effects of animal distribution on vegetation [55]. Overgrazing reduces the vegetation biomass and height, and restricts the regrowth ability of grassland. Our analysis shows that grazing activities in Qinghai notably decreased around 1998, coinciding with the implementation of national conservation policies (e.g., ecological compensation, restoration of degraded grassland). Grazing in Tibet was not active before 1995 and then rapidly increased, however grazing decreased after 2005 due to the late implementation of ecological conservation projects. The primary industry (mostly agriculture and animal husbandry) increased by nearly a factor of five in 1996–2015 compared with that in 1982–1994, which is also consistent with the EOS change turning points. The tertiary industry in Qinghai and Tibet quickly increased after the turning points, which indirectly reflects the intensification of human activities on the Qinghai-Tibetan Plateau.

4.2. Ecological Significance of the EOS and Its Turning Points

Phenological changes have great effects on the structure and function of ecosystems. At the community level, various species have different phenological responses to climate change, whereas the EOS can lead to a change in the competition for light and water conditions [17,56]. Moreover, plant species changes in the community introduced by the EOS can lead to phenological mismatches; for example, the period of high consumer demand for a resource does not match with the period of resource abundance [57]. At the ecosystem level, phenological grassland changes can modify certain land surface parameters (e.g., albedo, sensible heat flux, evaporation, boundary layer conductivity), which affects the regional carbon and water cycles [58]. For example, a later EOS may promote GPP and cause plants to close their stomates and increase water use efficiency if a soil deficit exists [59]. Moreover, the delayed EOS may also increase transpiration and partly offset the GPP, therefore leading to closer relationships between the net ecosystem productivities and EOS changes [60].

The existence of turning points indicates that the EOS trend over long-time periods does not remain unchanged, and the rates of EOS changes differ before and after these points. This observation has several advantages in ecosystem-related studies. First, climatic controls on the EOS in the Qinghai-Tibetan Plateau intersect with each other and follow non-linear relationships with the EOS. An analysis of the EOS before and after the turning points therefore helpful to evaluate the climatic driving mechanisms of the EOS. Second, the detection of spatial heterogeneity of the turning points is helpful for evaluating the large-scale implementation effects of ecological conservation projects. Third, an analysis of the turning points of the EOS relationships with ecosystem functions and services provide important guidelines for fine ecology planning and the development of protection policies.

4.3. Uncertainties, Challenges, and Future Directions

The uncertainties in this study arise from three aspects. First, although the EOS trends are consistent with the findings of MODIS NDVI and SPOT NDVI, some design shortcomings in the AVHRR sensor may potentially introduce noise into the GIMMS 3g NDVI dataset. Second, the human activities are difficult to quantify for lack of grazing data (intensity and boundary) and statistic data on the county levels for a long time. Third, there is a limited number of phenological stations on the Qinghai-Tibetan Plateau, and most are distributed in the east, which thus does not represent the EOS changes of the entire plateau. The results of the EOS extraction are not fully calibrated by observations owing to limited data availability.

We recommended the following perspectives for future studies. First, extreme climate events (e.g., cold, frost, drought) may have a more direct effect on vegetation phenology than gradual changes in mean climatic conditions [27,28]. Non-structural carbohydrate storage in plants is helpful to avoid damage caused by extreme events [61]. However, extreme climate conditions with variable frequencies and intensities in different seasons on the Qinghai-Tibetan Plateau require rigorous quantification. Second, although many studies have quantified the effects of climate variables in different seasons, spring phenology, growth season length, and human disturbances on the EOS changes, the joint contribution of these variables is low and the control mechanisms of the EOS and its turning points remain poorly understood. The strengthening and development of phenological observations stations are therefore necessary to explain the mechanism of phenology changes in the Qinghai-Tibetan Plateau. Third, ecosystem models are essential tools for simulating the carbon cycle in both historic and future climate scenarios however, their accuracies remain limited by the understanding of the EOS [62]. More reasonable algorithms and reliable observations are required to calibrate the ecosystem models, which will ultimately provide a new research direction but presently faces serious challenges.

5. Conclusions

This study applied multiple statistical methods and long-time series remote sensing data to determine the spatial heterogeneity and controls of autumn phenology on the Qinghai-Tibetan Plateau. The results are summarized as follows. (1) EOS turning points exhibit notable spatial heterogeneities. (2) The climatic controls of the EOS before and after the turning points varied in different subregions on the Qinghai-Tibetan Plateau. (3) Changes in the turning points are controlled by the joint effects of climate and human activities (grazing and economic development). This study is the first to demonstrate the spatial heterogeneity of turning points at a pixel scale and discuss their controls on the Qinghai-Tibetan Plateau, which is useful for exploring the mechanism of EOS changes and developing regional ecosystem conservation measures.

Author Contributions: Conceptualization, Y.Y. and N.Q.; methodology, N.Q.; validation, N.Q., J.Z., X.W. and L.K.; writing—original draft preparation, Y.Y. and N.Q.; writing—review and editing, N.M., Z.L., B.W., J.M. and R.L.; project administration, H.Z.; funding acquisition, H.Z. All authors have read and agreed to the published version of the manuscript.

Funding: This research was funded by The Second Tibetan Plateau Scientific Expedition and Research Program (STEP) (Grant No. 2019QZKK0307), the National Natural Science Foundation of China (41925005, 41701051, and 41801181) and Special Program of Research Center for Eco-environmental Sciences, CAS (RCEES-TDZ-2021-11).

Informed Consent Statement: Not applicable.

Data Availability Statement: The GIMMS 3g NDVI data presented in this study are openly available at the National Qinghai-Tibetan Plateau Data Center at [http://poles.tpdc.ac.cn/en/data/9775f2b4-7370-4e5e-a537-3482c9a83d88/. Accessed on 1 May 2021]. The climate data were downloaded from A Big Earth Data Platform for Three Poles with a spatial resolution of 0.1° and temporal resolution of 3 h [http://poles.tpdc.ac.cn/zh-hans/data/7a35329c-c53f-4267-aa07-e0037d913a21/. Accessed on 1 May 2021].

Conflicts of Interest: The authors declare no conflict of interest.

Appendix A

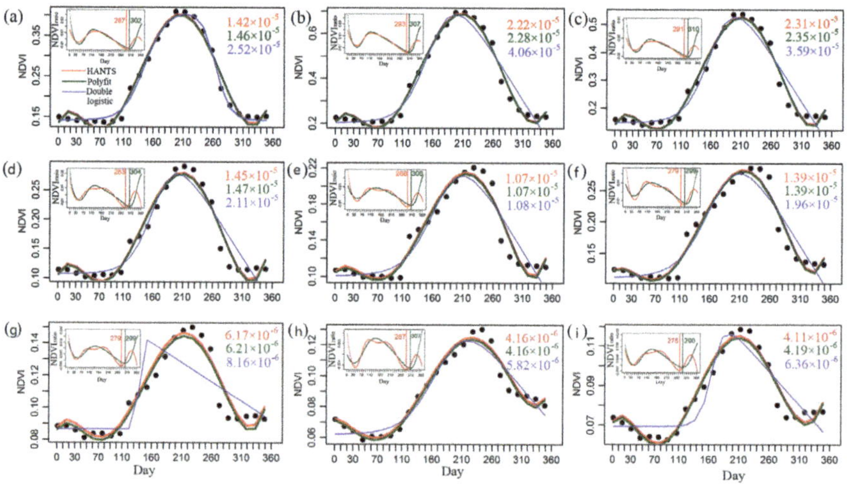

Figure A1. The NDVI fitting results of three equations, including HANTS (red lines and numbers), Polyfit (green lines and numbers), and Double logistic (red lines and numbers). (**a**–**i**) represent subregions I–IX, respectively. The upper-left panels are corresponding $NDVI_{ratio}$ changes and EOS with HANTS and Polyfit fitting methods. The upper-right numbers are RMESs for three fitting methods.

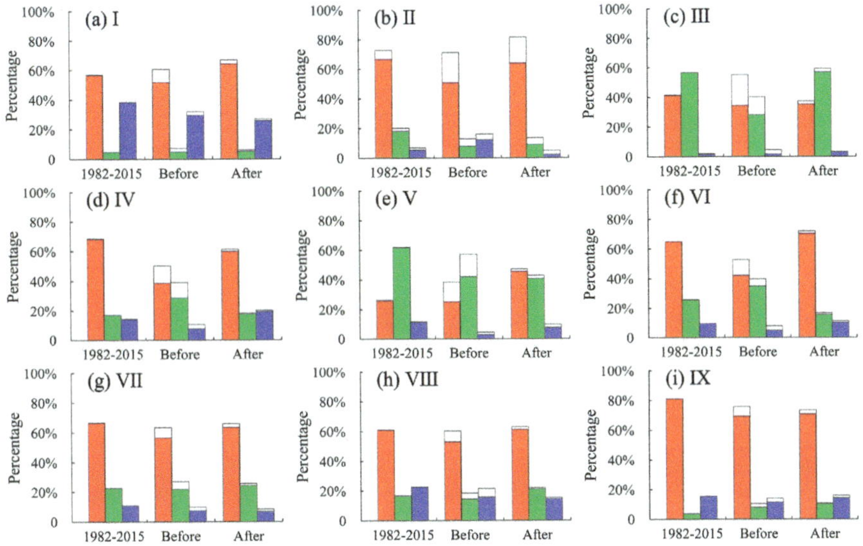

Figure A2. Relative proportions of the three climate variables that contributed to the EOS in each subregion. The legend coloring is same as in Figure 5. The white portions indicate that the proportion is not significant ($p > 0.05$).

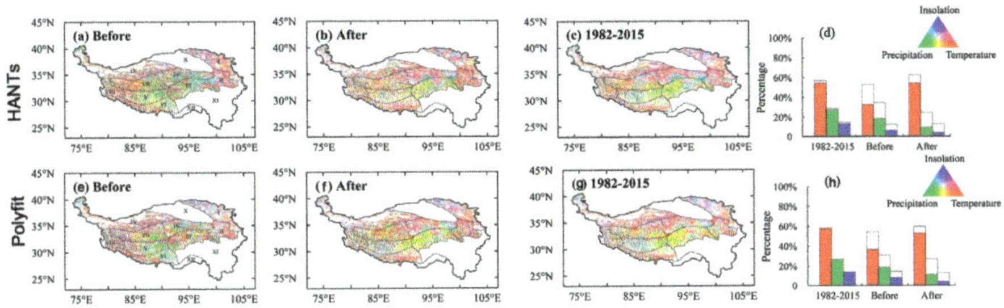

Figure A3. Relative influence of different climate variables (temperature, precipitation, and insolation) on EOS changes before (**a**,**e**) and after (**b**,**f**) the turning point, and over the entire study period (**c**,**g**) with the EOS values of HANTS and Polyfit methods respectively. (**d**,**h**) Area proportions controlled by the different dominant climate variables with results of HANTS and Polyfit methods.

References

1. Lieth, H. *Phenology and Seasonality Modeling*; Springer Science & Business Media: New York, NY, USA, 1974; Volume 8.
2. Wang, J.; Sun, H.; Xiong, J.; He, D.; Cheng, W.; Ye, C.; Yong, Z.; Huang, X. Dynamics and drivers of vegetation phenology in three-river headwaters region based on the Google Earth engine. *Remote Sens.* **2021**, *13*, 2528. [CrossRef]
3. Qiao, C.; Shen, S.; Cheng, C.; Wu, J.; Jia, D.; Song, C. Vegetation Phenology in the Qilian Mountains and its response to temperature from 1982 to 2014. *Remote Sens.* **2021**, *13*, 286. [CrossRef]
4. Bornez, K.; Richardson, A.D.; Verger, A.; Descals, A.; Penuelas, J. Evaluation of vegetation and PROBA-V phenology using PhenoCam and Eddy covariance data. *Remote Sens.* **2020**, *12*, 3077. [CrossRef]
5. Wang, S.Y.; Zhang, B.; Yang, Q.C.; Chen, G.S.; Yang, B.J.; Lu, L.L.; Shen, M.; Peng, Y.Y. Responses of net primary productivity to phenological dynamics in the Tibetan Plateau, China. *Agric. For. Meteorol.* **2017**, *232*, 235–246. [CrossRef]
6. An, S.; Chen, X.Q.; Zhang, X.Y.; Lang, W.G.; Ren, S.L.; Xu, L. Precipitation and minimum temperature are primary climatic controls of alpine grassland autumn phenology on the Qinghai-Tibet plateau. *Remote Sens.* **2020**, *12*, 431. [CrossRef]
7. Gonsamo, A.; Chen, J.M.; Ooi, Y.W. Peak season plant activity shift towards spring is reflected by increasing carbon uptake by extratropical ecosystems. *Glob. Chang. Biol.* **2018**, *24*, 2117–2128. [CrossRef]
8. Li, X.T.; Guo, W.; Li, S.H.; Zhang, J.Z.; Ni, X.N. The different impacts of the daytime and nighttime land surface temperatures on the alpine grassland phenology. *Ecosphere* **2021**, *12*, e03578. [CrossRef]

9. Lehikoinen, A.; Lindén, A.; Karlsson, M.; Andersson, A.; Crewe, T.L.; Dunn, E.H.; Gregory, G.; Karlsson, L.; Kristiansen, V.; Mackenzie, S.; et al. Phenology of the avian spring migratory passage in Europe and North America: Asymmetric advancement in time and increase in duration. *Ecol. Indic.* **2019**, *101*, 985–991. [CrossRef]
10. Zhu, W.Q.; Tian, H.Q.; Xu, X.F.; Pan, Y.Z.; Chen, G.S.; Lin, W.P. Extension of the growing season due to delayed autumn over mid and high latitudes in North America during 1982–2006. *Glob. Ecol. Biogeogr.* **2012**, *21*, 260–271. [CrossRef]
11. Bao, G.; Tuya, A.; Bayarsaikhan, S.; Dorjsuren, A.; Mandakh, U.; Bao, Y.H.; Li, C.L.; Vanchindorj, B. Variations and climate constraints of terrestrial net primary productivity over Mongolia. *Quat. Int.* **2020**, *537*, 112–125. [CrossRef]
12. Wu, C.; Gough, C.M.; Chen, J.M.; Gonsamo, A. Evidence of autumn phenology control on annual net ecosystem productivity in two temperate deciduous forests. *Ecol. Eng.* **2013**, *60*, 88–95. [CrossRef]
13. Sarvia, F.; De Petris, S.; Borgogno-Mondino, E. Exploring climate change effects on vegetation phenology by MOD13Q1 data: The piemonte region case study in the period 2001–2019. *Agronomy* **2021**, *11*, 555. [CrossRef]
14. Richardson, A.D.; Keenan, T.F.; Migliavacca, M.; Ryu, Y.; Sonnentag, O.; Toomey, M. Climate change, phenology, and phenological control of vegetation feedbacks to the climate system. *Agric. For. Meteorol.* **2013**, *169*, 156–173. [CrossRef]
15. Cheng, M.; Jin, J.X.; Jiang, H. Strong impacts of autumn phenology on grassland ecosystem water use efficiency on the Tibetan Plateau. *Ecol. Indic.* **2021**, *126*, 107682. [CrossRef]
16. Dong, M.; Jiang, Y.; Zheng, C.; Zhang, D. Trends in the thermal growing season throughout the Tibetan Plateau during 1960–2009. *Agric. For. Meteorol.* **2012**, *166–167*, 201–206. [CrossRef]
17. Shen, M.G.; Piao, S.L.; Dorji, T.; Liu, Q.; Cong, N.; Chen, X.Q.; An, S.; Wang, S.P.; Wang, T.; Zhang, G.X. Plant phenological responses to climate change on the Tibetan Plateau: Research status and challenges. *Natl. Sci. Rev.* **2015**, *2*, 454–467. [CrossRef]
18. Chen, X.Q.; An, S.; Inouye, D.W.; Schwartz, M.D. Temperature and snowfall trigger alpine vegetation green-up on the world's roof. *Glob. Chang. Biol.* **2015**, *21*, 3635–3646. [CrossRef]
19. Zhang, G.; Zhang, Y.; Dong, J.; Xiao, X. Green-up dates in the Tibetan Plateau have continuously advanced from 1982 to 2011. *Proc. Natl. Acad. Sci. USA* **2013**, *110*, 4309–4314. [CrossRef]
20. Li, P.; Peng, C.; Wang, M.; Luo, Y.; Li, M.; Zhang, K.; Zhang, D.; Zhu, Q. Dynamics of vegetation autumn phenology and its response to multiple environmental factors from 1982 to 2012 on Qinghai-Tibetan Plateau in China. *Sci. Total Environ.* **2018**, *637–638*, 855–864. [CrossRef]
21. Fu, Y.S.H.; Campioli, M.; Vitasse, Y.; De Boeck, H.J.; Van den Berge, J.; AbdElgawad, H.; Asard, H.; Piao, S.; Deckmyn, G.; Janssens, I.A. Variation in leaf flushing date influences autumnal senescence and next year's flushing date in two temperate tree species. *Proc. Natl. Acad. Sci. USA* **2014**, *111*, 7355–7360. [CrossRef]
22. Peng, J.; Wu, C.Y.; Wang, X.Y.; Lu, L.L. Spring phenology outweighed climate change in determining autumn phenology on the Tibetan Plateau. *Int. J. Climatol.* **2021**, *41*, 3725–3742. [CrossRef]
23. Cheng, M.; Jin, J.X.; Zhang, J.M.; Jiang, H.; Wang, R.Z. Effect of climate change on vegetation phenology of different land-cover types on the Tibetan Plateau. *Int. J. Remote Sens.* **2018**, *39*, 470–487. [CrossRef]
24. Chen, J.; Yan, F.; Lu, Q. Spatiotemporal Variation of Vegetation on the Qinghai–Tibet Plateau and the Influence of Climatic Factors and Human Activities on Vegetation Trend (2000–2019). *Remote Sens.* **2020**, *12*, 3150. [CrossRef]
25. Huang, K.; Zhang, Y.; Zhu, J.; Liu, Y.; Zu, J.; Zhang, J. The Influences of Climate Change and Human Activities on Vegetation Dynamics in the Qinghai-Tibet Plateau. *Remote Sens.* **2016**, *8*, 876. [CrossRef]
26. Horion, S.; Prishchepov, A.V.; Verbesselt, J.; de Beurs, K.; Tagesson, T.; Fensholt, R. Revealing turning points in ecosystem functioning over the Northern Eurasian agricultural frontier. *Glob. Chang. Biol.* **2016**, *22*, 2801–2817. [CrossRef]
27. Li, P.; Liu, Z.; Zhou, X.; Xie, B.; Li, Z.; Luo, Y.; Zhu, Q.; Peng, C. Combined control of multiple extreme climate stressors on autumn vegetation phenology on the Tibetan Plateau under past and future climate change. *Agric. For. Meteorol.* **2021**, *308–309*, 108571. [CrossRef]
28. Siegmund, J.F.; Wiedermann, M.; Donges, J.F.; Donner, R.V. Impact of temperature and precipitation extremes on the flowering dates of four German wildlife shrub species. *Biogeosciences* **2016**, *13*, 5541–5555. [CrossRef]
29. Bao, G.; Jin, H.; Tong, S.; Chen, J.; Huang, X.; Bao, Y.; Shao, C.; Mandakh, U.; Chopping, M.; Du, L. Autumn Phenology and Its Covariation with Climate, Spring Phenology and Annual Peak Growth on the Mongolian Plateau. *Agric. For. Meteorol.* **2021**, *298–299*, 108312. [CrossRef]
30. Zheng, D. *The Systematic Study of Ecogeographical Regions in China*; Commercial Press: Beijing, China, 2008.
31. Editorial Committee of Chinese Vegetation Map. *Vegetation Map of the People's Republic of China (1:1000000)*; Geological Publishing House: Beijing, China, 2006.
32. Pinzon, J.E.; Tucker, C.J. A Non-Stationary 1981–2012 AVHRR NDVI3g Time Series. *Remote Sens.* **2014**, *6*, 6929–6960. [CrossRef]
33. Deng, G.R.; Zhang, H.Y.; Yang, L.B.; Zhao, J.J.; Guo, X.Y.; Hong, Y.; Wu, R.H.; Dan, G. Estimating Frost during Growing Season and Its Impact on the Velocity of Vegetation Greenup and Withering in Northeast China. *Remote Sens.* **2020**, *12*, 1355. [CrossRef]
34. Wu, J.H.; Liang, S.L. Assessing Terrestrial Ecosystem Resilience using Satellite Leaf Area Index. *Remote Sens.* **2020**, *12*, 595. [CrossRef]
35. Wang, X.F.; Xiao, J.F.; Li, X.; Cheng, G.D.; Ma, M.G.; Zhu, G.F.; Arain, M.A.; Black, T.A.; Jassal, R.S. No trends in spring and autumn phenology during the global warming hiatus. *Nat. Commun.* **2019**, *10*, 10. [CrossRef]
36. Shen, M.G.; Sun, Z.Z.; Wang, S.P.; Zhang, G.X.; Kong, W.D.; Chen, A.P.; Piao, S.L. No evidence of continuously advanced green-up dates in the Tibetan Plateau over the last decade. *Proc. Natl. Acad. Sci. USA* **2013**, *110*, E2329. [CrossRef]

37. He, J.; Yang, K. China meteorological forcing dataset (1979–2015). In *A Big Earth Data Platform for Three Poles: 2016*; Northwest Institute of Eco-Environment and Resources: Lanzhou, China, 2016.
38. Jakubauskas, M.E.; Legates, D.R.; Kastens, J.H. Harmonic analysis of time-series AVHRR NDVI data. *Photogramm. Eng. Remote Sens.* **2001**, *67*, 461–470.
39. Piao, S.L.; Fang, J.Y.; Zhou, L.M.; Ciais, P.; Zhu, B. Variations in satellite-derived phenology in China's temperate vegetation. *Glob. Chang. Biol.* **2006**, *12*, 672–685. [CrossRef]
40. Liu, Q.; Fu, Y.S.H.; Zeng, Z.Z.; Huang, M.T.; Li, X.R.; Piao, S.L. Temperature, precipitation, and insolation effects on autumn vegetation phenology in temperate China. *Glob. Chang. Biol.* **2016**, *22*, 644–655. [CrossRef]
41. Stow, D.; Daeschner, S.; Hope, A.; Douglas, D.; Petersen, A.; Myneni, R.; Zhou, L.; Oechel, W. Variability of the seasonally integrated normalized difference vegetation index across the north slope of Alaska in the 1990s. *Int. J. Remote Sens.* **2003**, *24*, 1111–1117. [CrossRef]
42. Toms, J.D.; Lesperance, M.L. Piecewise regression: A tool for identifying ecological thresholds. *Ecology* **2003**, *84*, 2034–2041. [CrossRef]
43. Yu, H.; Luedeling, E.; Xu, J. Winter and spring warming result in delayed spring phenology on the Tibetan Plateau. *Proc. Natl. Acad. Sci. USA* **2010**, *107*, 22151–22156. [CrossRef]
44. Piao, S.L.; Wang, X.H.; Ciais, P.; Zhu, B.; Wang, T.; Liu, J. Changes in satellite-derived vegetation growth trend in temperate and boreal Eurasia from 1982 to 2006. *Glob. Chang. Biol.* **2011**, *17*, 3228–3239. [CrossRef]
45. Borcard, D.; Legendre, P.; Drapeau, P. Partialling out the spatial component of ecological variation. *Ecology* **1992**, *73*, 1045–1055. [CrossRef]
46. Oksanen, J.B.F.; Kindt, R.; Legendre, P.; O'hara, R.; Simpson, G.; Solymos, P.; Stevens, M.; Wagner, H. Multivariate Analysis of Ecological Communities. Version 1. Available online: http://cran.rproject.org/package=vegan (accessed on 1 May 2021).
47. Yang, Y.Z.; Wang, H.; Harrison, S.P.; Prentice, I.C.; Wright, I.J.; Peng, C.H.; Lin, G.H. Quantifying leaf-trait covariation and its controls across climates and biomes. *New Phytol.* **2019**, *221*, 155–168. [CrossRef]
48. Yamaura, Y.; Blanchet, F.G.; Higa, M. Analyzing community structure subject to incomplete sampling: Hierarchical community model vs. canonical ordinations. *Ecology* **2019**, *100*, e02759. [CrossRef]
49. Gill, A.L.; Gallinat, A.S.; Sanders-DeMott, R.; Rigden, A.J.; Gianotti, D.J.S.; Mantooth, J.A.; Templer, P.H. Changes in autumn senescence in northern hemisphere deciduous trees: A meta-analysis of autumn phenology studies. *Ann. Bot.* **2015**, *116*, 875–888. [CrossRef] [PubMed]
50. Zhang, Y.; Commane, R.; Zhou, S.; Williams, A.P.; Gentine, P. Light limitation regulates the response of autumn terrestrial carbon uptake to warming. *Nat. Clim. Chang.* **2020**, *10*, 739–743. [CrossRef]
51. Chen, H.; Zhu, Q.; Wu, N.; Wang, Y.; Peng, C.-H. Delayed spring phenology on the Tibetan Plateau may also be attributable to other factors than winter and spring warming. *Proc. Natl. Acad. Sci. USA* **2011**, *108*, E93. [CrossRef] [PubMed]
52. Sun, J.; Liu, M.; Fu, B.; Kemp, D.; Zhao, W.; Liu, G.; Han, G.; Wilkes, A.; Lu, X.; Chen, Y.; et al. Reconsidering the efficiency of grazing exclusion using fences on the Tibetan Plateau. *Sci. Bull.* **2020**, *65*, 1405–1414. [CrossRef]
53. Wu, G.-L.; Du, G.-Z.; Liu, Z.-H.; Thirgood, S. Effect of fencing and grazing on a Kobresia-dominated meadow in the Qinghai-Tibetan Plateau. *Plant. Soil* **2009**, *319*, 115–126. [CrossRef]
54. Li, G.Y.; Jiang, C.H.; Cheng, T.; Bai, J. Grazing alters the phenology of alpine steppe by changing the surface physical environment on the northeast Qinghai-Tibet Plateau, China. *J. Environ. Manag.* **2019**, *248*, 109257. [CrossRef]
55. Badingquiying; Smith, A.T.; Harris, R.B. Summer habitat use of plateau pikas (Ochotona curzoniae) in response to winter livestock grazing in the alpine steppe Qinghai-Tibetan Plateau. *Arct. Antarct. Alp. Res.* **2018**, *50*, e1447190. [CrossRef]
56. CaraDonna, P.J.; Iler, A.M.; Inouye, D.W. Shifts in flowering phenology reshape a subalpine plant community. *Proc. Natl. Acad. Sci. USA* **2018**, *115*, E9993. [CrossRef]
57. Visser, M.E.; Gienapp, P. Evolutionary and demographic consequences of phenological mismatches. *Nat. Ecol. Evol.* **2019**, *3*, 879–885. [CrossRef] [PubMed]
58. Teufel, B.; Sushama, L.; Arora, V.K.; Verseghy, D. Impact of dynamic vegetation phenology on the simulated pan-Arctic land surface state. *Clim. Dyn.* **2019**, *52*, 373–388. [CrossRef]
59. Jin, J.; Wang, Y.; Zhang, Z.; Magliulo, V.; Jiang, H.; Cheng, M. Phenology Plays an Important Role in the Regulation of Terrestrial Ecosystem Water-Use Efficiency in the Northern Hemisphere. *Remote Sens.* **2017**, *9*, 664. [CrossRef]
60. Liu, Y.; Wu, C. Understanding the role of phenology and summer physiology in controlling net ecosystem production: A multiscale comparison of satellite, PhenoCam and eddy covariance data. *Environ. Res. Lett.* **2020**, *15*, 104086. [CrossRef]
61. Palacio, S.; Maestro, M.; Montserrat-Marti, G. Seasonal dynamics of non-structural carbohydrates in two species of mediterranean sub-shrubs with different leaf phenology. *Environ. Exp. Bot.* **2007**, *59*, 34–42. [CrossRef]
62. Peng, J.; Wu, C.Y.; Zhang, X.Y.; Ju, W.M.; Wang, X.Y.; Lu, L.L.; Liu, Y.B. Incorporating water availability into autumn phenological model improved China's terrestrial gross primary productivity (GPP) simulation. *Environ. Res. Lett.* **2021**, *16*, 094012. [CrossRef]

remote sensing

MDPI

Article

Assessing Snow Phenology and Its Environmental Driving Factors in Northeast China

Hui Guo [1], Xiaoyan Wang [1,*], Zecheng Guo [1] and Siyong Chen [2]

[1] College of Earth and Environmental Sciences, Lanzhou University, Lanzhou 730030, China; hguo18@lzu.edu.cn (H.G.); 18709424424@163.com (Z.G.)

[2] School of Geography and Ocean Science, Nanjing University, Nanjing 210046, China; chensy18@lzu.edu.cn

* Correspondence: wangxiaoy@lzu.edu.cn; Tel.: +86-135-1964-0204

Abstract: Snow cover is an important water source and even an Essential Climate Variable (ECV) as defined by the World Meteorological Organization (WMO). Assessing snow phenology and its driving factors in Northeast China will help with comprehensively understanding the role of snow cover in regional water cycle and climate change. This study presents spatiotemporal variations in snow phenology and the relative importance of potential drivers, including climate, geography, and the normalized difference vegetation index (NDVI), based on the MODIS snow products across Northeast China from 2001 to 2018. The results indicated that the snow cover days (SCD), snow cover onset dates (SCOD) and snow cover end dates (SCED) all showed obvious latitudinal distribution characteristics. As the latitude gradually increases, SCD becomes longer, SCOD advances and SCED delays. Overall, there is a growing tendency in SCD and a delayed trend in SCED across time. The variations in snow phenology were driven by mean temperature, followed by latitude, while precipitation, aspect and slope all had little effect on the SCD, SCOD and SCED. With decreasing temperature, the SCD and SCED showed upward trends. The mean temperature has negatively correlation with SCD and SCED and positively correlation with SCOD. With increasing latitude, the change rate of the SCD, SCOD and SCED in the whole Northeast China were 10.20 d/degree, −3.82 d/degree and 5.41 d/degree, respectively, and the change rate of snow phenology in forested areas was lower than that in nonforested areas. At the same latitude, the snow phenology for different underlying surfaces varied greatly. The correlations between the snow phenology and NDVI were mainly positive, but weak correlations accounted for a large proportion.

Keywords: snow phenology; driving factors; spatiotemporal variations; Northeast China

Citation: Guo, H.; Wang, X.; Guo, Z.; Chen, S. Assessing Snow Phenology and Its Environmental Driving Factors in Northeast China. *Remote Sens.* **2022**, *14*, 262. https://doi.org/10.3390/rs14020262

Academic Editor: Alexander Kokhanovsky

Received: 2 November 2021
Accepted: 4 January 2022
Published: 7 January 2022

1. Introduction

Northeast China, one of the three major areas with snow cover in China [1,2], is an important agricultural production base. Snow cover can affect the scale and yield of agriculture by changing soil moisture, insultation from deep frost and prevention of microbial decomposition of organic matter. In addition, Snow accumulation and melting are important for storing and releasing water [3,4]. Snow phenology can describe seasonal snow cover variations and is also used to study the relationship between snow cover and climate change; parameters include the snow cover days (SCD), snow cover onset dates (SCOD) and snow cover end dates (SCED) [5,6]. According to the statistics, snow phenology generally has regular interannual variations [7]. Therefore, assessing snow phenology and its driving factors in Northeast China is essential for water resource management and agricultural development in this region.

However, in contrast with research on large-scale snow phenology, systematic studies of snow phenology in the region have been limited thus far. Chen et al. [8] used the MODIS snow products across Northeast China to study the spatiotemporal variations in snow cover. Ding and Gao [4] studied the SCD in Northeast China based on meteorological station data. Yang et al. [9] used MODIS data to explore only space-time distributions of SCD. Qiao et al. [10] used MODIS data to investigate the variations in snow phenology and their impact on vegetation growth in forested areas, which occupies 40% of the total area. Shi et al. [11] discussed the SCOD and SCED in the Mollisol areas across the northeastern plains of China based on snow depth data. In addition, Huang et al., Ke et al., and Ma et al. [5,6,12] used meteorological station data to analyze the snow phenology changes in snow-covered areas of China, including Northeast China.

When studying snow phenology based on meteorological observation stations, the stations give snow depth data for only the surrounding area [13]. Moreover, due to the scarcity and uneven distribution of stations, there are large discontinuities in the spatial distribution of the obtained snow phenology data, especially in forested areas and alpine regions [5,14–16]. The snow depth data retrieved by passive microwaves have a long time series, but the resolution is low [17], which is not appropriate for local regional research. The same is true of the Northern Hemisphere snow cover extent (NHSCE), which is more suitable for studying large-scale snow cover variations [18,19]. In contrast, MODIS data have not only higher spatial resolution but also higher temporal and spectral resolutions [20]. Therefore, MODIS is an ideal data source for studying continuous snow phenology, whether at the global or regional scale. For cloud contamination, many scholars have also studied and recovered data under clouds through a series of methods and achieved high cloud removal accuracy [21–26].

Based on the above situation, we analyzed the snow phenology of Northeast China from 2001 to 2018 in this study using the MODIS snow product. First, the daily cloud-free snow products were obtained through the conditional probability interpolation method based on a space-time cube, and accuracy was verified by ground observations. Based on this work, we explored the spatiotemporal variations in snow phenology and the relative importance of potential drivers, including climate, geography, and the NDVI, and we then discussed the roles of major factors in driving snow phenology.

2. Study Area

Northeast China is located at 115°30′~135°60′E, 38°42′~53°36′N, and the area is approximately 1.24 million km^2. The whole region is surrounded by the Daxingan Mountains, Xiaoxingan Mountains, and Chang-bai Mountains, with the Northeast Plain in the middle. It is a relatively independent and complete natural geographical area (Figure 1). The annual average temperature, precipitation, and elevation are −5~11 °C, 300~1000 mm, and 443 m, respectively [27–29]. Due to the existence of large-scale forests and climate factors, this region in China has a long snow period and thick snow cover [30]. The Daxingan Mountains, Xiaoxingan Mountains, and Changbai Mountains are the most important snowfall areas, with an average annual snowfall of more than 60 mm. The average snowfall is more than 90 mm in the northern Daxingan Mountains and Changbai Mountains, while in some plain areas, the average snowfall is less than 30 mm [31]. Here, the snowfall means the snow water equivalent.

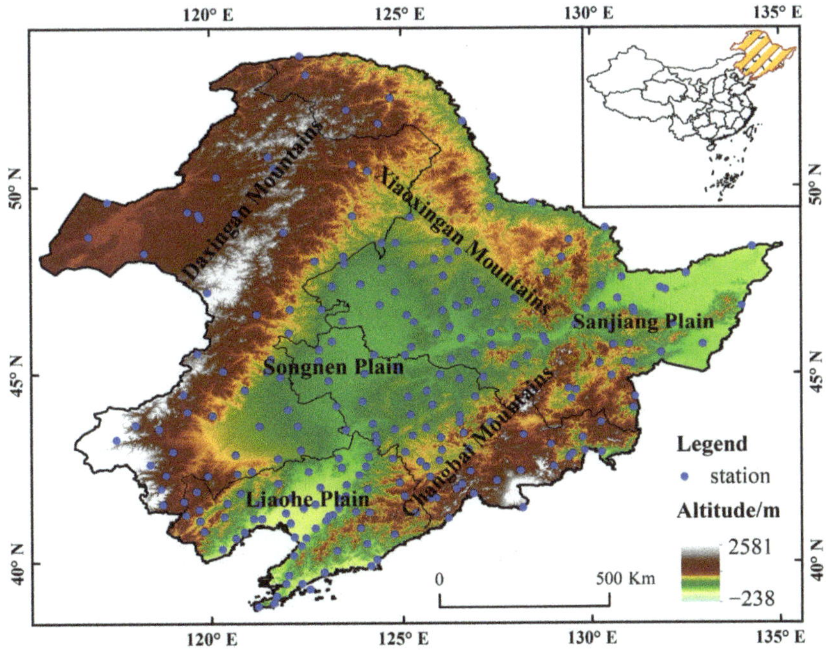

Figure 1. Topography and station distribution in the study area.

3. Methodology

3.1. Data Sources

3.1.1. MODIS Data

Daily MOD10A1 (V006) data with a 500 m resolution and containing 6574 scenes during 2001–2018 were downloaded with the help of the Google Earth Engine (https://code.earthengine.google.com (30 October 2021)). The product has four data types: NDSI data, quality assessment, albedo, and orbital information. In this research, only NDSI_Snow_Cover data were used as the main data source to extract snow products, and the NDSI of snow cover always greater than 0, but not all surface features with NDSI > 0 is snow cover. Detailed attribute information is shown in Table 1 [32].

Table 1. MODIS Snow Products (V006).

Value	Attributes	Value	Attributes
0–100	NDSI_Snow_Cover	239	ocean
200	missing data	250	cloud
201	no decision	254	detector saturated
211	night	255	fill
237	inland water		

3.1.2. Snow Depth Records

The snow depth records in Northeast China from 1 January 2013 to 31 December 2018 were collected from the National Cryosphere Desert Data Center (http://www.ncdc.ac.cn (30 October 2021)), and these records in snow season (October to April of the following year) were used to validate the cloud removal accuracy. The station distribution is shown in Figure 1. Generally, SD $\geq$ 1 cm indicates snow cover; otherwise, the station is regarded as snow-free [6,12,33].

3.1.3. Meteorological Data

Meteorological data with a 1 km resolution were acquired from the National Earth System Science Data Center, National Science and Technology Infrastructure of China (http://www.geodata.cn (30 October 2021)). The dataset is monthly data, and in this study, the temperature/precipitation in different months, the annual mean temperature (Temp-mean) and annual total precipitation (Pre-total) in hydrological years were used to explore the roles in snow phenology.

3.1.4. DEM

SRTM digital elevation data with a spatial resolution of 90 m was obtained from the Google Earth Engine (GEE). In addition, the data was resampled to 500 m in order to keep the resolution consistent with MODIS snow products. Based on this, the altitude, slope and aspect were derived.

3.1.5. Vegetation Data

To analyze the variations in snow phenology for different vegetation types in Northeast China, the MODIS vegetation product MOD13A1 normalized difference vegetation index (NDVI) from 2001 to 2018 was derived through the GEE cloud platform with spatial resolution and temporal resolution of 500 m and 16 d, respectively. In addition, the monthly/yearly NDVI maximum of all grids was then calculated by the maximum synthesis method to eliminate the interference of clouds and shadows.

3.1.6. Land Cover Data

MCD12Q1 is an annual land cover product with five classification standards. In this paper, the International Geosphere-Biosphere Program (IGBP) classification standard was chosen, and land cover types in this region were reclassified into water bodies, forested areas (including broadleaf forest, coniferous forest, mixed forest and shrub), and nonforested areas (except water and forested areas), and then we explore the differences in snow cover phenology for various underlying surfaces at the same latitude.

3.2. Methodology

3.2.1. MODIS Snow Product Cloud Removal

Due to cloud contamination, we used a noval cloud removal algorithm to obtain daily cloud-free MODIS snow products. First, a threshold of 10 (expand 100 times) was used to delineate between snow and snow-free conditions in the NDSI_Snow_Cover data layer [34], and the rest attributes were classified as clouds, except for inland water and ocean. Second, the conditional probability interpolation method based on a space-time cube was used to remove the clouds. In addition, snow probability of the cloud pixels was calculated by using the conditional probability of the central pixel and every neighboring pixel in a space-time cube of 5 × 5 × 5 under the same snow condition as the weight. Finally, the snow condition of pixels covered with clouds was recovered according to the snow probability [26]. The formulas are as follows:

$$P(C_{x,y}|C_n) = \frac{\sum 1 - ABS(C_{x,y,t} - C_{n,t'})}{N_{x,y}} \quad (1)$$

$$P(x_0, y_0, t_0) = \frac{\sum P(C_{x_0,y_0}|C_n) \times S_n}{\sum P(C_{x_0,y_0}|C_n) \times V_{n'}} \quad (2)$$

$$C(x_0, y_0, t_0) = \begin{cases} snow, & P(x_0, y_0, t_0) \geq 0.5 \\ snowfree, & P(x_0, y_0, t_0) < 0.5 \end{cases} \quad (3)$$

Here, $P(C_{x,y}|C_n)$ is the conditional probability having the same snow condition for the central and n-th adjacent pixels in the space-time cube. $C_{x,y,t}$ and $C_{n,t'}$ represent snow (C = 1) or snow-free (C = 0) conditions for days t and t', respectively. $N_{x,y}$ are cloud-free days for the central pixel and n-th neighboring pixels within the study time; $P(x_0, y_0, t_0)$ is the snow probability of the cloud gaps. S_n means that the n-th pixel has snow (S_n = 1) or is snow-free (S_n = 0); $V_{n'}$ indicates whether the n-th pixel is cloudless ($V_{n'}$ = 1) or covered by clouds ($V_{n'}$ = 0); and $C(x_0, y_0, t_0)$ is the snow condition.

3.2.2. Snow Phenology Calculation

Snow phenology mainly includes snow cover days (SCD), snow cover onset dates (SCOD) and snow cover end dates (SCED). In this study, daily cloud-free snow products were obtained through cloud removal algorithm above, and snow phenological parameters in hydrological year, which was defined from 1 September to 31 August of the following year, were calculated pixel by pixel. The SCD was the total days when a pixel is snow in a hydrological year. The SCOD was the first day when pixel was covered with snow lasting at least five days for the first time, and SCED was the last day of at least 5 days of continuous snow. This avoided the influence of instantaneous snowfall [10,16,35].

3.2.3. Cloud-Free Snow Product Accuracy Assessment

Currently, station data from meteorological observatories are usually regarded as "truth" data and used to evaluate the cloud removal accuracy. The accuracy assessment metrics include overall accuracy (*OA*), underestimation error (*UE*) and overestimation error (*OE*) based on the confusion matrix (Table 2), which are defined as follows:

$$OA = \frac{a+d}{a+b+c+d} \tag{4}$$

$$UE = \frac{b}{a+b+c+d} \tag{5}$$

$$OE = \frac{c}{a+b+c+d} \tag{6}$$

Table 2. Confusion matrix.

		MODIS	
		Snow	**Snow-Free**
Truth	snow	a	b
	snow-free	c	d

The definitions of a, b, c and d are given in Table 2. *OA* represents the proportion that pixels are consistent with the truth and MODIS classification; *UE* is the proportion that pixels are snow-free in MODIS, but the corresponding pixels in the truth are covered with snow; and *OE* refers to the proportion that pixels are covered with snow in MODIS, but the corresponding pixels in the truth are snow-free.

3.2.4. Trend Analysis

The Mann–Kendall test and Theil–Sen median analysis were used to explore the variation trend in snow phenology from hydrological years 2001–2018. In the Mann–Kendall test, to calculate whether the trend of snow phenology was increasing or decreasing at the 0.05 confidence level, the Z value was divided into five types: significant decrease (Z < 1.96), slow decrease ($-1.96 \le Z < 0$), nonsignificant change (Z = 0), slow increase ($0 > Z \ge 1.96$) and significant increase (Z > 1.96). The formulas are as follows:

$$Z = \begin{cases} \frac{S-1}{\sqrt{var(S)}}, S > 0 \\ 0 \quad , S = 0 \\ \frac{S+1}{\sqrt{var(S)}}, S < 0 \end{cases} \tag{7}$$

where,

$$var(S)=\frac{n(n-1)(2n+5)}{18} \tag{8}$$

$$S=\sum_{i=1}^{n-1}\sum_{j=i+1}^{n} sgn(S_j-S_i) \tag{9}$$

$$sgn(S_j-S_i)=\begin{cases}1, S_j-S_i>0\\0, S_j-S_i=0\\-1, S_j-S_i<0\end{cases} \tag{10}$$

When $Z>0$, the trend is upward, and when $Z<0$, it is downward. S_i/S_j represent the value in years *i/j*, n is the length of the time series. When $|Z|>Z_{1-\alpha/2}$ (α is the significance level), the trend is significant in the time series. In this paper, $\alpha = 0.05$ was used.

In the Theil–Sen median analysis, to explore the details of the variations in snow phenology, S_{snow} was divided into seven types: <−4 d/a, -<−4–−2 d/a, −2–0 d/a, 0 d/a, 0–2 d/a, 2–4 d/a and >4 d/a. The formula is as follows:

$$S_{snow}= Median\left(\frac{S_j-S_i}{j-i}\right), \quad \forall j>i \tag{11}$$

where $S_{snow}>0$ and $S_{snow}<0$ represent upward and downward trends, respectively.

3.2.5. Relative Importance of Multiple Factors to Snow Phenology

The use of geodetector is a common statistical approach that can analyze spatial variations and reveal the driving factors behind them [36]. A geodetector contains four subdetectors: factor detector, risk detector, ecological detector and interaction detector [37]. In this research, we employed a factor detector to quantify the relative contributions of vegetation and geographical (altitude, slope, aspect, latitude and longitude) and meteorological (temperature and precipitation) factors to snow phenology variations, and the dominant driving factor was then the highest contributor to snow phenology variations.

The factor detector is calculated by the following *q*-statistic:

$$q= 1-\frac{\sum_{h=1}^{L} N_h\sigma_h^2}{N\sigma^2}=1-\frac{SSW}{SST} \tag{12}$$

$$SSW=\sum_{h=1}^{L} N_h\sigma_h^2,\ SST=\ N\sigma^2 \tag{13}$$

where $0\le q\le 1$, and the larger the value, the greater the influence of the factor. h is the number of strata for variables or factors, N represents the number of units in stratum h, and σ^2 and σ_h^2 denote the variance in the entire study area and stratum h, respectively. SW and SST are the sum of squares within the data and the total sum of squares, respectively.

3.2.6. Correlation Analysis

To investigate the influence of major driving factors on snow phenology, we calculated the correlation coefficients between these factors and snow phenology:

$$r_{xy}=\frac{\sum_{i=1}^{n}(x_i-\overline{x})(y_i-\overline{y})}{\sqrt{\sum_{i=1}^{n}(x_i-\overline{x})^2\sum_{i=1}^{n}(y_i-\overline{y})^2}} \tag{14}$$

where x_i and y_i represent the values in the *i*-th year and $\overline{x}$ and $\overline{y}$ are the average values for all years. If $r>0$, two variables are positively correlated, and if $r<0$, it is negatively correlated. When $|r|\le 0.3$, two variables are weakly correlated or have no correlation; $0.3<|r|\le 0.5$ indicates that there is a moderate correlation between two variables; $0.5<|r|\le 1$ indicates strong correlation [38].

4. Results

4.1. Validation of the Daily Cloud-Free MODIS Snow Products

Based on above accuracy assessment metrics, the available station data in Northeast China were screened to quantitatively evaluate the cloud removal results. The accuracies of the three evaluation indicators were obtained (Table 3), and Figure 2 summarizes the monthly accuracy evaluation results for ground observation stations. Overall, the *OA* of the MODIS cloud-free product was more than 0.9, the *UE* and *OE* were less than 0.1, and the *UE* values were greater than the *OE* values. Generally, the accuracies in the snow stabilization season were higher than that in the snow accumulation and melting season. However, in the October and April, the *OA* was the highest, and the *UE* and *OE* were the lowest, which may be attributed to the lack of snow during this period. There seems odd relative to others in the February, which is due to the lack of 9 days MODIS data (19 February 2016–27 February 2016). In the cloud removal processing, the missing data are thought to be cloud. Therefore, the cloud removal accuracy is very low for these data because there is no spatial and temporal neighbor data available. The analysis reveals that the cloud removal product had good reliability and met the needs of the snow phenology analysis.

Table 3. Accuracy evaluation results for ground observation stations.

Time	*OA*	*UE*	*OE*
1 October 2013–30 April 2014	0.95	0.03	0.02
1 October 2014–30 April 2015	0.94	0.05	0.01
1 October 2015–30 April 2016	0.92	0.06	0.03
1 October 2016–30 April 2017	0.94	0.04	0.02
1 October 2017–30 April 2018	0.93	0.05	0.02

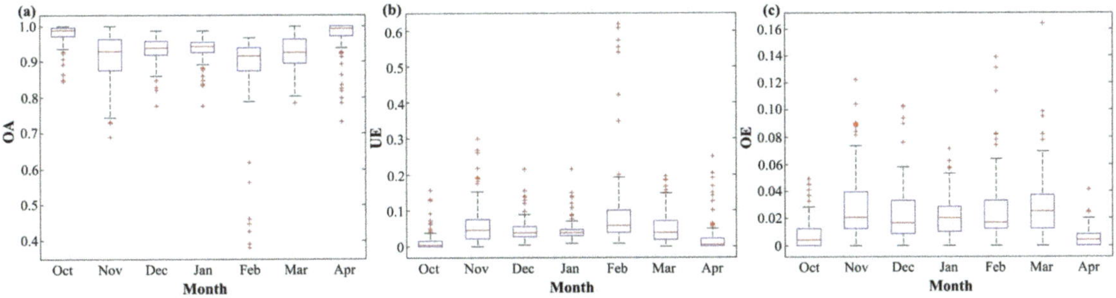

Figure 2. Boxplots of the three evaluation indices (**a**) OA, (**b**) UE and (**c**) OE in different months.

4.2. Spatiotemporal Variations and Trends in Snow Phenology

4.2.1. SCD

The distribution of SCD in Northeast China was visually consistent with those of topography and latitude, showing the characteristics of "high SCD in the mountains and low SCD in the plains" and "high SCD at high latitudes and low SCD at low latitudes". The SCD values in the northern Daxingan Mountains were more than 150 d, those of the Xiaoxingan Mountains were mainly within 120–150 d, and the SCD values of the plain area were significantly less than those of the mountainous area. The SCD values of the Sanjiang Plain were within 90–150 d, and the Liaohe Plain had the shortest SCD values of less than 60 d (Figure 3).

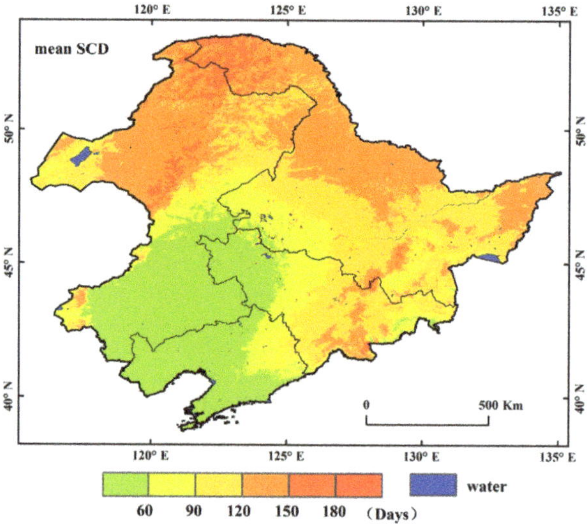

Figure 3. Spatial distribution of annual average SCD in Northeast China from HY2001 to HY2017.

Figure 4 showed the histogram of annual average SCD computed for a 10-day interval from HY2001 to HY2017. The average SCD in Northeast China ranged from 0 to 230 d, with an obvious bimodal distribution, and the annual average SCD was 93 d. The frequency of SCD in the range from 10–20 d accounted for 9.95%, and that in the range from 120–130 d accounted for 13.92%.

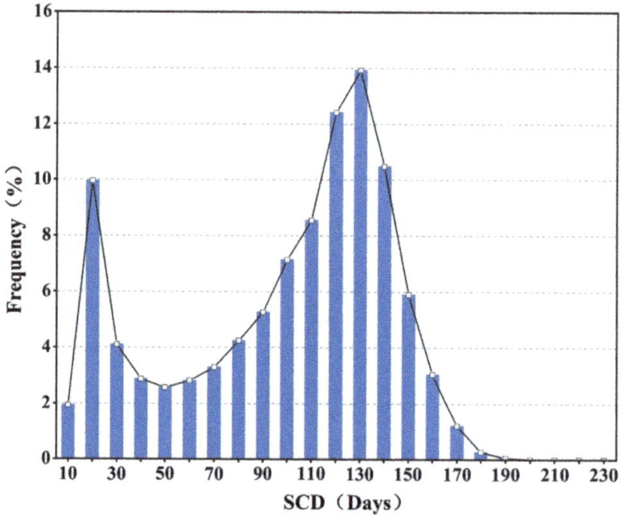

Figure 4. Histogram of annual average SCD from HY2001 to HY2017 in Northeast China.

Figure 5a indicates that the SCD in Northeast China mainly increased, accounting for 62.27% of the total, among which only 4.73% increased significantly. The regions where the SCD increased were mainly in the southern Daxingan Mountains, Xiaoxingan Mountains and Changbai Mountains. SCD decreased mainly in the northern Daxingan Mountains and the southwestern Changbai Mountains, accounting for 20.25% of the total area, and only 0.24% of the total area recorded significant decreases. Figure 5b showed the area with an

SCD trend >0 d/a accounted for 60.61%, which was similar to the spatial distribution of SCD with an increasing trend; the area—revealing no trend was concentrated in the Liaohe Plain and Songnen Plain, accounting for 16.80%; the areas with trends of less than 0 d/a accounted for 18.65% and were mainly located in the northern Daxingan Mountains and the southwestern Changbai Mountains. The results of the Theil–Sen trend and Mann–Kendall significance test were consistent, which further verified the accuracy of these trends.

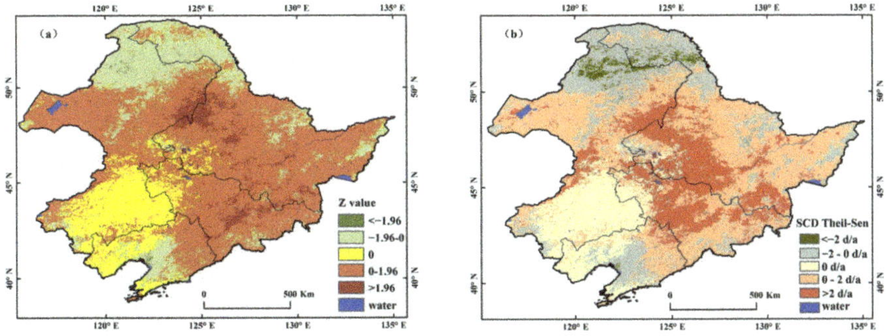

Figure 5. Trend of SCD in Northeast China. (**a**) Significance test and (**b**) trend of SCD from HY2001 to HY2017.

4.2.2. SCOD

Figure 6 shows that the SCOD was delayed from north to south, which meant an obvious increase with latitude. At high latitudes, the SCOD were mainly concentrated within 60–90 d, that is, in November of the current year, and in the northern part of the Daxingan Mountains, the SCOD were at the end of September or early October of the current year. In low-latitude areas, the SCOD were concentrated within 90–120 d, that is, December of the current year. The SCOD were later in some areas of the southern Songnen Plain, and the latest was in January of the next year. The annual average SCOD in Northeast China were mainly concentrated within 60–100 d, from November to the middle of January of the next year, accounting for 84.78% of the total area. Among them, the peak was 70–80 d, which was approximately the middle of November of the current year, with a proportional area of 30.22% (Figure 7).

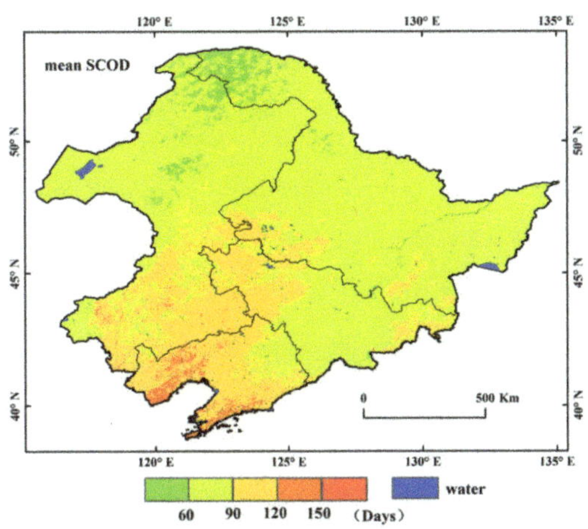

Figure 6. Spatial distribution of annual average SCOD in Northeast China from HY2001 to HY2017.

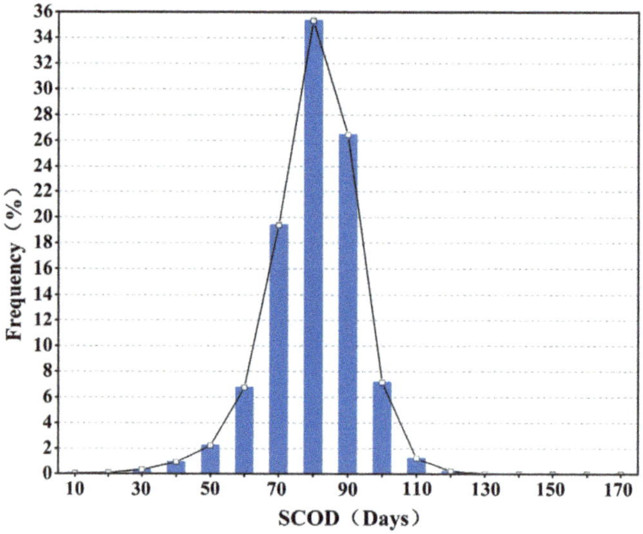

Figure 7. Histogram of annual average SCOD from HY2001 to HY2017 in Northeast China.

Figure 8 shows the significance tests and the trend of SCOD from HY2001 to HY2017. The results of the Theil–Sen trend and Mann–Kendall significance test were consistent. The area in which the SCOD showed an early trend accounted for 31.93%, while the delayed trend accounted for 32.80%. The trends of significant advance and delay were relatively small, only 0.24% and 0.12%, respectively (Figure 8a). Overall, the area with an increase in Figure 8a corresponded to an SCOD trend > 0 d/a in Figure 8b, and that the area with a decrease corresponded to an SCOD trend < 0 d/a.

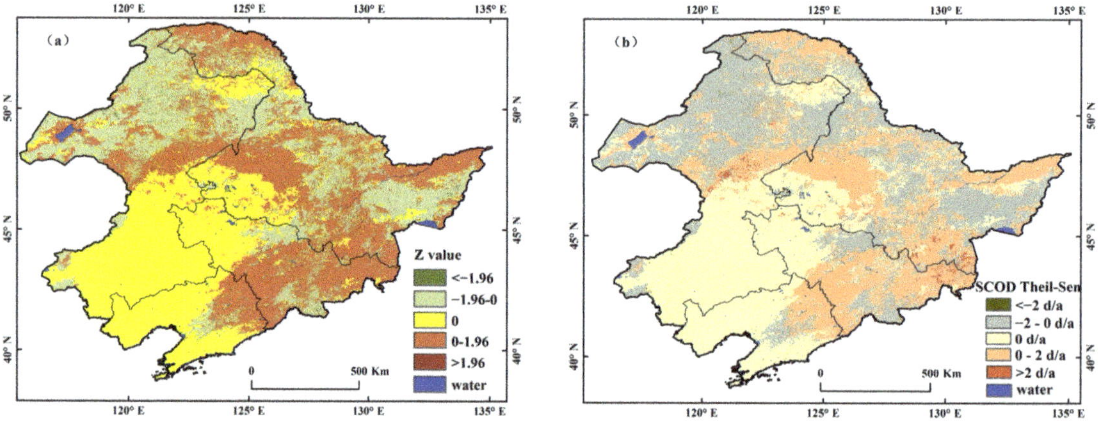

Figure 8. Trend of SCOD in Northeast China. (**a**) Significance test and (**b**) trend of SCOD from HY2001 to HY2017.

4.2.3. SCED

The distribution of SCED was consistent with that of SCD, presenting "high SCED in the mountains and low SCED in the plains" and "high SCED at high latitudes and low SCED at low latitudes" trends. From the perspective of spatial distribution, the SCED in some areas of the Daxingan Mountains, Xiaoxingan Mountains and Changbai Mountains were mainly in the range of 210 d to 240 d in approximately April of the next year. The SCED of the Songnen Plain and Liaohe Plain were relatively early, probably between January and February of the next year (Figure 9). The annual average SCED distribution in Northeast China also had two peaks, which were 130–140 d and 200–210 d, corresponding to the middle of January and late April of the next year, accounting for 5.54% and 20.50% of the area, respectively. The area with an SCED of less than 240 d accounted for 99.59%; that is, at the end of April of the next year, the snow in Northeast China almost disappeared (Figure 10).

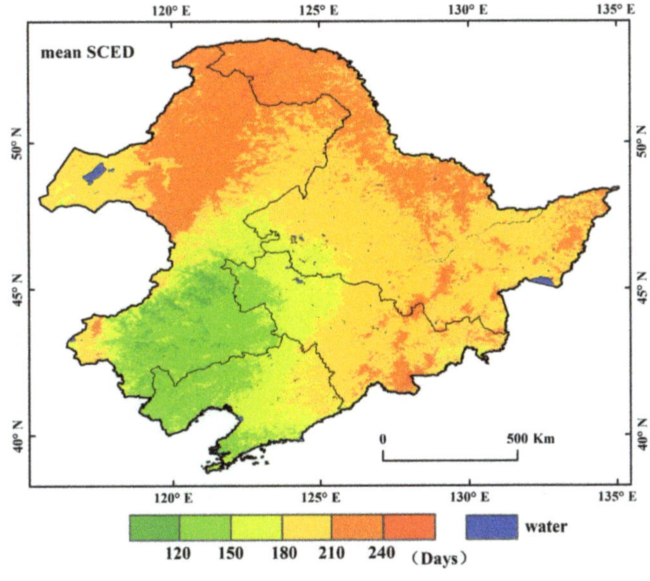

Figure 9. Spatial distribution of annual average SCED in Northeast China from HY2001 to HY2017.

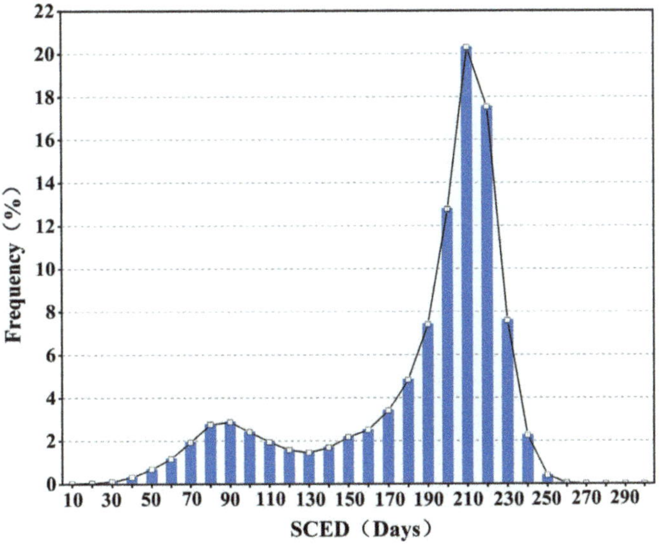

Figure 10. Histogram of annual average SCED from HY2001 to HY2017 in Northeast China.

For the SCED, 29.44% of the total area showed an early trend, while 36.70% had a delayed trend. In addition, only 1.49% and 2.53% of the area were significantly early and delayed, respectively (Figure 11a). The delayed areas were mainly distributed in the Sanjiang Plain, Daxingan Mountains, Xiaoxingan Mountains and Changbai Mountains; the areas with an advancing trend of SCED were mainly concentrated in the middle of the region and the transition area between plain and mountainous areas. Figure 11b shows the trend of SCED in Northeast China. Compared with the significance test, the overall spatial distribution characteristics and the increasing/decreasing trends of SCED were highly consistent. The area with an SCED trend > 0 d/a accounted for 36.36%; that with an SCED trend < 0 d/a accounted for 29.11%.

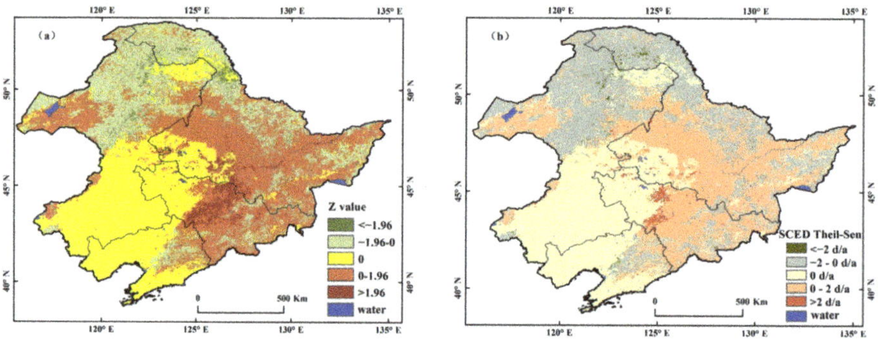

Figure 11. Trend of SCEDs in Northeast China. (**a**) Significance test and (**b**) trend of SCED from HY2001 to HY2017.

In general, the correlation analysis of the SCD, SCOD, and SCED revealed significant relationships between the SCED and SCD (r = 0.70). The correlation between the SCOD and SCD was −0.14, and the increasing trend of the SCD was determined by the advancement of the SCOD and the delay of the SCED. Considering that there was no significant change in the SCOD across time, the snow phenology variations in Northeast China from HY2001 to HY2017 were attributed mostly to the changes in SCED.

4.3. Roles of Multiple Factors in Snow Phenology

Figure 12 shows that the geographical and meteorological factors and the NDVI all affected the SCD, SCOD and SCED. Annual mean temperature had the greatest impact on the SCD, SCOD and SCED, followed by latitude. Precipitation, aspect and slope all had little effect on the SCD, SCOD and SCED, and all these q values were less than 0.1. Compared with the SCOD, the NDVI and longitude both had a greater impact on the SCED and SCD, with q values of 0.35 (0.15) and 0.30 (0.13), respectively. However, altitude was an important factor affecting the SCOD compared with the SCD and SCED.

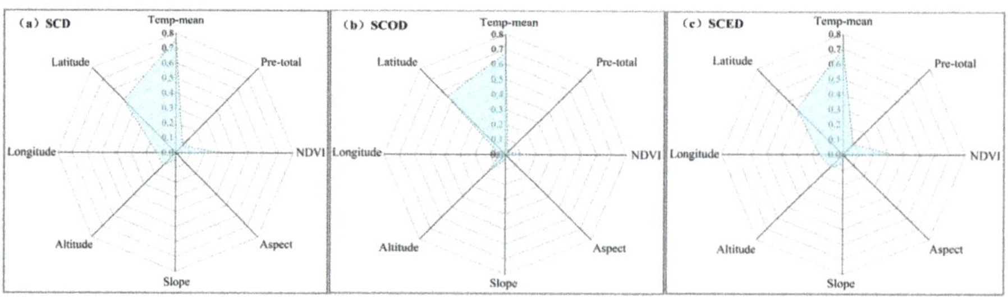

Figure 12. Effects of meteorological and geographic factors and the NDVI on snow phenology. (**a**) SCD, (**b**) SCOD, (**c**) SCED. Note: $p < 0.01$.

We further analyzed the relative importance of monthly temperature and precipitation to snow phenology, and the results are shown in Figure 13. Obviously, the influence of monthly temperature on snow phenology is much greater than that of monthly precipitation. In terms of the roles of different months, the temperature from January to May has a stronger impact on snow phenology, while the impact from June to August is relatively weak. Precipitation has a greater impact on snow phenology in May, September and snow stable period, which the snow completely melted in May, and snowfall occurred in some areas in September.

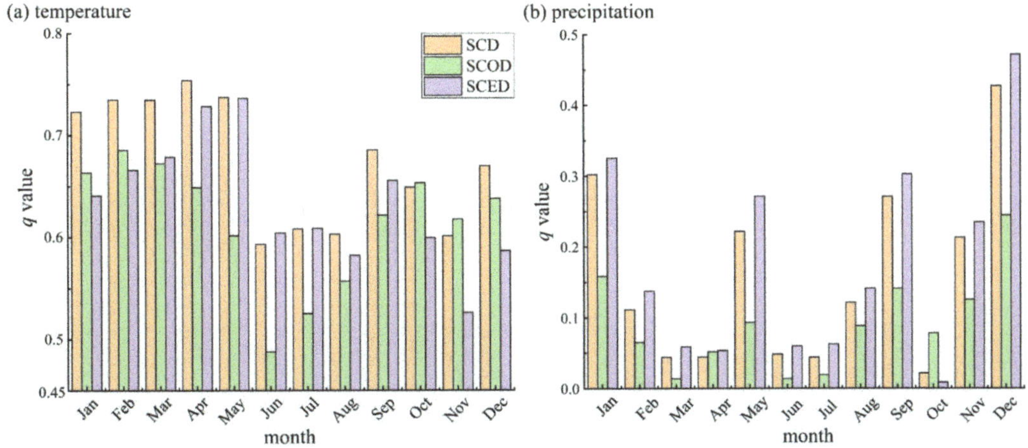

Figure 13. Roles of temperature and precipitation in different months in snow phenology. (**a**) temperature, (**b**) precipitation.

5. Discussions

5.1. Response of Snow Phenology to Climate

Figure 14 shows the annual variations in the mean temperature and snow phenology from HY2001 to HY2017. Except for the SCOD, there were clear increases in both the SCD and SCED and a decrease in temperature. According to statistics, SCD and SCED were both strongly negatively correlated with the mean temperature; at the 95% confidence level, the correlation coefficients were −0.73 and −0.57, respectively. The correlation with the SCOD was moderate (r = 0.41), which can be explained that besides mean temperature, the effect of latitude on SCOD was also great (q = 0.55).

The spatial pattern of correlation between the mean temperature and snow phenology is presented in Figure 15. For 99.91% of the pixels, the mean temperature was negatively correlated with the SCD, which means the lower the mean temperature was, the longer the SCD in the whole study area; 59.40% of the pixels showed strong negative correlations, and areas with weak negative correlation were mainly distributed in the northern Daxingan Mountains and Xiaoxingan Mountains (Figure 15a). The correlations between the SCOD and mean temperature were mainly positive, and the weak positive correlations accounted for a large proportion (40.81%). Areas with strong positive correlations accounted for only 8.19% and were distributed mainly in the Xiaoxingan Mountains and Changbai Mountains (Figure 15b). The correlations between the mean temperature and SCED were similar to those of the SCD, and most regions had negative correlations (93.86%). The difference was that the proportion of strong negative correlations was relatively small (33.06%), and the average correlation of the SCED was −0.38, while that of the SCD was −0.52 (Figure 15c).

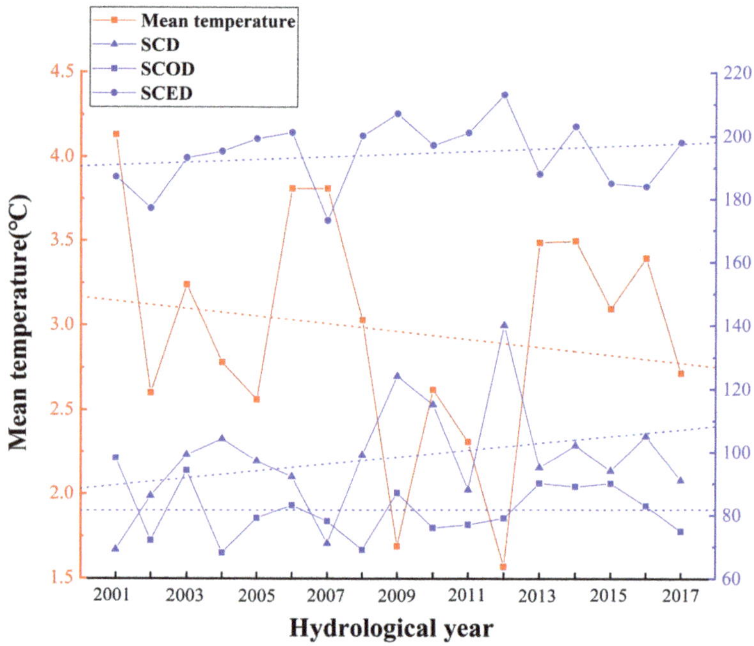

Figure 14. Annual variations in the mean temperature and snow phenology from HY2001 to HY2017.

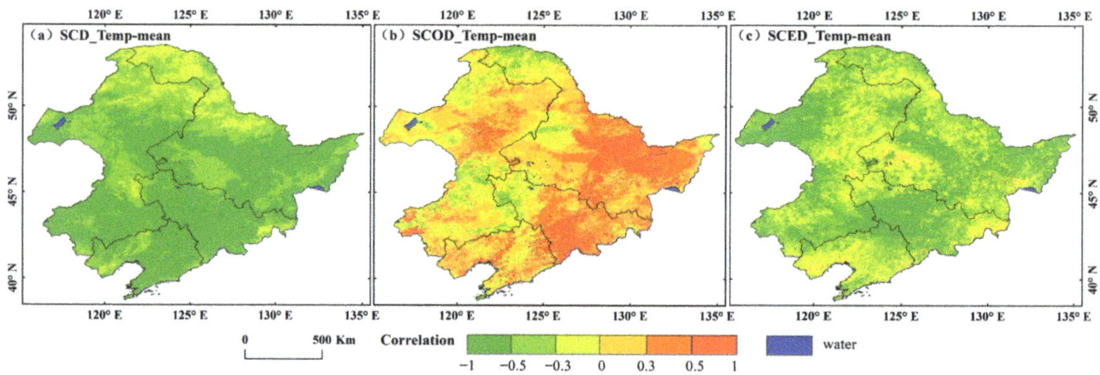

Figure 15. Spatial pattern of the correlations between the mean temperature and snow phenology. (**a**) SCD, (**b**) SCOD, (**c**) SCED.

5.2. Geographical and Vegetation Controls on Snow Phenology

To quantitatively investigate the latitudinal zonation of snow phenology further, statistics were analyzed in combination with the underlying surface conditions. The slope and the regularity between snow phenology and latitude in the nonforested and forested areas further quantitatively proved the latitudinal zonation of snow phenology, as shown in Table 4 and Figure 16. In Northeast China, as the latitude increased by 1 degree, the average SCD increased by 10.2 d, the SCOD advanced by 3.82 d and the SCED was delayed by 5.41 d. Generally, the change rate in forested areas was slower. The change rate in the SCD, SCOD and SCED were 5.41 d/degree, −2.02 d/degree and 2.91 d/degree, respectively, which were closely related to the closed environments of the forested areas themselves.

Table 4. Linear slope and R^2 of the mean SCD, SCOD, and SCED with latitudes from 39.22°N to 53.22°N in the nonforested and forested areas in Northeast China.

	SCD		SCOD		SCED	
	Slope	R^2	Slope	R^2	Slope	R^2
Northeast China	10.2	0.97	−3.82	0.89	5.41	0.96
Nonforested areas	7.8	0.76	−3.41	0.74	4.12	0.71
Forested areas	5.41	0.98	−2.02	0.93	2.91	0.97

Note: The slope represents the change rate in the days when the latitude increased by 1° (d/degree).

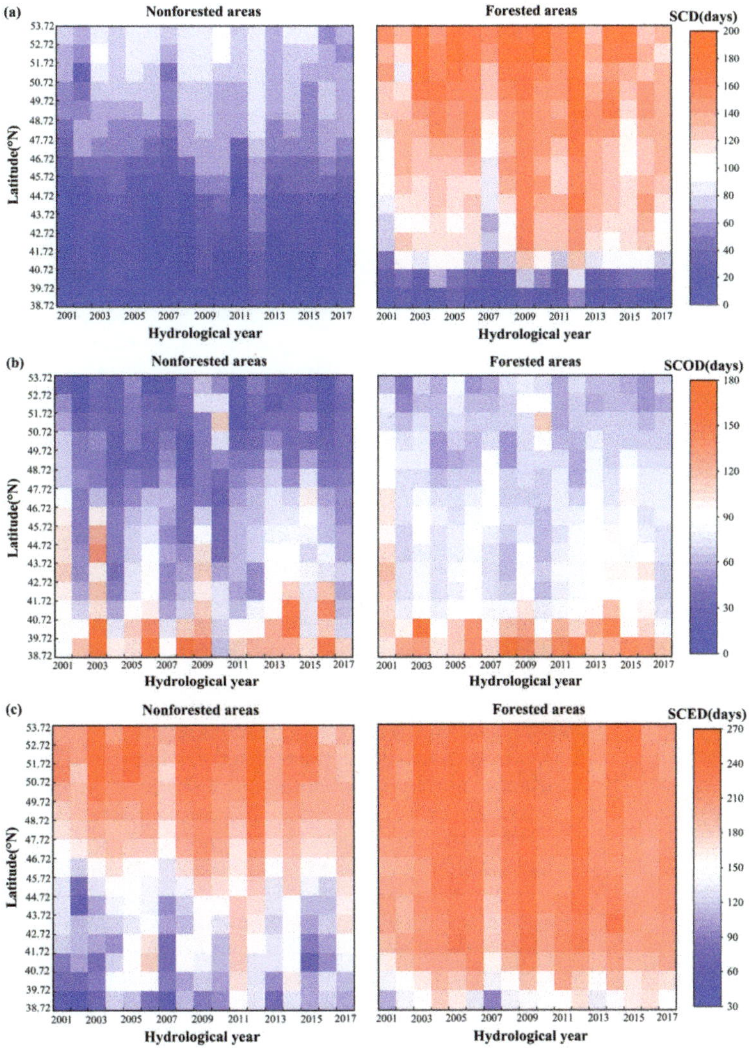

Figure 16. Snow phenology related to latitude in the nonforested and forested areas. (**a**) SCD, (**b**) SCOD, (**c**) SCED.

The higher the latitude was, the longer the SCD, the earlier the SCOD and the later the SCED, in both nonforested and forested areas. At the same latitude, the SCD of forested areas was higher than that of nonforested areas. At high latitudes, the SCOD of forested areas was later than that of nonforested areas. For the SCED, at low latitudes, the snow melting time in forested areas was later than that in nonforested areas, and the SCED was basically the same at high latitudes; these changes can be explained by the observations that snow melting in high-latitude areas was mainly affected by temperature and that the SCED in this region was in approximately April, so the rapidly increasing temperature led to the melting of snow in the forested and nonforested areas.

In contrast to the mean temperature, the NDVI denoting vegetation greenness was positively correlated with the SCD, SCOD and SCED, accounting for 68.69%, 59.22% and 58.68%, respectively. However, the weak correlation ($-0.3 < r < 0.3$) between snow phenology and the NDVI was dominant, with 69.85%, 74.99% and 72.65% of pixels (Figure 17a–c), which illustrates that the NDVI was not the main factor affecting snow phenology.

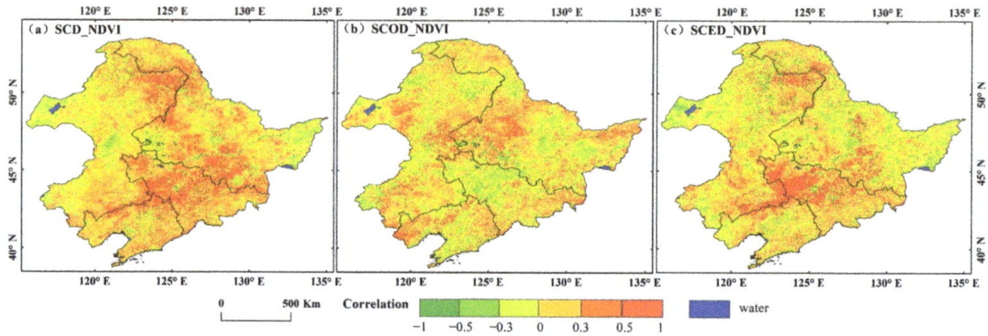

Figure 17. Spatial pattern of correlations between the NDVI and snow phenology. (a) SCD, (b) SCOD, (c) SCED.

5.3. Comparison with Previous Results

Shi et al. [11] reported strong spatial heterogeneity in snow phenology in the Mollisol region of Northeast China. The SCD increased from southwest to northeast gradually, and snow cover began to accumulate in mid-November and completely melted in late March during 1978–2016. In the western Changbai Mountains and northern Daxingan Mountains, the SCOD always began in early October, and the SCED in the western Daxingan Mountains, Xiaoxingan Mountains, and Changbai Mountains always occurred in May [8]; which was similar to the results of this study. In the forested areas of Northeast China, the SCOD were later, and the SCED were earlier in the plains. The SCOD occurred between late November and mid-November in the regions with high altitudes, and the SCED occurred later with increasing latitude and altitude. The earlier SCOD and later SCED led to an increase in the SCD, especially in high mountain areas. In most areas, the SCOD experienced an advancing trend, the SCED exhibited an obvious delaying trend, and the SCD showed an opposing trend from south to north in Northeast China from 2004 to 2018 [10].

This paper found that SCD and SCED increased, and SCOD basically did not change in Northeast China from 2001 to 2018. However, previous studies on snow phenology in China found that due to the increase in temperature, the SCOD in most areas were delayed, and SCED were advanced [5,6,12]. These contradictions may have been caused by the inconsistency of the time span. Although snow cover will decrease under the background of global warming, the decrease of mean temperature in a short time has led to the increase of snow cover. Studies have shown that there are strong correlations between meteorological and geographic factors and snow cover [39–41]. Vegetation change was also closely related to snow phenology [10,42,43]. Temperature and precipitation could affect snow cover variations [44,45]. However, in this research, the results showed that temperature was the main factor affecting the variations in snow phenology, and precipitation had little effect. In the three basins of Songhua River in Northeast China, temperature, precipitation and altitude were considered to be the three most important factors [9]. Huang et al. [12] studied the snow cover variations across China from 1951 to 2018 based on snow depth dataset and used model to analyze the driving effect of multiple factors on snow cover phenology and found that the most important factors influencing the SCD, SCOD and SCED were annual coldest monthly minimum temperature, altitude and annual mean temperature, respectively. This difference may have been caused by inconsistencies in time scales, study areas, data and research methods. In addition, we both found that precipitation had little effect on snow phenology.

6. Conclusions

Snow cover is one of the most active features on the land surface [46], and it is very important to quantitatively explore and understand snow phenology variations for local climate change. The snow phenology variations and their driving factors in Northeast China from 2001 to 2018 were assessed based on daily cloud-free snow products generated in this research. The conclusions were as follows.

(1) The SCD, SCOD and SCED all showed the characteristics of latitudinal zonal distribution, and the SCED and SCD distributions had obvious consistency. With increasing latitude, the SCD was longer, the SCOD began earlier, and the SCED appeared later. Overall, the SCD showed mainly an increasing trend, which was mostly distributed in the southern Daxing'an Mountains, Xiaoxing'an Mountains and Changbai Mountains. The SCOD showed advanced and delayed trends that accounted for 31.93% and 32.80%, respectively. The corresponding proportions of the SCED accounted for 29.44% and 36.70%, respectively, which meant that the SCED showed a delayed trend overall. On the Liaohe Plain and Songnen Plain, the snow phenology basically did not change.
(2) For snow phenology, the mean temperature was identified as the most important driver, followed by latitude. In terms of the roles of temperature in different months, the snow phenology is mainly affected by the temperature in winter of current year and spring of the next year. The decrease in temperature directly led to the extension of SCD, the advancement of SCOD and the delay in SCED. Precipitation, aspect and the slope all had little effect on snow phenology. Compared with the SCOD, the NDVI and longitude both had a greater impact on the SCED and SCD, while SCOD showed a greater impact from altitude.
(3) The mean temperature was mainly negatively correlated with the SCD and SCED and mostly positively correlated with the SCOD. As the latitude increased, the snow phenology changed gradually, and the change rate in the SCD, SCOD and SCED in the whole Northeast China were 10.20 d/degree, −3.82 d/degree and 5.41 d/degree, respectively. The change rate in the snow phenology in forested areas and nonforested areas were inconsistent, and it was slower in forested areas than nonforested areas. Snow phenology was mainly positively correlated with the NDVI, but weak correlations with the NDVI accounted for a large proportion.

Author Contributions: Conceptualization, H.G. and Z.G.; methodology, H.G. and X.W.; software, H.G. and Z.G.; validation, H.G., S.C.; writing—original draft preparation, H.G.; writing—review and editing, H.G., X.W., Z.G. and S.C.; supervision, X.W. All authors have read and agreed to the published version of the manuscript.

Funding: This research was funded by the Science and Technology Basic Resources Investigation Program of China grant number 2017FY100500 and the National Natural Science Foundation of China grant number 41771373, 41971325.

Acknowledgments: We gratefully acknowledge the Google Earth Engine, the National Earth System Science Data Center and the National Cryosphere Desert Data Center for obtaining data freely. Moreover, we would like to thank the two reviewers for their valuable comments and suggestions.

Conflicts of Interest: The authors declare no conflict of interest.

References

1. Shi, Y.; Gao, X.J.; Wu, J.; Giorgi, F. Changes in snow cover over China in the 21st century as simulated by a high resolution regional climate model. *Environ. Res. Lett.* **2011**, *6*, 045401. [CrossRef]
2. Peng, S.S.; Piao, S.L.; Ciais, P.; Fang, J.Y.; Wang, X.H. Change in winter snow depth and its impacts on vegetation in China. *Glob. Change Biol.* **2010**, *16*, 3004–3013. [CrossRef]
3. Qi, W.; Feng, L.; Liu, J.G.; Yang, H. Snow as an Important Natural Reservoir for Runoff and Soil Moisture in Northeast China. *J. Geophys. Res. Atmos.* **2020**, *125*, e2020JD033086. [CrossRef]
4. Ding, T.; Gao, H. Relationship between winter snow cover days in Northeast China and rainfall near the Yangtze River basin in the following summer. *J. Meteorol. Res.* **2015**, *29*, 400–411. [CrossRef]
5. Ke, C.-Q.; Li, X.-C.; Xie, H.; Ma, D.-H.; Liu, X.; Kou, C. Variability in snow cover phenology in China from 1952 to 2010. *Hydrol. Earth Syst. Sci.* **2016**, *20*, 755–770. [CrossRef]
6. Ma, N.; Yu, K.; Zhang, Y.; Zhai, J.; Zhang, Y.; Zhang, H. Ground observed climatology and trend in snow cover phenology across China with consideration of snow-free breaks. *Clim. Dyn.* **2020**, *55*, 2867–2887. [CrossRef]
7. Sun, Y.; Zhang, T.; Liu, Y.; Zhao, W.; Huang, X. Assessing Snow Phenology over the Large Part of Eurasia Using Satellite Observations from 2000 to 2016. *Remote Sens.* **2020**, *12*, 2060. [CrossRef]
8. Chen, S.; Yang, Q.; Xie, H.; Zhang, H.; Lu, P.; Zhou, C. Spatiotemporal variations of snow cover in northeast China based on flexible multiday combinations of moderate resolution imaging spectroradiometer snow cover products. *J. Appl. Remote Sens.* **2014**, *8*, 084685. [CrossRef]
9. Yang, Q.; Song, K.; Hao, X.; Chen, S.; Zhu, B. An Assessment of Snow Cover Duration Variability Among Three Basins of Song hua River in Northeast China Using Binary Decision Tree. *Chin. Geogr. Sci.* **2018**, *28*, 946–956. [CrossRef]
10. Qiao, D.; Zhou, J.; Liang, S.; Fu, X. Combined Effects of Precipitation and Temperature on the Responses of Forest Spring Phenology to Winter Snow Cover Dynamics in Northeast China. *IEEE Access* **2019**, *7*, 138950–138962. [CrossRef]
11. Shi, H.; Zhou, L.L.; Xu, X.Q.; Fan, H.M. Snow Cover Distribution's Correlation with Climatic Factors in Northeast China's Mollisol Region. *Appl. Ecol. Environ. Res.* **2020**, *18*, 6539–6554. [CrossRef]
12. Huang, X.; Liu, C.; Zheng, Z.; Wang, Y.; Li, X.; Liang, T. Snow cover variations across China from 1951–2018. *Cryosphere* **2020**, 1–20. [CrossRef]
13. Wang, Y.; Zheng, Z. Spatial Representativeness Analysis for Snow Depth Measurements of Meteorological Stations in North east China. *J. Hydrometeorol.* **2020**, *21*, 791–805. [CrossRef]
14. Woo, M.K.; Thorne, R. Snowmelt contribution to discharge from a large mountainous catchment in subarctic Canada. *Hydrol. Processes* **2006**, *20*, 2129–2139. [CrossRef]
15. Tang, Z.G.; Wang, J.; Li, H.Y.; Yan, L.L. Spatiotemporal changes of snow cover over the Tibetan plateau based on cloud-re moved moderate resolution imaging spectroradiometer fractional snow cover product from 2001 to 2011. *J. Appl. Remote Sens.* **2013**, *7*, 073582. [CrossRef]
16. Wang, X.; Zhu, Y.; Chen, Y.; Zheng, H.; Liu, H.; Huang, H.; Liu, K.; Liu, L. Influences of forest on MODIS snow cover map ping and snow variations in the Amur River basin in Northeast Asia during 2000–2014. *Hydrol. Processes* **2017**, *31*, 3225–3241. [CrossRef]
17. Liang, S.; Li, X.; Zheng, X.; Jiang, T.; Li, X.; Qiao, D. Effects of Winter Snow Cover on Spring Soil Moisture Based on Remote Sensing Data Product over Farmland in Northeast China. *Remote Sens.* **2020**, *12*, 2716. [CrossRef]
18. Brown, R.D.; Robinson, D.A. Northern Hemisphere spring snow cover variability and change over 1922–2010 including an assessment of uncertainty. *Cryosphere* **2011**, *5*, 219–229. [CrossRef]
19. Chen, X.; Liang, S.; Cao, Y.; He, T.; Wang, D. Observed contrast changes in snow cover phenology in northern middle and high latitudes from 2001–2014. *Sci. Rep.* **2015**, *5*, 16820. [CrossRef]
20. Liang, H.; Huang, X.D.; Sun, Y.H.; Wang, Y.L.; Liang, T.G. Fractional Snow-Cover Mapping Based on MODIS and UAV Data over the Tibetan Plateau. *Remote Sens.* **2017**, *9*, 1332. [CrossRef]
21. Paudel, K.P.; Andersen, P. Monitoring snow cover variability in an agropastoral area in the Trans Himalayan region of Ne pal using MODIS data with improved cloud removal methodology. *Remote Sens. Environ.* **2011**, *115*, 1234–1246. [CrossRef]
22. Hall, D.K.; Riggs, G.A.; Foster, J.L.; Kumar, S.V. Development and evaluation of a cloud-gap-filled MODIS daily snow-cover product. *Remote Sens. Environ.* **2010**, *114*, 496–503. [CrossRef]
23. Da Ronco, P.; De Michele, C. Cloud obstruction and snow cover in Alpine areas from MODIS products. *Hydrol. Earth Syst. Sci.* **2014**, *18*, 4579–4600. [CrossRef]
24. Huang, X.D.; Deng, J.; Ma, X.F.; Wang, Y.L.; Feng, Q.S.; Hao, X.H.; Liang, T.G. Spatiotemporal dynamics of snow cover based on multi-source remote sensing data in China. *Cryosphere* **2016**, *10*, 2453–2463. [CrossRef]

25. Huang, Y.; Liu, H.X.; Yu, B.L.; We, J.P.; Kang, E.L.; Xu, M.; Wang, S.J.; Klein, A.; Chen, Y.N. Improving MODIS snow prod ucts with a HMRF-based spatio-temporal modeling technique in the Upper Rio Grande Basin. *Remote Sens. Environ.* **2018**, *204*, 568–582. [CrossRef]
26. Chen, S.Y.; Wang, X.Y.; Guo, H.; Xie, P.Y.; Wang, J.; Hao, X.H. A Conditional Probability Interpolation Method Based on a Space-Time Cube for MODIS Snow Cover Products Gap Filling. *Remote Sens.* **2020**, *12*, 3577. [CrossRef]
27. Chen, S.Y.; Wang, X.Y.; Guo, H.; Xie, P.Y.; Sirelkhatim, A.M. Spatial and Temporal Adaptive Gap-Filling Method Producing Daily Cloud-Free NDSI Time Series. *IEEE J. Sel. Top. Appl. Earth Obs. Remote Sens.* **2020**, *13*, 2251–2263. [CrossRef]
28. Wang, X.Y.; Chen, S.Y.; Wang, J. An Adaptive Snow Identification Algorithm in the Forests of Northeast China. *IEEE J. Sel. Top. Appl. Earth Obs. Remote Sens.* **2020**, *13*, 5211–5222. [CrossRef]
29. Li, X.; Cheng, G.D.; Jin, H.J.; Kang, E.; Che, T.; Jin, R.; Wu, L.Z.; Nan, Z.T.; Wang, J.; Shen, Y.P. Cryospheric change in China. *Glob. Planet Change* **2008**, *62*, 210–218. [CrossRef]
30. Qian, A.; Yi, S.; Chang, L.; Sun, G.; Liu, X. Using GRACE Data to Study the Impact of Snow and Rainfall on Terrestrial Water Storage in Northeast China. *Remote Sens.* **2020**, *12*, 4166. [CrossRef]
31. Sun, X.Z.; Sun, Z.B.; Luo, Y. Characteristics of snowfall from 1960 to 2005 in northeast China. *J. Meteorol. Environ.* **2010**, *26*.
32. Riggs, G.A.; Hall, D.K.; Román, M.O. *MODIS Snow Products Collection 6 User Guide*; National Snow and Ice Data Center: Boulder, CO, USA, 2015.
33. Xu, W.; Ma, H.; Wu, D.; Yuan, W. Assessment of the Daily Cloud-Free MODIS Snow-Cover Product for Monitoring the Snow-Cover Phenology over the Qinghai-Tibetan Plateau. *Remote Sens.* **2017**, *9*, 585. [CrossRef]
34. Zhang, H.B.; Zhang, F.; Zhang, G.Q.; Che, T.; Yan, W.; Ye, M.; Ma, N. Ground-based evaluation of MODIS snow cover prod uct V6 across China: Implications for the selection of NDSI threshold. *Sci. Total Environ.* **2019**, *651*, 2712–2726. [CrossRef] [PubMed]
35. Wang, X.; Wu, C.; Wang, H.; Gonsamo, A.; Liu, Z. No evidence of widespread decline of snow cover on the Tibetan Plateau over 2000–2015. *Sci. Rep.* **2017**, *7*, 14645. [CrossRef] [PubMed]
36. Wang, J.-F.; Li, X.-H.; Christakos, G.; Liao, Y.-L.; Zhang, T.; Gu, X.; Zheng, X.-Y. Geographic detectors-based health risk assessment and its application in the neural tube defects study of the Heshun region, China. *Int. J. Geogr. Inf. Sci.* **2010**, *24*, 107–127. [CrossRef]
37. Wei, W.; Guo, Z.C.; Zhou, L.; Xie, B.B.; Zhou, J.J. Assessing environmental interference in northern China using a spatial dis tance model: From the perspective of geographic detection. *Sci. Total Environ.* **2020**, *709*, 136170. [CrossRef]
38. Cohen, J. *Statistical Power Analysis for the Behavioral Sciences*, 2nd ed.; Lawrence Erlbaum Associates: Hillsdale, NJ, USA, 1988; p. 83.
39. Ye, K.H. Interannual variability of March snow mass over Northern Eurasia and its relation to the concurrent and preceding surface air temperature, precipitation and atmospheric circulation. *Clim. Dyn.* **2019**, *52*, 2813–2836. [CrossRef]
40. Zhong, X.Y.; Zhang, T.J.; Kang, S.C.; Wang, K.; Zheng, L.; Hu, Y.T.; Wang, H.J. Spatiotemporal variability of snow depth across the Eurasian continent from 1966 to 2012. *Cryosphere* **2018**, *12*, 227–245. [CrossRef]
41. Litaor, M.I.; Williams, M.; Seastedt, T.R. Topographic controls on snow distribution, soil moisture, and species diversity of herbaceous alpine vegetation, Niwot Ridge, Colorado. *J. Geophys. Res.-Biogeosci.* **2008**, *113*. [CrossRef]
42. Qiao, D.; Wang, N. Relationship between Winter Snow Cover Dynamics, Climate and Spring Grassland Vegetation Phenology in Inner Mongolia, China. *ISPRS Int. J. Geo-Inf.* **2019**, *8*, 42. [CrossRef]
43. Jost, G.; Weiler, M.; Gluns, D.R.; Alila, Y. The influence of forest and topography on snow accumulation and melt at the wa tershed-scale. *J. Hydrol.* **2007**, *347*, 101–115. [CrossRef]
44. Qin, D.H.; Liu, S.Y.; Li, P.J. Snow cover distribution, variability, and response to climate change in western China. *J. Clim.* **2006**, *19*, 1820–1833.
45. Hantel, M.; Ehrendorfer, M.; Haslinger, A. Climate sensitivity of snow cover duration in Austria. *Int. J. Climatol.* **2000**, *20*, 615–640. [CrossRef]
46. Li, L.-Y.; Ke, C.-Q. Analysis of spatiotemporal snow cover variations in Northeast China based on moderate-resolution-imag ing spectroradiometer data. *J. Appl. Remote Sens.* **2014**, *8*, 084695. [CrossRef]

remote sensing

Article

Remote Sensing Phenology of the Brazilian Caatinga and Its Environmental Drivers

Rodolpho Medeiros [1], João Andrade [2], Desirée Ramos [3], Magna Moura [4], Aldrin Martin Pérez-Marin [5], Carlos A. C. dos Santos [1], Bernardo Barbosa da Silva [1] and John Cunha [6,*]

1 Academic Unity of Atmospheric Sciences, Technology and Natural Resources Center, Federal University of Campina Grande, Campina Grande 58109-970, Brazil; rodolpholuiz42@gmail.com (R.M.); carlos.santos@ufcg.edu.br (C.A.C.d.S.); bernardo.silva@ufcg.edu.br (B.B.d.S.)
2 Centre for Technology and Geosciences, Department of Civil Engineering, Federal University of Pernambuco, Recife 50740-550, Brazil; joao.jma@ufpe.br
3 Department of Biodiversity, São Paulo State University—UNESP, Rio Claro, Jaboticabal 14884-900, Brazil; dm.ramos@unesp.br
4 Empresa Brasileira de Pesquisa Agropecuária, Embrapa Semiárido, Petrolina 56302-970, Brazil; magna.moura@embrapa.br
5 National Institute of Semi-Arid, Campina Grande 58429-970, Brazil; aldrin.perez@insa.gov.br
6 Centre for the Sustainable Development of the Semi-Arid, Federal University of Campina Grande, Campina Grande 58429-900, Brazil
* Correspondence: john.brito@ufcg.edu.br

Citation: Medeiros, R.; Andrade, J.; Ramos, D.; Moura, M.; Pérez-Marin, A.M.; dos Santos, C.A.C.; da Silva, B.B.; Cunha, J. Remote Sensing Phenology of the Brazilian Caatinga and Its Environmental Drivers. *Remote Sens.* **2022**, *14*, 2637. https://doi.org/10.3390/rs14112637

Academic Editors: Xuanlong Ma, Jiaxin Jin, Xiaolin Zhu, Yuke Zhou and Qiaoyun Xie

Received: 30 April 2022
Accepted: 30 May 2022
Published: 31 May 2022

Abstract: The Caatinga is the largest nucleus of Seasonally Dry Tropical Forests (SDTF) in the Neotropics. The leafing patterns of SDTF vegetation are adapted to the current environmental and climate variability, but the impacts of climate change tend to alter plants' phenology. Thus, it is necessary to characterise phenological parameters and evaluate the relationship between vegetation and environmental drivers. From this information, it is possible to identify the dominant forces in the environment that trigger the phenological dynamics of the Caatinga. In this way, remote sensing represents an essential tool to investigate the phenology of vegetation, particularly as it has a long series of vegetation monitoring and allows relationships with different environmental drivers. This study has two objectives: (i) estimate phenological parameters using an Enhanced Vegetation Index (EVI) time-series over 20 years, and (ii) characterise the relationship between phenologic dynamics and environmental drivers. TIMESAT software was used to determine four phenological parameters: Start Of Season (SOS), End Of Season (EOS), Length Of Season (LOS), and Amplitude (AMPL). Boxplots, Pearson's, and partial correlation coefficients defined relationships between phenologic dynamics and environmental drivers. The non-parametric test of Fligner–Killeen was used to test the interannual variability in SOS and EOS. Our results show that the seasonality of vegetation growth in the Caatinga was different in the three experimental sites. The SOS was the parameter that presented the greatest variability in the days of the year (DOY), reaching a variation of 117 days. The sites with the highest SOS variability are the same ones that showed the lowest EOS variation. In addition, the values of LOS and AMPL are directly linked to the annual distribution of rainfall, and the longer the rainy season, the greater their values are. The variability of the natural cycles of the environmental drivers that regulate the ecosystem's phenology and the influence on the Caatinga's natural dynamics indicated a greater sensitivity of the phenologic dynamics to water availability, with precipitation being the limiting factor of the phenologic dynamics. Highlights: The EVI time series was efficient in estimating phenological parameters. The high variability of the start of season (SOS) occurred in sites with low variability of end of the season (EOS) and vice versa. The precipitation and water deficit presented a higher correlation coefficient with phenological dynamics. Length of Season (LOS) and amplitude (AMPL) are directly linked to the annual distribution of rainfall.

Keywords: land surface phenology; vegetation indexes; seasonally dry tropical forest

1. Introduction

The Seasonally Dry Tropical Forests (SDTF) are characterised by a strong seasonal and irregular distribution of rainfall throughout the year, resulting in long dry season periods [1,2]. The Brazilian Caatinga, with an area of approximately 900,000 km^2, is the largest nucleus of SDTF in the Neotropics [The tropical New World biogeographic region comprises Central America, the Caribbean, and South America] [3]. Moreover, with 3347 plant species, of which 526 are endemic, the Caatinga is the richest nucleus of SDTF in the Neotropics [4]. Its remarkable floristic diversity makes the Caatinga twice as rich as the Amazon rainforest when considering the species/area relationship [4]. Phenology studies recurring life-cycle events such as bird migration, flower blooming, or leaf emergence and senescence and the causes of their timing by biotic and abiotic forces [5,6]. The leafing patterns of Caatinga vegetation are adapted to the intense climate and water seasonality, being highly dependent on the climate interannual variability [7,8]. Most leaves fall during the dry season, and the first rainfall pulses trigger a quick leaf flush in the wet season [7–9].

Precipitation and soil moisture are the leading environmental drivers for the leaf changes in the Caatinga [8,9]. Still, the temperature can also be a driver for species presenting scheduled phenology in dry ecosystems [10]. The Intergovernmental Panel on Climate Change [11] forecasts an increase of 1.5 °C in the global mean air temperature for the next two decades in an optimistic scenario, considering the reduction of current emissions of CO_2. Besides the temperature increases, climate changes are likely to alter the precipitation regimes in the following decades [12,13]. The consequences of these changes are, among others, the projected drying out of surface soils [12] and prolonged dry seasons, with an increase of 47% of the area of the Northeast Region of Brazil (NEB) subjected to extreme drought events until 2070 [13]. Thus, understanding the past and current vegetation's response to the environmental drivers is paramount to predicting its behaviour in climate change scenarios, allowing the detection of changes in the timing of leaf patterns and their causes.

The ability to monitor global vegetation phenology, or Land Surface Phenology (LSP), has increased with the validated Remote Sensing (RS) and modelling approaches to mapping phenology [14,15]. Long-term data from satellite products are useful tools for understanding the phenological responses of vegetation to current environmental drivers using Vegetation Indices (VIs), allowing it to predict its responses to climate change scenarios. VIs time series has received the attention given its potential to characterise interactions between climate and vegetation with broad applications in different ecosystems [16–18]. Several VIs are calculated based upon different spectral bands and, therefore, evidence of different components of the environment [19]. The Enhanced Vegetation Index (EVI) has been widely used to characterise vegetation phenology [20,21] due to its sensitivity to high biomass and reduced atmospheric and soil effects. EVI is calculated from the near-infrared (NIR), Red, and Blue bands and can be derived for different satellite platforms, such as Landsat, Sentinel, and MODIS. The use of algorithms to determine the main phenological metrics from the VIs time series has favoured the representation of the phenological stages of each cropping system, allowing a crop-type classification based on their phenological metrics [22]. However, the studies driving this on a global scale have been primarily focused on forest ecosystems, associating phenological changes in vegetation with climate patterns, particularly with rainfall data [23]. These studies showed that the phenological dynamics strongly depend on the seasonality of rainfall [21,24]. Still, the studies on a regional scale indicate that other environmental drivers also trigger phenological changes [25–28].

The LSP applied to an ecosystem scale seems to offer the best opportunities to advance understanding of environmental triggers and determinants for phenological dynamics, given the possibility to understand it on a broad scale, encompassing areas in a range of contrasting environmental conditions. For instance, the early greening or pre-rain green-up, a phenomenon where trees produce leaves before the rain starts, was registered in the woodlands and savannas of southern Africa through RS satellite techniques [29,30]. Furthermore, the application of the LSP at a continental scale and using long-term time-series

(2002–2014) allowed us to measure the variability in leaf flushing (i.e., greening) among years and to identify the photoperiod as the environmental cue for early greening [30].

The use of LSP and their drivers will be significant for SDTFs where interannual rainfall variability and rainy season duration change on a spatial and temporal scale [31,32], factors that are expected to influence the phenological strategies of plant communities in this vegetation [10]. There was also great regional variability and interannual fluctuation in vegetation phenology, and the overall phenological trends shifted later [21]. By following the studies for dry forests, Tong et al. [15] reported that the interannual rainfall variability was a more dominant force than fire events and land-use change in the phenological trend in tropical areas. For example, Jesus et al. [25] noticed changes in the phenological patterns for dense and open vegetation areas of the Caatinga, suggesting that factors that vary at spatial scales, such as the vegetation structure, would also be necessary for the phenological responses of the vegetation. In an experimental area in the Caatinga, the phenological response was directly related to soil water availability [9].

Despite advances in the analysis of phenological patterns and their associated environmental drivers, mainly observed in studies at the ecosystem scale, the application of long-term time series of vegetation indices in studies of the Caatinga vegetation is scarce. In addition, there are limitations to the diversity of sites studied for the Caatinga. For example, when analysing the dominant environmental drivers for the phenology of seasonally dry ecosystems (Caatinga, Cerrado), Alberton et al. [33] observed that the dominant drivers in these ecosystems were distinct, with light (measured as day-length) being more relevant in explaining leafing patterns in Cerrado communities than rainfall for Caatinga communities. Therefore, comparing sites of the same ecosystem can better define the environmental drivers associated with the phenological dynamics. There are also limitations to the number of environmental drivers analysed in the studies carried out for the Caatinga. Analysis with more environmental drivers could reveal meaningful soil–plant–atmosphere interactions, which may occur to a lesser extent.

Given the above, this study proposes to evaluate the Caatinga phenological sensitivity to environmental drivers in three Caatinga ecoregions. The seasonality of vegetation will be observed from the EVI time series over 20 years (2000–2019) and environmental drivers (precipitation, air temperature, soil moisture, and water deficit) from global databases. This study has two objectives: (i) estimate phenological parameters using an EVI time-series over 20 years, and (ii) characterise the relationship between phenologic dynamics and environmental drivers. The results will also be expected to serve as a baseline against which to compare future changes in Caatinga phenology due to natural or anthropogenic causes.

2. Material and Methods

2.1. Study Areas

The semi-arid climate (Köppen's BSh) [34], where the Caatinga is located, presents accumulated precipitation below 600 mm and an average air temperature of 25 °C to 30 °C throughout the year [35]. However, seasonally, the average annual rainfall varies widely from 300 mm to 1000 mm annually, mainly concentrated in a period of 3 to 4 months, during summer and autumn, followed by a prolonged dry season lasting 8 to 9 months during winter and spring [36]. The Caatinga is characterised by high interannual precipitation variability, with droughts that can last for years and high potential evapotranspiration rates of between 1500 mm and 2000 mm annually [37]. The Caatinga region has different soil categories, ranging from shallow, rocky, and relatively fertile to deep with high natural fertility and sedimentary or sandy, deep, and non-fertile [35]. In the Caatinga, the typical vegetation comprises deciduous and thorny forests or small forests that grow mainly in exposed crystalline rocky terrains, such as the Depression Sertaneja [38]. The region's geomorphology ranges from lowlands of 300 m, mountains, plateaus, and plateaus with an altitude of up to 1000 m [35].

The study was carried out at three SDTF protected sites, each area being in a different ecoregion (Figure 1): (i) Meridional Sertaneja Depression (MSD) ecoregion experimental site located Embrapa Semi-arid, Petrolina, Pernambuco; (ii) Sertanejas Depressions (SD) ecoregion experimental site located at Seridó Ecological Station, Serra Negra do Norte, Paraíba; (iii) Borborema Plateau (BP) ecoregion experimental site located at Semi-arid National Institute, Campina Grande, Paraíba. The MSD experimental site's geographic coordinates are 9°02′47′′ S, 40°19′16′′ W, 395 m altitude, and its vegetation is dry xerophilous forest, with trees and shrubs that average five metres in height [39]. The average annual temperature is 26.2 °C, and the rainfall is 387 mm [39], concentrated mainly from January to April. The predominant soil type in the experimental site is Red-Yellow Argisol, followed by Haplossol and Hydromorphic Vertisol [40]. The SD experimental site's geographic coordinates are 6°34′42′′ S, 37°15′05′′ W, and 203 m altitude. Its area is characterised by dry xerophilous forest with deciduous plant species and the predominance of small, widely dispersed trees and shrubs up to 10 metres high, which develop and grow only in the rainy season between January and May [41]. The average annual precipitation varies between 400 and 700 mm, and the average yearly temperature ranges from 28 to 30 °C. The dominant soil is Neosol Litolic, shallow (about 40 cm), stony, and low fertility [42]. The BP experimental site's geographic coordinates are 7°16′49′′ S, 35°58′34′′ W, and 492 m altitude. The vegetation present in the experimental site is composed of dense Caatinga, with tree heights between three and seven metres. The region experiences a rainy season starting in March/April and extending until August, while the dry season lasts for five to six months [43]. The average annual rainfall of Campina Grande is about 750 mm, and the average annual air temperature is 23.3 °C [44]. The soil is nitric planosol [45].

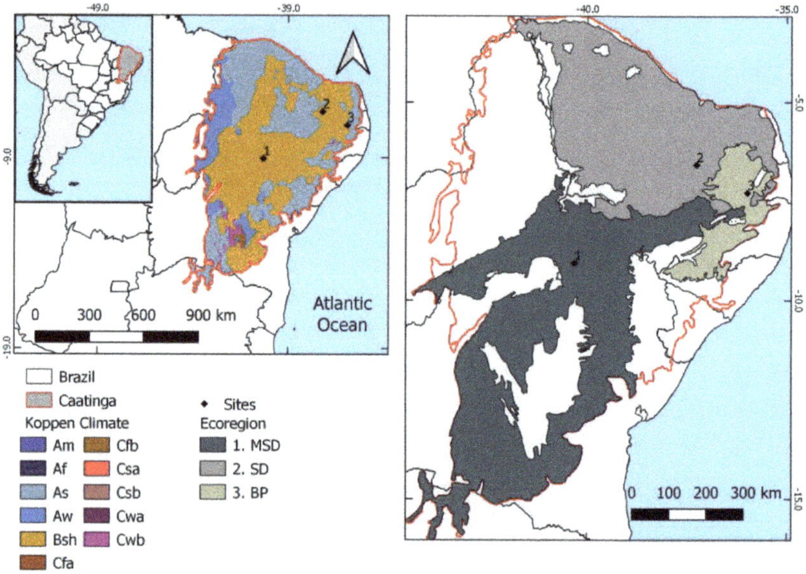

Figure 1. Map of the Caatinga with the physical boundaries of the ecoregions and the locations under study. Koppen's climate classification for Brazil from ALVARES et al. [34] emphasizes the semi-arid (BSh) climate of Caatinga SDTF in orange. Location of sites in ecoregions: 1. MSD (Meridional Sertaneja Depression ecoregion); 2. SD (Sertanejas Depressions ecoregion); 3. BP (Borborema Plateau ecoregion).

2.2. Data Processing

The processes for obtaining geospatial images and data were performed using the Google Earth Engine (GEE) tool [46]. Cloud computing platforms, such as GEE, facilitate the processing of satellite images globally. With access and analysis of several petabytes of images and a broad set of geospatial data, dating from 40 years of Earth observation data, updated and expanded daily, the tool allows viewing, manipulating, editing, and creating spatial data [46]. The EVI vegetation indices derived from the Moderate Resolution Imaging Spectroradiometer (MODIS) sensor reflectance data were used to represent the variability of native vegetation, investigating some biophysical mechanisms that involve soil–plant–atmosphere relationships in the natural dynamics of the Caatinga. The MODIS sensor images and the TerraClimate and CHIRPS data sets' environmental drivers were evaluated from 2000 to 2019. The environmental drivers were made up of monthly precipitation, air temperature, soil moisture, and water deficit time series for the three experimental areas.

2.2.1. MODIS Data

This research used the surface reflectance data of the MODIS 16-day nadir BRDF-Adjusted Reflectance product (MCD43A4); the dataset is produced daily using 16 days of Terra and Aqua MODIS data at 500 metres resolution. The reflectance bands were used to obtain the EVI vegetation index [47]. The first available images of MODIS date from February 2000, and this study had its period limited to including images until December 2019. The EVI series were assembled for each studied site, with the value of the MCD43A4 product pixel corresponding to the geographic coordinates of each testing location. After that, the EVI time series was converted into a monthly EVI time series, using the mean value of each month.

2.2.2. TerraClimate e CHIRPS Data

The air temperature, precipitation, soil moisture, and water deficit are some of the potential environmental drivers for the leaf exchange strategies of Caatinga plant species [33]. These possible environmental drivers acting on the Caatinga vegetation were observed in three different sites of the Caatinga. All the variables mentioned were obtained monthly from the TerraClimate data set [48], except for precipitation data. Monthly precipitation data were obtained from the Climate Hazards Group InfraRed Precipitation with Stations (CHIRPS) database [49]. The exact centre pixels used to construct the EVI time series were used to obtain the environmental drivers' time series.

2.2.3. Phenological Metrics

The phenological metrics represent the characteristics of the vegetation within its phenological cycle, or phenophases, corresponding to dimensionless output parameters and can be calculated based on the EVI time series. In this study, the TIMESAT software [50,51] was used to analyse 20 years of EVI time series (from 2000 to 2019) and to compute 4 phenological metrics: Start of Season (SOS), End Of Season (EOS), Length of the season (LOS), and Amplitude (AMPL) difference between the peak and the base level value. After applying the Savitzky-Golay filter [52] in TIMESAT, a seasonality parameter per year was chosen, representing a phenological cycle with a start and end level of 20% of the seasonal amplitude. This threshold value was used in several studies and is known to be accurate in registering the plant's phenological transitions [53–58]. The four phenological metrics are shown in boxplot graphics for each ecoregion studied and used for correlations with environmental drivers.

2.2.4. Seasonal Variability Analysis

Interannual variability in phenological metrics across sites—To test if the interannual variability in SOS and EOS (i.e., the variances in phenological transition dates) differs among the three sites, we used the non-parametric test Fligner–Killeen [59]. The Fligner–Killeen test compares the homogeneity of variances among samples [59]. The same analysis was performed to test if the variability of SOS differs from the EOS within each site. To test if the LOS and the AMPL of EVI differ among sites, we performed a one-way analysis of variance (ANOVA) followed by the Tukey post-hoc test. In addition, we used box plots to compare the three experimental sites' seasonal data on EVI, phenological metrics, and environmental drivers for 20 years.

Influence of environmental drivers on phenological metrics—The environmental drivers were also observed monthly to assess their influence on phenological parameters. This study analysed these relationships through scatter plots with smoothed (r) correlation curves from Pearson's classification (P) between the EVI and environmental drivers. Pearson's correlation between EVI time series and monthly environmental drivers was submitted to different monthly lag periods (0–3). The lag monthly of each environmental driver with the highest Pearson's correlation was applied to the partial correlation method. Then the partial correlation method was used to analyse the environmental drivers in the monthly EVI time series [60]. Data analysis was carried out in R [61].

3. Results

3.1. Seasonal Profiles and Phenology of SDTF Studied Sites

The interannual variability did not differ among sites for the SOS ($X^2 = 1.0$; df = 2; p-value = 0.606) nor the EOS ($X^2 = 1.3$; df = 2; p-value = 0.520). The variability was higher for the SOS than the EOS at MSD ($X^2 = 10.6$; df = 1; p-value = 0.001) and BP ($X^2 = 11.1$; df = 1; p-value = 0.0008) but not for SD ($X^2 = 0.8$; df = 1; p-value = 0.363). The LOS differed among sites ($F_{2,54} = 4.1$; p-value = 0.02), with SD presenting a shorter LOS than BP (p-value = 0.01) but with no differences between MSD vs. BP (p-value = 0.37) and MSD vs. SD (p-value = 0.28). The amplitude of EVI also differed among sites ($F_{2,54} = 5.2$; p-value = 0.008), with MSD presenting lower amplitude than BP (p-value = 0.006) but with no differences between SD vs. BP (p-value = 0.52) and MSD vs. SD (p-value = 0.10).

The monthly values of the seasonal EVI profile for each ecoregion studied using 20 years of time series are presented in the box plot of Figure 2. The continuous line on the boxplot indicates the median of the monthly values for 20 years of the EVI time series. It can be seen in Figure 2 that the median values are always between 0.2 and 0.5. Despite the similarity in the amplitude of the EVI values, the graph of the median of the monthly values has different temporal behaviour for the different ecoregions studied. The maximum values on the MSD experimental site are similar in February and March. The monthly median presents a well-defined maximum value for the SD experimental site, occurring in April. The maximum EVI values occur in May and June at the BP experimental site. For MSD and SD studied ecoregions, the minimum values occur between six and seven months after the maximum values are observed, while for BP, it happens after four to five months.

Although the graphs with the monthly EVI values presented in Figure 2 understand the temporal behaviour, they do not reveal as many characteristics about the vegetation as the graphs with the phenological metrics presented in Figure 3. When analysing the SOS, it is noticed that the BP and MSD experimental sites present a greater interquartile range when compared to the SD experimental site. For most observations, the SOS from the SD experimental site has taken place between January and February. On the MSD experimental site, it is observed that SOS occurs most frequently between November and January. At BP, it is observed that the highest frequency of SOS occurs between January and May (117 days).

Interestingly, the interquartile range observed for the SOS is not for the EOS and LOS metrics for the MSD and BP experimental sites. For the EOS and LOS, it was observed that the MSD and BP sites have a smaller interquartile range than the SD experimental site, with interquartile range values of less than 30 days for both situations. For the SD experimental site, the variation in the interquartile range in the EOS is similar to that observed for SOS, with a variation slightly greater than 30 days interval and the LOS reaching an interquartile range of 60 days. A relationship of the amplitude with the other metrics, SOS, EOS, and LOS, was not observed. The amplitude's highest values were observed for the BP and SD experimental sites, with 25% of the observations above 0.46. The amplitude values for the MSD experimental area had 75% of the observations below 0.31.

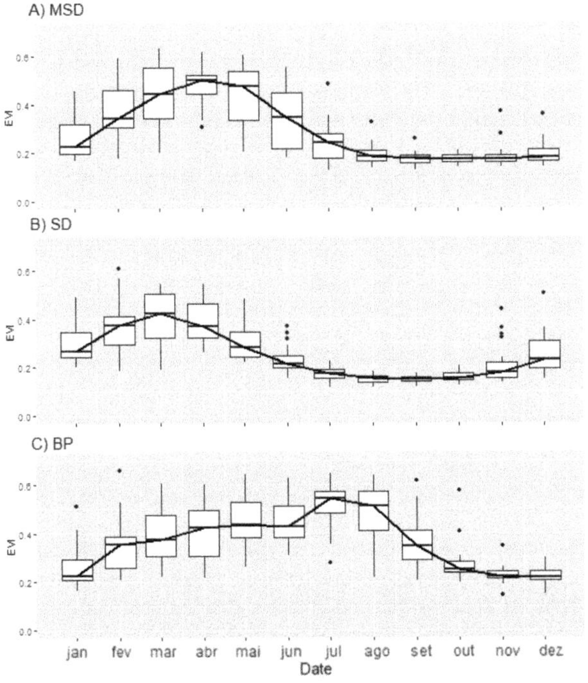

Figure 2. Monthly EVI values for 20 years of time series in (**A**) MSD ecoregion experimental, (**B**) SD ecoregion experimental, and (**C**) BP ecoregion experimental.

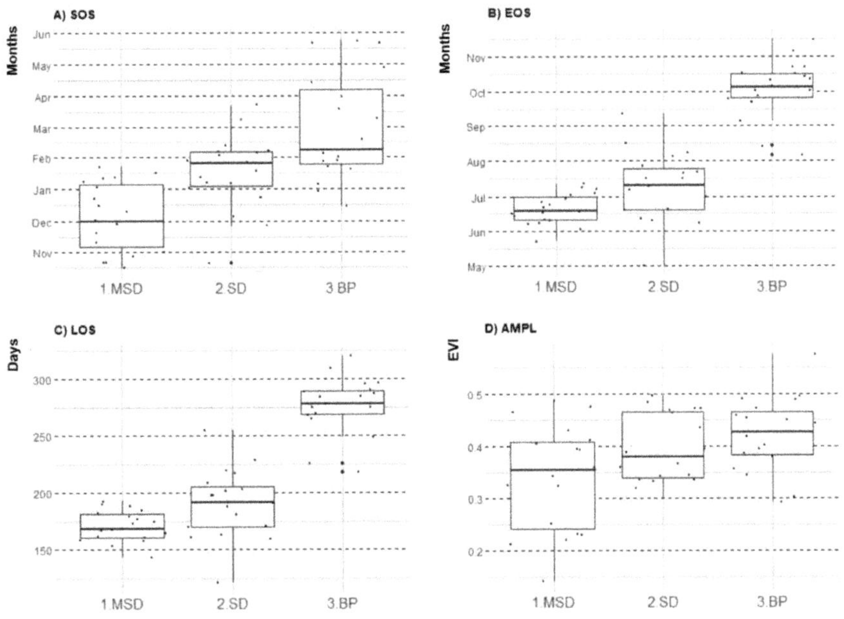

Figure 3. Caatinga phenological metrics of the three ecoregions studied, (**A**) SOS, (**B**) EOS, (**C**) LOS, and (**D**) Amplitude.

3.2. Environmental Drivers

Due to the high number of deciduous species, the SDTF, such as the Caatinga, presents high variability in plant biomass in the annual cycle. Figures 4–6 present the environmental drivers' boxplots that may influence the triggers of phenological changes in the Caatinga. When observing the shape of the graphs of the median behaviour of each environmental driver, it can be seen that there is a similarity in the water deficit, precipitation, and soil moisture behaviour. Pearson's correlation coefficient for the environmental drivers and time series EVI is presented in Figure 7. Figure 7 only presents the lag (0–3 months) environmental drivers with the highest Pearson's correlation. With the application of Pearson's correlation, this study allowed for analysis of each environmental driver and identified their action times on vegetation. These results reinforce the importance of precipitation as the environmental driver that best reflects plant biomass production in the Caatinga. Thus, it is the environmental driver that presents the highest coefficient correlation among all: MSD (r = 0.7258; $p < 0.05$, lag = 1), SD (r = 0.8267; $p < 0.05$, lag = 1), and BP (r = 0.7546; $p < 0.05$, lag = 1). Subsequently, water deficit and soil moisture had the highest correlation values. The water deficit's correlations were: MSD (r = −0.6, $p < 0.01$, lag = 0); SD (r = −0.79, $p < 0.01$, lag = 0); BP (r = −0.69, $p < 0.01$, lag = 1). Soil moisture is one of the main environmental triggers of the Caatinga, and in shallow soils, there is a tendency to saturate and also dry out more quickly, not allowing, in many cases, a long-term response from the vegetation: MSD (r = 0.52; $p < 0.05$, lag = 0), SD (r = 0.69; $p < 0.05$, lag = 0) and BP (r = 0.52; $p < 0.05$, lag = 0). Pearson's correlation coefficient between the EVI and air temperature time series showed the weakest but most significant correlation. While MSD and SD presented positive relations between EVI and temperature (r = 0.6; $p < 0.05$, lag = 2) and (r = 0.59; $p < 0.05$, lag = 3), BP (r = −0.56; $p < 0.05$, lag = 0) showed a negative association.

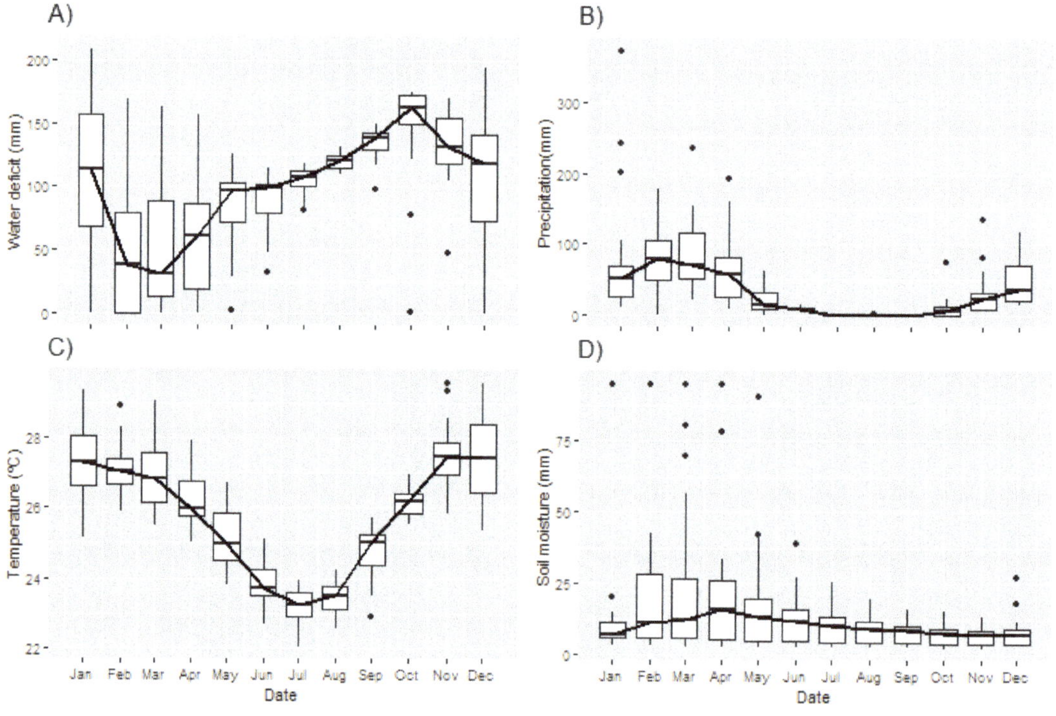

Figure 4. Environment drivers for MSD in (**A**) Water Deficit, (**B**) Precipitation, (**C**) Air Temperature, and (**D**) Soil moisture.

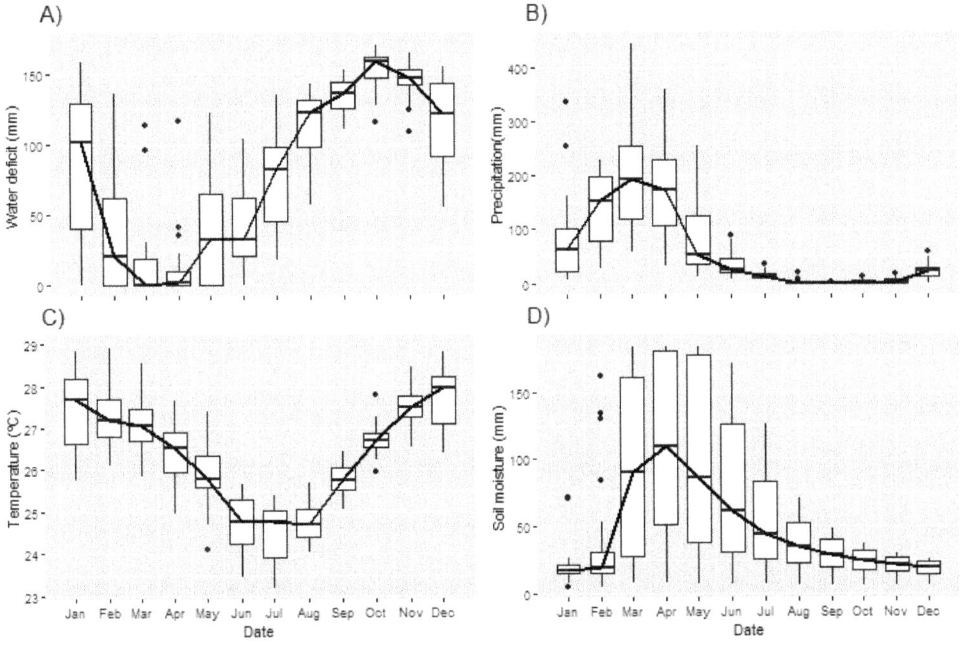

Figure 5. Environment drivers for SD in (**A**) Water Deficit, (**B**) Precipitation, (**C**) Air Temperature, and (**D**) Soil moisture.

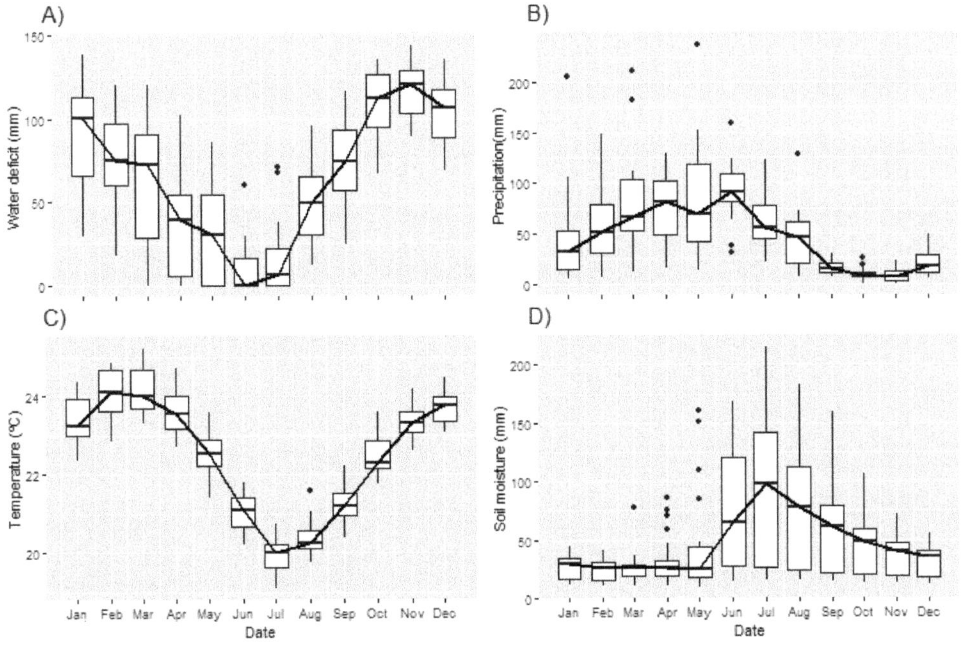

Figure 6. Environment drivers for BP in (**A**) Water Deficit, (**B**) Precipitation, (**C**) Air Temperature, and (**D**) Soil moisture.

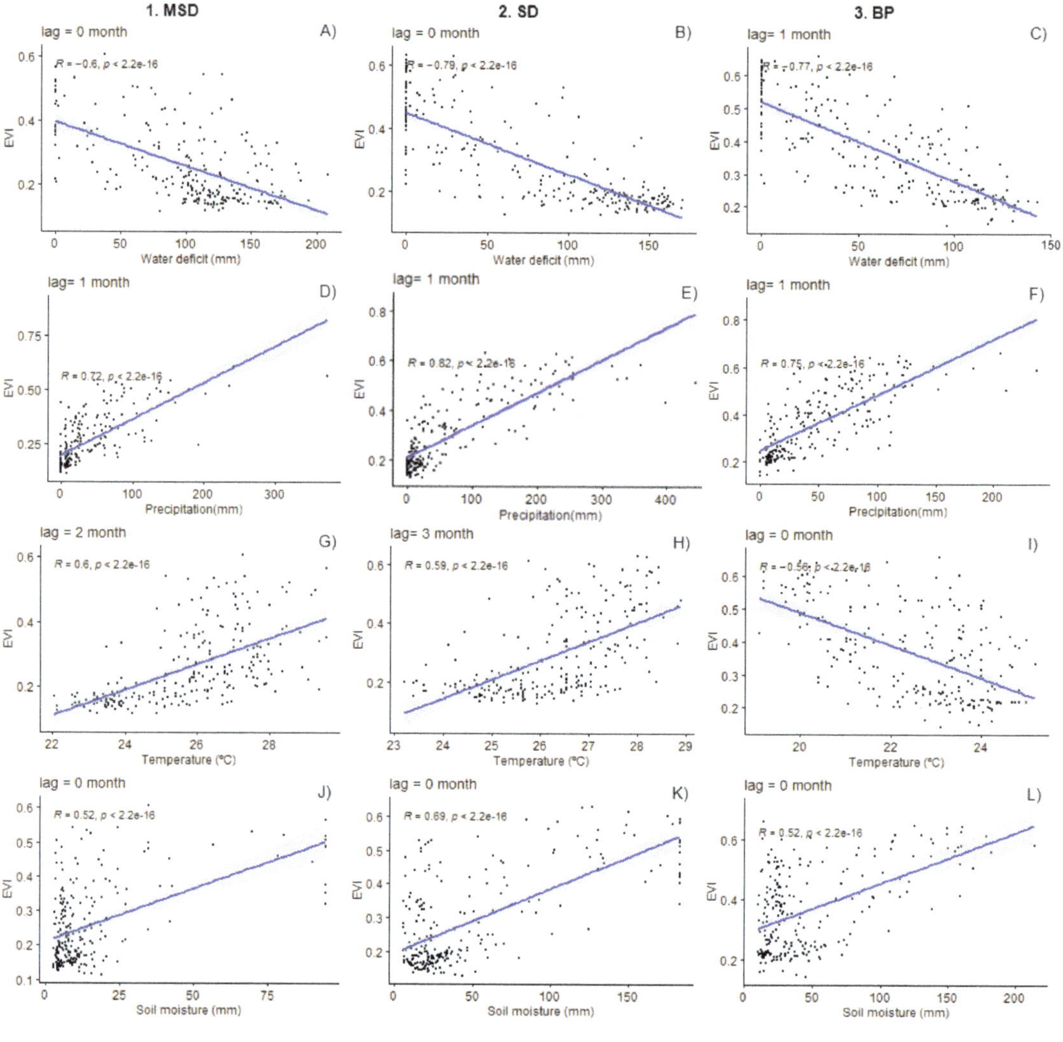

Figure 7. Correlation between EVI and environment drivers for MSD, SD, and BP in water deficit (**A–C**), precipitation (**D–F**), air temperature (**G–I**), and (**D**) Soil moisture (**J–L**). The lag months have the highest (r^2) between the environmental drivers and the EVI monthly time series.

In Pearson's correlation, environmental drivers were analysed one by one. However, the observed behaviour may not reflect the actual effects on vegetation represented by the EVI time series. There may be an association of impact between the analysed environmental drivers. Partial correlation analysis is a way to solve this problem, allowing the analysis of multiple variables. The partial correlation analysis between the environmental drivers and EVI is presented in Table 1 with five scenarios. Scenario 1 considers all environmental drivers to calculate partial correlations, and in the other scenarios, the effect of one of the drivers is retained. Precipitation and soil moisture always showed a positive partial correlation. In contrast, the water deficit showed negative correlations, with r values ranging from −0.25 to −0.62. Only at the BP site, the partial correlation with temperature was negative. In scenario 1, with all environmental drivers, precipitation was the environmental driver that presented the highest correlation, with r values ranging from 0.48 to 0.51. The other environmental drivers are better perceived when precipitation is removed from the analysis (scenario 2). In scenario 2, the highest correlations were for temperature in MSD (0.46), water deficit and soil moisture for SD (0.40), and water deficit for BP (−0.62). This dependence relationship becomes evident when the drivers with the highest partial correlation identified in scenario 2 are removed from the analysis. Thus, the highest values of r for precipitation are seen in scenarios 5, 3, and 4 for MSD, SD, and BP sites, respectively.

Table 1. Partial correlation coefficients between the EVI and environment drivers across MSD, SD, and BP sites. Scenario 1 considers all environmental drivers to calculate partial correlations. For the other scenarios, the effect of one of the variables is removed: scenarios 2 (without precipitation), 3 (without soil moisture), 4 (without water deficit), and 5 (without temperature). "-" indicates that the environment driver was not used to calculate the partial correlation. Only for statistical significance ($p < 0.05$). Missing estimates (NS) are not significant.

Site	Scenarios	Air Temperature (°C)	Water Deficit (mm)	Soil Moisture (mm)	Precipitation (mm)
1. MSD	1	0.28	−0.25	NS	0.48
	2	0.46	−0.27	0.26	-
	3	0.26	−0.34	-	0.52
	4	0.34	-	0.26	0.49
	5	-	−0.32	NS	0.58
2. SD	1	0.09	−0.41	0.11	0.51
	2	0.26	−0.40	0.40	-
	3	0.06	−0.50	-	0.61
	4	0.36	-	0.33	0.51
	5	-	−0.52	0.08	0.55
3. BP	1	NS	−0.29	0.18	0.48
	2	NS	−0.62	0.16	-
	3	−0.14	−0.32	-	0.47
	4	−0.28	-	0.23	0.69
	5	-	−0.39	0.22	0.48

4. Discussion

Compared to the following season, the greater rainfall variability at the beginning of the wet season may be the primary driver for the interquartile range of the SOS to be greater than that observed for the EOS and LOS (Figure 3). The higher frequency and accumulation are the main characteristics of rainfall contributing to the higher AMPL and LOS values at the BP experimental site. The partial coefficient of air temperature in the BP experimental site was always lower. In tropical regions, the air temperature variability is low, with about four degrees of annual amplitude difference from the monthly normal. However, in the MSD experimental site, there was a greater partial correlation between air temperature (scenario 2). The climatological water deficit is calculated from the reference evapotranspiration and precipitation, providing an efficient measure of the water availability and demand of the environment. Air temperature, directly related to water vapour-pressure deficit, is one of the climatic factors contributing to the variability of reference evapotranspiration [62]. The average monthly precipitation is quite variable in the annual cycle, which has a greater impact on the water deficit.

In the Caatinga, vegetation's phenology and photosynthetic activity are associated with water availability [9]. In drier regions such as MSD and SD, intra-annual water availability is lower, leading to shorter growing season periods. The water stress makes the environmental conditions (Figures 2 and 3) harsh for leaf maintenance for vegetation. In contrast, BP with more regular water availability (Figures 2 and 3) shows larger LOS, which means more water and carbon exchange time at the soil–vegetation–atmosphere interface [63]. Carbon assimilation was related to EVI at Caatinga by Mendes et al. [64], demonstrating that the vegetation presents more significant photosynthetic activity and productivity (seasonal and total) during high EVI value periods. The vegetation at BP showed a higher senescence rate than in the other areas. The phenological parameter is related to leaf loss in the dry season and the maximum use of water availability for its recovery after the dry period with accelerated regrowth and increased metabolic processes in the rainy season [65]. These vegetation strategies are adaptive to optimise the phenological, vegetative, and reproductive processes [10]. The study areas showed similarities in the seasonality of phenological parameters, corroborating the high association level with water availability.

Many studies have shown that rainfall seasonality regulates SDTF canopy seasonality [24,66], but there is a complex relationship between environmental drivers and the vegetation response [9,28], mainly for the Caatinga where the plant physiology is adapted to drought and elevated temperatures [67]. As expected for the Caatinga vegetation, the months with the highest water deficit have the lowest leaf cover (lowest EVI values), a result previously found by other studies in the region [7,33,68]. About 70% of the year has a water deficit [69]. According to Flerchinger et al. [70], about 90% of the rainwater in arid and semi-arid regions returns to the atmosphere through evapotranspiration. As there is not such a significant variability in the average air temperature between the tropics, the annual constancy provides a high evaporative demand from the atmosphere throughout the year [69], conditioning the water deficit to the seasonal fluctuation of the rainfall, adjusting the Caatinga phenological cycle to the water availability. Phenological transitions are an excellent indicator of climate change [71], and future scenarios estimate greater water demand from plants and the occurrence of droughts, providing a more significant water deficit [72,73], and with that, being able to alter the phenology of the plants. The water deficit slows down plants' growth, causes leaves, fruits, and flowers to fall, and, in the short term, tends to anticipate flowering and the beginning of fruiting, reducing plant cycles. In contrast, they tend to extend or even prevent the regular completion of the plant cycle [74].

The months with the highest spectral response of vegetation occurred in the rainy season when rain and soil moisture were predominantly distributed. The peak of the EVI was preceded by the month of greatest precipitation at SD and MSD. The rain at the BP experimental site shows a smoother and more regular distribution than in other areas, with the EVI closely following seasonal fluctuations in precipitation. The ecosystem accumulates sufficient water reserves in the soil and biomass for both sites under study, resulting in slower leaf fall during the dry season. The EVI followed the monthly rainfall distributions linearly, as shown in Figure 4. Likewise, the air temperature is observed after the maximum peak of the EVI. The air temperature also decreases the vegetation response, resulting in greater water stress for the research areas. Because of the high diversity of species at the Caatinga, the studied sites could present plants at different stages of adaptation or with physiological aspects related to water uptake or leaf abscission. The fact that MSD and SD present shorter LOS than BP could be related to vegetation strategies to use the water more efficiently during the available period. The peak of EVI values was higher at MSD and SD, and this would be related to more biomass production resulting from the water use efficiency (WUE) of the species composition.

5. Conclusions

This study observed that using the EVI time series efficiently estimated phenological parameters. The determination of the phenological parameters for 20 years allowed the evaluation of their seasonality. The seasonality of Caatinga vegetation growth differed between the three experimental sites. The SOS was the parameter that presented the greatest variability in the days of the year (DOY), reaching a variation of 117 days. The high variability of SOS occurred in sites with low variability of EOS and vice versa. The relationship between air temperature and vegetation was insignificant for the two experimental sites and presented a significant value for the site drier (MSD). At the experimental site where the highest vegetative peak co-occurred with the low air temperature, the highest LOS and AMPL were observed. In addition, the values of LOS and AMPL are directly linked to the annual distribution of rainfall, and the longer the rainy season, the greater their values are. The variability of the natural cycles of the environmental drivers that regulate the ecosystem's phenology and the influence on the Caatinga's natural dynamics indicated a greater sensitivity of the vegetation to water availability, with the water deficit being the limiting factor of the vegetation. There is a greater need for new research and studies aiming to reduce the impact of climatic risks by characterising and quantifying species' response patterns to likely climate change. There will be changes in phenological patterns, production, and spatial distribution. Quantifying these impacts and devising appropriate actions to minimise such adversities represent significant research challenges in the coming decades.

Author Contributions: Formal analysis, R.M., J.A., M.M., A.M.P.-M. and C.A.C.d.S.; Investigation, R.M., J.A., M.M. and B.B.d.S.; Methodology, D.R. and J.C.; Software, R.M., J.A. and D.R.; Supervision, J.C.; Visualization, A.M.P.-M., C.A.C.d.S. and B.B.d.S.; Writing—original draft, R.M., J.A., D.R. and B.B.d.S.; Writing—review & editing, D.R., M.M., C.A.C.d.S. and J.C. All authors have read and agreed to the published version of the manuscript.

Funding: The study was financed by the Pró-Reitoria de Pós-Graduação e Pesquisa of the Federal University of Campina Grande (PRPG-UFCG).

Acknowledgments: The authors thank the Coordenação de Aperfeiçoamento de Pessoal de Nível Superior-Brazil (CAPES)-Finance Code 001 for the as a Brazilian CAPES scholarship to the first and second author, the project Nova geração de modelos para estimativas de evapotranspiração para o Semiárido Brasileiro (CNPq Proc. 409341/2021-5). This work forms part of the project Evolução na representação do balanço de energia por meio da integração de dados de campo e satélite aplicados a computação em nuvem para o semiárido brasileiro (FAPESQ under startup grant number #010/2021). The authors are also grateful to CNPq for the PQ (Productivity and Research) grants to the sixth autor (Proc. 304493/2019-8).

Conflicts of Interest: The authors declare no conflict of interest.

References

1. Pennington, R.T.; Lehmann, C.; Rowland, L.M. Tropical savannas and dry forests. *Curr. Biol.* **2018**, *28*, R541–R545. [CrossRef] [PubMed]
2. Moro, M.F.; Nic Lughadha, E.; de Araújo, F.S.; Martins, F.R. A Phytogeographical Metaanalysis of the Semiarid Caatinga Domain in Brazil. *Bot. Rev.* **2016**, *82*, 91–148. [CrossRef]
3. De Queiroz, L.P.; Cardoso, D.; Fernandes, M.F.; Moro, M.F. Diversity and Evolution of Flowering Plants of the Caatinga Domain. *Caatinga* **2017**, 23–63. [CrossRef]
4. Fernandes, M.F.; Cardoso, D.; de Queiroz, L.P. An updated plant checklist of the Brazilian Caatinga seasonally dry forests and woodlands reveals high species richness and endemism. *J. Arid Environ.* **2020**, *174*, 104079. [CrossRef]
5. Leith, H. Phenology and Seasonality Modeling. *Soil Sci.* **1974**, *120*, 461. [CrossRef]
6. Morisette, J.T.; Richardson, A.D.; Knapp, A.K.; Fisher, J.I.; Graham, E.A.; Abatzoglou, J.; Wilson, B.E.; Breshears, D.D.; Henebry, G.M.; Hanes, J.M.; et al. Tracking the rhythm of the seasons in the face of global change: Phenological research in the 21st century. *Front. Ecol. Environ.* **2009**, *7*, 253–260. [CrossRef]
7. Machado, I.C.S.; Barros, L.M.; Sampaio, E.V.S.B. Phenology of Caatinga Species at Serra Talhada, PE, Northeastern Brazil. *Biotropica* **1997**, *29*, 57–68. [CrossRef]
8. Alberton, B.; Torres, R.D.S.; Cancian, L.F.; Borges, B.D.; Almeida, J.; Mariano, G.C.; dos Santos, J.; Morellato, P. Introducing digital cameras to monitor plant phenology in the tropics: Applications for conservation. *Perspect. Ecol. Conserv.* **2017**, *15*, 82–90. [CrossRef]
9. Paloschi, R.; Ramos, D.; Ventura, D.; Souza, R.; Souza, E.; Morellato, L.; Nóbrega, R.; Coutinho, I.A.C.; Verhoef, A.; Körting, T.; et al. Environmental Drivers of Water Use for Caatinga Woody Plant Species: Combining Remote Sensing Phenology and Sap Flow Measurements. *Remote Sens.* **2020**, *13*, 75. [CrossRef]
10. Vico, G.; Thompson, S.E.; Manzoni, S.; Molini, A.; Albertson, J.D.; Almeida-Cortez, J.S.; Fay, P.A.; Feng, X.; Guswa, A.J.; Liu, H.; et al. Climatic, ecophysiological, and phenological controls on plant ecohydrological strategies in seasonally dry ecosystems. *Ecohydrology* **2015**, *8*, 660–681. [CrossRef]
11. Riahi, K.; Schaeffer, R.; Arango, J.; Calvin, K.; Guivarch, C.; Hasegawa, T.; Jiang, K.; Kriegler, E.; Matthews, R.; Peters, G.P.; et al. Mitigation pathways compatible with long-term goals. In *IPCC, 2022: Climate Change 2022: Mitigation of Climate Change. Contribution of Working Group III to the Sixth Assessment Report of the Intergovernmental Panel on Climate Change*; Shukla, P.R., Skea, J.,

Slade, R., Al Khourdajie, A., van Diemen, R., McCollum, D., Pathak, M., Some, S., Vyas, P., Fradera, R., et al., Eds.; Cambridge University Press: Cambridge, UK; New York, NY, USA, 2022. [CrossRef]

12. Torres, R.R.; Lapola, D.M.; Gamarra, N.L.R. Future Climate Change in the Caatinga. *Caatinga* **2017**, 383–410. [CrossRef]
13. Marengo, J.A.; Cunha, A.P.M.A.; Nobre, C.A.; Neto, G.G.R.; Magalhaes, A.R.; Torres, R.R.; Sampaio, G.; Alexandre, F.; Alves, L.M.; Cuartas, L.A.; et al. Assessing drought in the drylands of northeast Brazil under regional warming exceeding 4 °C. *Nat. Hazards* **2020**, *103*, 2589–2611. [CrossRef]
14. Xu, L.; Myneni, R.; Iii, F.S.C.; Callaghan, T.V.; Pinzon, J.E.; Tucker, C.J.; Zhu, Z.; Bi, J.; Ciais, P.; Tømmervik, H.; et al. Temperature and vegetation seasonality diminishment over northern lands. *Nat. Clim. Chang.* **2013**, *3*, 581–586. [CrossRef]
15. Tong, X.; Tian, F.; Brandt, M.; Liu, Y.; Zhang, W.; Fensholt, R. Trends of land surface phenology derived from passive microwave and optical remote sensing systems and associated drivers across the dry tropics 1992–2012. *Remote Sens. Environ.* **2019**, *232*. [CrossRef]
16. Brando, P.M.; Goetz, S.J.; Baccini, A.; Nepstad, D.C.; Beck, P.S.A.; Christman, M.C. Seasonal and interannual variability of climate and vegetation indices across the Amazon. *Proc. Natl. Acad. Sci. USA* **2010**, *107*, 14685–14690. [CrossRef]
17. Olmos-Trujillo, E.; González-Trinidad, J.; Júnez-Ferreira, H.; Pacheco-Guerrero, A.; Bautista-Capetillo, C.; Avila-Sandoval, C.; Galván-Tejada, E. Spatio-Temporal Response of Vegetation Indices to Rainfall and Temperature in A Semiarid Region. *Sustainability* **2020**, *12*, 1939. [CrossRef]
18. Andrade, J.; Cunha, J.; Silva, J.; Rufino, I.; Galvão, C. Evaluating single and multi-date Landsat classifications of land-cover in a seasonally dry tropical forest. *Remote Sens. Appl. Soc. Environ.* **2021**, *22*, 100515. [CrossRef]
19. Huete, A.; Didan, K.; Miura, T.; Rodriguez, E.P.; Gao, X.; Ferreira, L.G. Overview of the radiometric and biophysical performance of the MODIS vegetation indices. *Remote Sens. Environ.* **2002**, *83*, 195–213. [CrossRef]
20. Adole, T.; Dash, J.; Atkinson, P. A systematic review of vegetation phenology in Africa. *Ecol. Inform.* **2016**, *34*, 117–128. [CrossRef]
21. Suepa, T.; Qi, J.; Lawawirojwong, S.; Messina, J. Understanding spatio-temporal variation of vegetation phenology and rainfall seasonality in the monsoon Southeast Asia. *Environ. Res.* **2016**, *147*, 621–629. [CrossRef]
22. Htitiou, A.; Boudhar, A.; Lebrini, Y.; Hadria, R.; Lionboui, H.; Elmansouri, L.; Tychon, B.; Benabdelouahab, T. The Performance of Random Forest Classification Based on Phenological Metrics Derived from Sentinel-2 and Landsat 8 to Map Crop Cover in an Irrigated Semi-arid Region. *Remote Sens. Earth Syst. Sci.* **2019**, *2*, 208–224. [CrossRef]
23. Pastor-Guzman, J.; Dash, J.; Atkinson, P. Remote sensing of mangrove forest phenology and its environmental drivers. *Remote Sens. Environ.* **2018**, *205*, 71–84. [CrossRef]
24. De Jesus, J.B.; Kuplich, T.M.; Barreto, D.D.C.; da Rosa, C.N.; Hillebrand, F.L. Temporal and phenological profiles of open and dense Caatinga using remote sensing: Response to precipitation and its irregularities. *J. For. Res.* **2021**, *32*, 1067–1076. [CrossRef]
25. Songsom, V.; Koedsin, W.; Ritchie, R.J.; Huete, A. Mangrove Phenology and Environmental Drivers Derived from Remote Sensing in Southern Thailand. *Remote Sens.* **2019**, *11*, 955. [CrossRef]
26. Wang, G.; Huang, Y.; Wei, Y.; Zhang, W.; Li, T.; Zhang, Q. Inner Mongolian grassland plant phenological changes and their climatic drivers. *Sci. Total Environ.* **2019**, *683*, 1–8. [CrossRef]
27. Huang, J.-G.; Ma, Q.; Rossi, S.; Biondi, F.; Deslauriers, A.; Fonti, P.; Liang, E.; Mäkinen, H.; Oberhuber, W.; Rathgeber, C.B.K.; et al. Photoperiod and temperature as dominant environmental drivers triggering secondary growth resumption in Northern Hemisphere conifers. *Proc. Natl. Acad. Sci. USA* **2020**, *117*, 20645–20652. [CrossRef]
28. Godoy-Veiga, M.; Cintra, B.B.L.; Stríkis, N.M.; Cruz, F.W.; Grohmann, C.H.; Santos, M.S.; Regev, L.; Boaretto, E.; Ceccantini, G.; Locosselli, G.M. The value of climate responses of individual trees to detect areas of climate-change refugia, a tree-ring study in the Brazilian seasonally dry tropical forests. *For. Ecol. Manag.* **2021**, *488*, 118971. [CrossRef]
29. Whitecross, M.; Witkowski, E.; Archibald, S. No two are the same: Assessing variability in broad-leaved savanna tree phenology, with watering, from 2012 to 2014 at Nylsvley, South Africa. *S. Afr. J. Bot.* **2016**, *105*, 123–132. [CrossRef]
30. Ryan, C.M.; Williams, M.; Grace, J.; Woollen, E.; Lehmann, C.E.R. Pre-rain green-up is ubiquitous across southern tropical Africa: Implications for temporal niche separation and model representation. *New Phytol.* **2016**, *213*, 625–633. [CrossRef]
31. Sampaio, E.V. Overview of the Brazilian caatinga. *Seas. Dry Trop. For.* **1995**, *1*, 35–63. [CrossRef]
32. Gutiérrez, A.P.A.; Engle, N.L.; De Nys, E.; Molejón, C.; Martins, E.S. Drought preparedness in Brazil. *Weather Clim. Extremes* **2014**, *3*, 95–106. [CrossRef]
33. Alberton, B.; Torres, R.D.S.; Silva, T.S.F.; da Rocha, H.R.; Moura, M.S.B.; Morellato, L.P.C. Leafing Patterns and Drivers across Seasonally Dry Tropical Communities. *Remote Sens.* **2019**, *11*, 2267. [CrossRef]
34. Alvares, C.A.; Stape, J.L.; Sentelhas, P.C.; Moraes, G.J.L.; Sparovek, G. Köppen's climate classification map for Brazil. *Meteorol. Z.* **2013**, *22*, 711–728. [CrossRef]
35. Sampaio, E.V.S.B. Características e potencialidades. In *Uso Sustentável e Conservação dos Recursos Florestais da Caatinga*; Gariglio, M.A., Sampaio, E.V.S.B., Cestaro, L.A., Kageyama, P.Y., Eds.; Serviço Florestal Brasileiro: Brasília, Brazil, 2010; pp. 29–48.
36. Prado, D. As caatingas da América do Sul. In *Ecologia e Conservação da Caatinga*; Leal, I.R., Tabarelli, M., Silva, J.M.C., Eds.; Universitária da UFPE: Recife, Brazil, 2003; pp. 3–73.
37. Nimer, E. Climatologia da região Nordeste do Brasil. Introdução à climatologia dinâmica. *Rev. Bras. Geogr.* **1972**, *34*, 3–51.
38. Velloso, A.L.; Sampaio, E.V.S.B.; Giulietti, A.M.; Barbosa, M.R.V.; Castri, A.A.J.F.; Queiroz, L.P.; Fernandes, A.; Oren, D.C.; Cestaro, L.A.; Carvalho, A.J.E.; et al. Ecorregiões: Propostas para o Bioma Caatinga. In *Associação Plantas do Nordeste*; The Nature Conservancy do Brasil: Recife, Brazil, 2002; p. 75.

39. Souza, R.; Feng, X.; Antonino, A.; Montenegro, S.; Souza, E.; Porporato, A. Vegetation response to rainfall seasonality and interannual variability in tropical dry forests. *Hydrol. Process.* **2016**, *30*, 3583–3595. [CrossRef]
40. Kill, L.H.P. Caracterização da vegetação da Reserva Legal da Embrapa Semiárido. *Embrapa Semiárido Pet.* **2017**, *1*, 1–27.
41. Tavares-Damasceno, J.P.; Silveira, J.L.G.D.S.; Câmara, T.; Stedile, P.D.C.; Macario, P.; Toledo-Lima, G.S.; Pichorim, M. Effect of drought on demography of Pileated Finch (Coryphospingus pileatus: Thraupidae) in northeastern Brazil. *J. Arid Environ.* **2017**, *147*, 63–70. [CrossRef]
42. Althoff, T.D.; Menezes, R.; de Carvalho, A.L.; Pinto, A.D.S.; Santiago, G.A.C.F.; Ometto, J.; Von Randow, C.; Sampaio, E.V.D.S.B. Climate change impacts on the sustainability of the firewood harvest and vegetation and soil carbon stocks in a tropical dry forest in Santa Teresinha Municipality, Northeast Brazil. *For. Ecol. Manag.* **2016**, *360*, 367–375. [CrossRef]
43. Cunha, J.E.B.L.; Rufino, I.A.A.; Ideião, S.M.A. Determinação da temperatura da superfície na cidade de Campina Grande-PB a partir de imagens do satélite Landsat 5-TM. In *Anais XIV Simpósio Brasileiro de Sensoriamento Remoto*; INPE: Natal, Brazil, 2009.
44. INMET, National Institute of Meteorology of Brazil, 2021. Available online: https://bdmep.inmet.gov.br/# (accessed on 1 October 2021).
45. Embrapa, Empresa Brasileira de Pesquisa Aagropecuária. Centro Nacional de Pesquisa de Solos. *Sistema Brasileiro de Classificação de Solo*, 2nd ed.; Embrapa Solos: Rio de Janeiro, Brazil, 2006.
46. Gorelick, N.; Hancher, M.; Dixon, M.; Ilyushchenko, S.; Thau, D.; Moore, R. Google Earth Engine: Planetary-scale geospatial analysis for everyone. *Remote Sens. Environ.* **2017**, *202*, 18–27. [CrossRef]
47. Guerschman, J.P.; Scarth, P.F.; McVicar, T.R.; Renzullo, L.J.; Malthus, T.J.; Stewart, J.B.; Rickards, J.E.; Trevithick, R. Assessing the effects of site heterogeneity and soil properties when unmixing photosynthetic vegetation, non-photosynthetic vegetation and bare soil fractions from Landsat and MODIS data. *Remote Sens. Environ.* **2015**, *161*, 12–26. [CrossRef]
48. Abatzoglou, J.T.; Dobrowski, S.; Parks, S.A.; Hegewisch, K.C. TerraClimate, a high-resolution global dataset of monthly climate and climatic water balance from 1958–2015. *Sci. Data* **2018**, *5*, 170191. [CrossRef] [PubMed]
49. Funk, C.; Peterson, P.; Landsfeld, M.; Pedreros, D.; Verdin, J.; Shukla, S.; Husak, G.; Rowland, J.; Harrison, L.; Hoell, A.; et al. The climate hazards infrared precipitation with stations—A new environmental record for monitoring extremes. *Sci. Data* **2015**, *2*, 150066. [CrossRef] [PubMed]
50. Jönsson, P.; Eklundh, L. TIMESAT—A program for analyzing time-series of satellite sensor data. *Comput. Geosci.* **2004**, *30*, 833–845. [CrossRef]
51. Jönsson, P.; Eklundh, L. *TIMESAT 3.1—Software Manual*; Lund University: Lund, Sweden, 2012; p. 82. Available online: http://web.nateko.lu.se/timesat/timesat.asp (accessed on 1 May 2021).
52. Savitzky, A.; Golay, M.J.E. Smoothing and Differentiation of Data by Simplified Least Squares Procedures. *Anal. Chem.* **1964**, *36*, 1627–1639. [CrossRef]
53. Kong, F.; Li, X.; Wang, H.; Xie, D.; Li, X.; Bai, Y. Land Cover Classification Based on Fused Data from GF-1 and MODIS NDVI Time Series. *Remote Sens.* **2016**, *8*, 741. [CrossRef]
54. Streher, A.S.; Sobreiro, J.F.F.; Morellato, P.; Silva, T. Land Surface Phenology in the Tropics: The Role of Climate and Topography in a Snow-Free Mountain. *Ecosystems* **2017**, *20*, 1436–1453. [CrossRef]
55. Diem, P.K.; Pimple, U.; Sitthi, A.; Varnakovida, P.; Tanaka, K.; Pungkul, S.; Leadprathom, K.; LeClerc, M.Y.; Chidthaisong, A. Shifts in Growing Season of Tropical Deciduous Forests as Driven by El Niño and La Niña during 2001–2016. *Forests* **2018**, *9*, 448. [CrossRef]
56. Wang, Y.; Zang, S.; Tian, Y. Mapping paddy rice with the random forest algorithm using MODIS and SMAP time series. *Chaos Solitons Fractals* **2020**, *140*. [CrossRef]
57. Ramírez-Cuesta, J.; Minacapilli, M.; Motisi, A.; Consoli, S.; Intrigliolo, D.; Vanella, D. Characterization of the main land processes occurring in Europe (2000–2018) through a MODIS NDVI seasonal parameter-based procedure. *Sci. Total Environ.* **2021**, *799*, 149346. [CrossRef]
58. Doussoulin-Guzmán, M.-A.; Pérez-Porras, F.-J.; Triviño-Tarradas, P.; Ríos-Mesa, A.-F.; Porras, A.G.-F.; Mesas-Carrascosa, F.-J. Grassland Phenology Response to Climate Conditions in Biobio, Chile from 2001 to 2020. *Remote Sens.* **2022**, *14*, 475. [CrossRef]
59. Conover, W.J.; Johnson, M.E.; Johnson, M.M. A Comparative Study of Tests for Homogeneity of Variances, with Applications to the Outer Continental Shelf Bidding Data. *Technometrics* **1981**, *23*, 351–361. [CrossRef]
60. Zhu, W.; Zheng, Z.; Jiang, N.; Zhang, D. A comparative analysis of the spatio-temporal variation in the phenologies of two herbaceous species and associated climatic driving factors on the Tibetan Plateau. *Agric. For. Meteorol.* **2018**, *248*, 177–184. [CrossRef]
61. R Development Core Team. *R: A Language and Environment for Statistical Computing*; R Foundation for Statistical Computing: Vienna, Austria, 2019; Available online: https://www.r-project.org (accessed on 20 February 2019).
62. De Souza, L.S.B.; Silva, M.T.L.; Alba, E.; de Moura, M.S.B.; Neto, J.F.D.C.; de Souza, C.A.A.; da Silva, T.G.F. New method for estimating reference evapotranspiration and comparison with alternative methods in a fruit-producing hub in the semi-arid region of Brazil. *Arch. Meteorol. Geophys. Bioclimatol. Ser. B* **2022**, 1–10. [CrossRef]
63. Dos Santos, C.A.C.; Mariano, D.A.; Nascimento, F.D.C.A.D.; Dantas, F.R.D.C.; de Oliveira, G.; Silva, M.T.; da Silva, L.L.; da Silva, B.B.; Bezerra, B.G.; Safa, B.; et al. Spatio-temporal patterns of energy exchange and evapotranspiration during an intense drought for drylands in Brazil. *Int. J. Appl. earth Obs. Geoinf. ITC J.* **2019**, *85*, 101982. [CrossRef]

64. Mendes, K.R.; Campos, S.; Da Silva, L.L.; Mutti, P.R.; Ferreira, R.R.; Medeiros, S.S.; Perez-Marin, A.M.; Marques, T.V.; Ramos, T.M.; Vieira, M.M.D.L.; et al. Seasonal variation in net ecosystem CO_2 exchange of a Brazilian seasonally dry tropical forest. *Sci. Rep.* **2020**, *10*, 1–16. [CrossRef] [PubMed]
65. Bezerra, M.V.C.; Da Silva, B.B.; Bezerra, B.G. Avaliação dos efeitos atmosféricos no albedo e NDVI obtidos com imagens de satélite. *Rev. Bras. Eng. Agrícola Ambient.* **2011**, *15*, 709–717. [CrossRef]
66. Allen, K.; Dupuy, J.M.; Gei, M.G.; Hulshof, C.; Medvigy, D.; Pizano, C.; Salgado-Negret, B.; Smith, C.M.; Trierweiler, A.; Van Bloem, S.J.; et al. Will seasonally dry tropical forests be sensitive or resistant to future changes in rainfall regimes? *Environ. Res. Lett.* **2017**, *12*, 023001. [CrossRef]
67. Trovão, D.M.D.B.M.; Fernandes, P.D.; De Andrade, L.A.; Neto, J.D. Variações sazonais de aspectos fisiológicos de espécies da Caatinga. *Rev. Bras. Eng. Agrícola Ambient.* **2007**, *11*, 307–311. [CrossRef]
68. Miranda, R.D.Q.; Nóbrega, R.L.B.; de Moura, M.S.B.; Raghavan, S.; Galvíncio, J.D. Realistic and simplified models of plant and leaf area indices for a seasonally dry tropical forest. *Int. J. Appl. Earth Obs. Geoinf. ITC J.* **2020**, *85*, 101992. [CrossRef]
69. Marengo, J.A.; Torres, R.R.; Alves, L.M. Drought in Northeast Brazil—past, present, and future. *Theor. Appl. Climatol.* **2016**, *129*, 1189–1200. [CrossRef]
70. Flerchinger, G.N.; Hanson, C.L.; Wight, J.R. Modeling Evapotranspiration and Surface Energy Budgets across a Watershed. *Water Resour. Res.* **1996**, *32*, 2539–2548. [CrossRef]
71. Langvall, O.; Löfvenius, M.O. Long-term standardized forest phenology in Sweden: A climate change indicator. *Int. J. Biometeorol.* **2021**, *65*, 381–391. [CrossRef] [PubMed]
72. Marin, F.R.; Assad, E.D.; Barbarisi, B.F.; Pilau, F.G.; Pacheco, L.R.F.; Zullo, J.J.; Pinto, H.S. Efeito das mudanças climáticas sobre a aptidão climática para cana-de-açúcar no Estado de São Paulo. In *Congresso Brasileiro de Agrometeorologia*; Embrapa/SBAgro: Aracaju, Brazil, 2007; Volume 15.
73. Nepomuceno, A.L.; Farias, J.R.B.; Salinet, L.H.; Polizel, A.M.; Neumaier, N.; Beneventi, M.A.; Stolf, R.; Rolla, A.A.P. Engenharia genética como ferramenta no desenvolvimento de plantas de soja adaptadas a cenários futuros de mudanças climáticas. In *Congresso Brasileiro de Agrometeorologia*; Embrapa/SBAgro: Aracaju, Brazil, 2007.
74. Bergamaschi, H. O clima como fator determinante da fenologia das plantas. In *Fenologia: Ferramenta para Conservação, Melhoramento e Manejo de Recursos Vegetais Arbóreos*; Embrapa Florestas: Colombo, Brazil, 2007; Volume 1, pp. 291–310.

remote sensing

Communication

Comparison of Vegetation Phenology Derived from Solar-Induced Chlorophyll Fluorescence and Enhanced Vegetation Index, and Their Relationship with Climatic Limitations

Cong Wang [1], Yijin Wu [1], Qiong Hu [1,*], Jie Hu [2], Yunping Chen [2], Shangrong Lin [3] and Qiaoyun Xie [4]

1 Key Laboratory for Geographical Process Analysis & Simulation of Hubei Province & School of Urban and Environmental Sciences, Central China Normal University, Wuhan 430079, China; wangcong@ccnu.edu.cn (C.W.); wuyijin@ccnu.edu.cn (Y.W.)
2 Macro Agriculture Research Institute, College of Plant Science and Technology, Huazhong Agricultural University, Wuhan 430070, China; hu_jie@webmail.hzau.edu.cn (J.H.); yunping.chen@webmail.hzau.edu.cn (Y.C.)
3 Southern Marine Science and Engineering Guangdong Laboratory (Zhuhai), School of Atmospheric Sciences, Sun Yat-sen University, Zhuhai 519082, China; linsr@radi.ac.cn
4 School of Life Sciences, Faculty of Science, University of Technology Sydney, Sydney, NSW 2007, Australia; qiaoyun.xie@uts.edu.au
* Correspondence: huqiong@ccnu.edu.cn

Citation: Wang, C.; Wu, Y.; Hu, Q.; Hu, J.; Chen, Y.; Lin, S.; Xie, Q. Comparison of Vegetation Phenology Derived from Solar-Induced Chlorophyll Fluorescence and Enhanced Vegetation Index, and Their Relationship with Climatic Limitations. *Remote Sens.* **2022**, *14*, 3018. https://doi.org/10.3390/rs14133018

Academic Editor: Sofia Bajocco

Received: 26 April 2022
Accepted: 20 June 2022
Published: 23 June 2022

Publisher's Note: MDPI stays neutral with regard to jurisdictional claims in published maps and institutional affiliations.

Abstract: Satellite-based vegetation datasets enable vegetation phenology detection at large scales, among which Solar-Induced Chlorophyll Fluorescence (SIF) and Enhanced Vegetation Index (EVI) are widely used proxies for detecting phenology from photosynthesis and greenness perspectives, respectively. Recent studies have revealed the divergent performances of SIF and EVI for estimating different phenology metrics, i.e., the start of season (SOS) and the end of season (EOS); however, the underlying mechanisms are unclear. In this study, we compared the SOS and EOS of natural ecosystems derived from SIF and EVI in China and explored the underlying mechanisms by investigating the relationships between the differences of phenology derived from SIF and EVI and climatic limiting factors (i.e., temperature, water and radiation). The results showed that the differences between phenology generated using SIF and EVI were diverse in space, which had a close relationship with climatic limitations. The increasing climatic limitation index could result in larger differences in phenology from SIF and EVI for each dominant climate-limited area. The phenology extracted using SIF was more correlated with climatic limiting factors than that using EVI, especially in water-limited areas, making it the main cause of the difference in phenology from SIF and EVI. These findings highlight the impact of climatic limitation on the differences of phenology from SIF and EVI and improve our understanding of land surface phenology from greenness and photosynthesis perspectives.

Keywords: vegetation phenology; climatic limitation; solar-induced chlorophyll fluorescence; enhanced vegetation index

1. Introduction

Vegetation phenology is the study of the timing of recurring biological events of plants and their interactions among periodic changes in the natural environment [1]. It indicates the response and adaptation of vegetation ecosystems to seasonal and interannual environmental change [2,3]. Since the industrial revolution, climate change (e.g., global warming) induced by human activities has had a profound impact on vegetation phenology; at the same time, changes in vegetation phenology have been regarded as a sensitive indicator of climate change and the carbon cycle [4]. Information on vegetation phenology is playing an increasingly important role in global change monitoring, ecological environment simulation and climate change response [2,4].

Remote sensing provides a useful approach to characterizing seasonal and interannual changes in land surface vegetation from regional to global scales [5–7]. Land surface phenology is mainly extracted based on satellite vegetation datasets using the key phenological metrics, i.e., the start of season (SOS) and the end of season (EOS) [8], to characterize the timing of vegetation dynamics during the growing season. Vegetation indices (VIs), calculated from land surface reflectance, are widely used to extract land surface phenology and analyze its response to climate change in various studies from the leaf and canopy greenness perspective. For example, the global land surface phenology product MCD12Q2 was generated by Zhang et al. [6,8] using the EVI (Enhanced Vegetation Index) time series, which is the only global land surface phenology product available in recent years. However, as VIs are not capable of providing us with a direct proxy of physiological processes, they cannot be perfectly applied to modeling frameworks [9]. In this case, some studies have explored the potential of vegetation phenology extraction from an photosynthetic perspective. Solar-induced chlorophyll fluorescence (SIF), as a new physiological proxy for photosynthesis activity [10], presenting a weak signal emitted by green plants during photosynthesis [11–14]. Compared to traditional VIs, SIF provides a direct indicator for monitoring vegetation physiological functioning [15,16], and has a close relationship with carbon uptake of vegetation. Some studies have indicated that satellite-based SIF observations are highly correlated with in situ Gross Primary Productivity (GPP) over flux towers, and thus have the potential to reveal GPP dynamics under environmental changes over a large scale [15,17].

Some studies have reported that phenology derived from SIF and EVI were different across various vegetation types [18,19], such as coniferous forests, deciduous forests, grasslands and croplands [15–17,20,21]. For example, Wang et al. [14] revealed that EVI-based EOS could be later than SIF-based EOS for more than two weeks in grasslands in Australia, and such differences would be larger when plants are stressed with decreasing soil moisture. Moreover, for different phenological metrics, i.e., SOS and EOS, SIF and EVI also performed differently. For example, Walther et al. [15] indicated that the EVI-based SOS of boreal evergreen coniferous forest was much later (about a month) than the SIF-based SOS, but the EVI-based EOS was slightly advanced (about 1 to 2 weeks) to the SIF-based EOS. Although some studies have revealed differences in phenology derived from SIF and EVI among land cover types, the driving factors and underlying mechanisms are less known.

Except for croplands, which could be largely affected by human activities, the dynamics of land surface phenology are driven by the physical characteristics of the vegetation itself and the external climate environment [22]. The external climate factors that affect vegetation phenology mainly include temperature, precipitation and radiation, which interact to promote or limit natural vegetation growth [23,24]. For example, Ma et al. [25] revealed that 80% of EVI-based phenology dynamics in dryland ecosystems are driven by the variability of annual precipitation. In contrast, recent studies have indicated that SIF has quicker responses to external environmental stress information (e.g., water stress) than EVI did [26,27], as SIF contains additional information on stress conditions that reflects fluorescence efficiency [11]. However, climate controls on EVI-based phenology (greenness) and SIF-based phenology (photosynthesis) have not been compared, and a comprehensive analysis across different climatic conditions is still scarce.

In this study, we defined the climatic limiting controls on vegetation growth as climatic limitations, which include temperature-limiting, water-limiting, and radiation-limiting factors. We focus on naturally vegetated areas in China and divide them into climate-limited areas (i.e., temperature, water, and radiation limitations). We then employed SIF and EVI to extract phenology from photosynthesis and greenness perspectives, respectively, and compared their characteristics across climate-limited areas. We further explored the underlying mechanisms by investigating the relationships between the differences in phenology derived from SIF and EVI and climatic limiting factors. This work can provide insights into the mechanistic differences between SIF and EVI in characterizing land surface phenology to improve our understanding of vegetation dynamics from greenness and photosynthesis perspectives and their interactions with climate conditions.

2. Materials and Methods

2.1. Data Sources and Reprocessing

2.1.1. SIF Datasets

The GOSIF (Global OCO-2 SIF) is a reconstructed SIF product based on Orbiting Carbon Observatory-2 (OCO-2) observations, Moderate Resolution Imaging Spectroradiometer (MODIS) vegetation data, and meteorological reanalysis data. The GOSIF datasets from 2003 to 2016 were downloaded from http://globalecology.unh.edu/ (accessed on 10 February 2021), which were globally spatio-temporal continuous at 0.05° and 8-day resolution derived with a machine learning algorithm trained with OCO-2 SIF [28]. The datasets had a good performance validated by original SIF observations (RMSE = 0.07 W m^{-2} μm^{-1} sr^{-1}) and also showed a good correlation with the in-situ GPP over flux sites ($R^2 = 0.73$, $p < 0.001$) [28].

2.1.2. EVI Datasets

The MODIS Terra/Aqua Vegetation Indices (MOD13C1/MYD13C1, V006) were combined to generate EVI time series from 2003 to 2016 at 8-day interval and 0.05° spatial resolution, which were available at https://ladsweb.modaps.eosdis.nasa.gov/ (accessed on 15 March 2021). Global MOD13C1 and MYD13C1 are cloud-free spatial composites of MOD13A2 and MYD13A2 at 16-day intervals and 1 km spatial resolutions, respectively.

2.1.3. Land Cover Map

We utilized the global land cover product (GLC), freely available at http://data.ess.tsinghua.edu.cn/index.html (accessed on 22 September 2021) to map the natural vegetated areas and mask croplands that are vulnerable to human interference [29]. This product consists of 17 land cover types, among which the developed land types and non-vegetated land types were masked to generate natural vegetated areas. The accuracy for 2010, 2015 and 2020 are 86.39% ± 9.05%, 86.44% ± 8.99% and 84.83% ± 10.19%, respectively [29]. We aggregated the original land cover dataset from 2015 to 0.05° to match the spatial resolution of the SIF and EVI datasets in this study.

2.1.4. Meteorological Datasets

The reanalysis meteorological datasets from 2003 to 2016 were obtained from the ERA-Interim global reanalysis data (https://apps.ecmwf.int/datasets/data/interim-full-daily/levtype=sfc/, accessed on 13 December 2020). Here, we selected 2 m temperature, total precipitation and photosynthetically active radiation (PAR) from ERA-5. All of these variables were obtained from the surface-level fields of this reanalysis at 0.125° for a daily interval. The potential evapotranspiration (PET) was downloaded from the gridded Climatic Research Unit (CRU) datasets (https://catalogue.ceda.ac.uk/uuid/10d3e3640f004c578403419aac167d82, accessed on 27 January 2021). The original climate datasets from 2003 to 2016 were aggregated to a monthly time scale at a 0.05° spatial resolution to calculate climatic limitation indices.

2.2. Methods

2.2.1. Phenology Extraction

Due to cloud, atmosphere and snow contamination, we used Savitzky–Golay filtering to smooth the time series of GOSIF and MODIS EVI. Then, a double logistic function was fitted based on the smoothed time series to generate continuous curves. The double logistic is a flexible model for monitoring seasonal and inter-annual land surface dynamics based on satellite data, which has been widely used for various vegetation types at global or regional scales [30,31]. The double logistic function can be written as follows:

$$V(t) = V_{\min} + (V_{\max} - V_{\min}) \times (\frac{1}{1 + e^{-mS\times(t-S)}} + \frac{1}{1 + e^{mA\times(t-A)}} - 1) \tag{1}$$

where $V(t)$ is the value of vegetation proxies (i.e., GOSIF or MODIS EVI) at day of the year (DOY) t, $V_{\max}$ is the maximum vegetation proxies in the year, $V_{\min}$ is the minimum vegetation proxies in the year, mS and mA are the maximum slope of the curve in green up and in senescence, respectively, S and A are their corresponding DOYs. Finally, the SOS and EOS were estimated as follows [32]:

$$SOS = \frac{2\ln(\sqrt{3} - \sqrt{2})}{mS} + S \tag{2}$$

$$EOS = \frac{2\ln(\sqrt{3} - \sqrt{2})}{mA} + A \tag{3}$$

2.2.2. Determination of Climate-Limited Area

We used long-term monthly average climate data to develop scaling factors (0–1) (refer to the climatic limitation index) [24]. The temperature limitation index, radiation limitation index and water limitation index were calculated using the criteria proposed by Nemani et al. [24] as follows:

$$iT = \begin{cases} 1 - \frac{T_{\min} - TM_{\min}}{TM_{\max} - TM_{\min}}, TM_{\min} < T_{\min} < TM_{\max} \\ 1, T_{\min} < TM_{\min} \\ 0, T_{\min} > TM_{\max} \end{cases} \tag{4}$$

where iT is the temperature limitation index, $T_{\min}$ is the daily minimum temperature, $TM_{\min}$ and $TM_{\max}$ are the thresholds of the daily minimum temperature, which were set as -5 °C and 5 °C in this study, respectively.

$$iR = \begin{cases} 1 - \frac{R_{mean} - RM_{\min}}{RM_{\max} - RM_{\min}}, RM_{\min} < R_{mean} < RM_{\max} \\ 1, R_{mean} < RM_{\min} \\ 0, R_{mean} > RM_{\max} \end{cases} \tag{5}$$

where iR is the radiation limitation index, R_{mean} is the daily mean PAR, $RM_{\min}$ and $RM_{\max}$ are the thresholds of the daily mean PAR, which were set as 75 W and 150 W, respectively. In addition, we used the ratio of precipitation to potential evapotranspiration (P/PET) as an indicator of water-limited conditions, as below:

$$iW = \begin{cases} 1 - \frac{P}{0.75*PET}, \frac{P}{PET} < 0.75 \\ 0, \frac{P}{PET} \geq 0.75 \end{cases} \tag{6}$$

The spatial patterns of the three climatic limitation indices are shown in Figure 1a. For classification, we define the pixels as the dominant temperature-limited area if: (1) iT is higher than iR and iW, and (2) iT is larger than 0.25. Radiation-limited areas and water-limited areas were determined by the same criteria. We determined the pixels as having no climatic limitation where iT, iR and iW are all lower than 0.25 (Figure 1b).

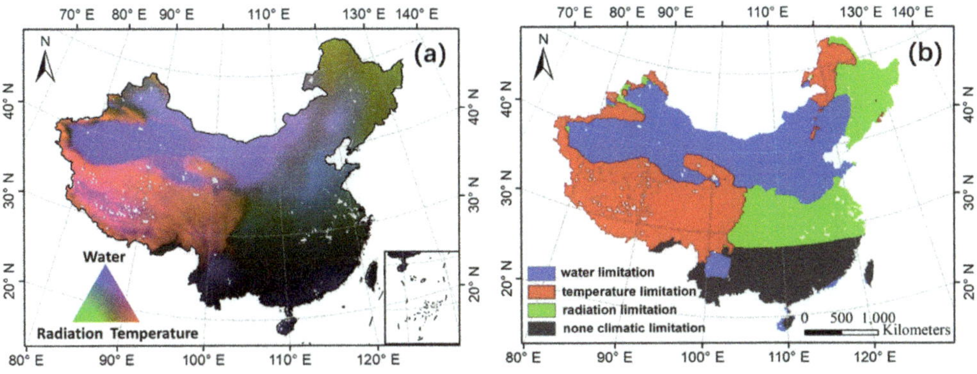

Figure 1. Spatial pattern of climatic limitations in China (**a**) and dominant climatic limitations (**b**).

2.2.3. Relationship of Phenology Derived from SIF and EVI and Climatic Limitations

We randomly sampled 5000 pixels for each climate-limited area and adopted linear correlation regression analysis to explore the relationship between phenology generated using SIF and EVI and the dominant climatic limiting factors. Furthermore, we adopted the *C*-index proposed by Garonna et al. [33] to quantify the relative contributions of phenology derived from SIF and EVI to their differences with climatic limitation indices [34], which were calculated as follows:

$$C = \frac{|S_{\rm SIF}| - |S_{\rm EVI}|}{|S_{\rm SIF}| + |S_{\rm EVI}|} \quad (7)$$

where $S_{\rm SIF}$ or $S_{\rm EVI}$ is the gradient (i.e., slope) of linear regression relationships between SOS/EOS generated using SIF or EVI and climatic limitation indices. As the *C*-index is unitless, ranging from −1 to 1, the contribution ratio based on the *C*-index (*Cr*) can be calculated as Equation (8). If the *Cr* of phenology from SIF or EVI is larger than 50%, this means that this factor is mostly attributable to the difference of phenology between SIF and EVI under climatic limitations.

$$C_r = \frac{1 + C}{2} \times 100\% \quad (8)$$

3. Results

In general, the multi-year average phenology of natural vegetation from SIF and EVI has consistent spatial patterns in China, with a delaying pattern in SOS and an advanced pattern in EOS from southeast to northwest (Figure 2a–d), which is consistent with previous studies reported by Wang et al. [35]. Furthermore, we found substantial differences between the SIF and the EVI in the derived phenological metrics (Figure 2e,f). Specifically, the SOS derived from SIF is generally later than that from EVI, which accounts for 70% of the total natural vegetated area, except for those areas of evergreen forest in the south. The EOS from SIF is generally earlier than that from EVI, accounting for 87% of the total natural vegetated area.

Then, the differences in phenology derived from SIF and EVI were presented statistically across different climate-limited areas (Figure 3). We found that the differences in SOS generated using SIF and EVI (here denoted as ΔSOS) mainly ranged from 0 to 20 days (Figure 3a), while the differences in EOS generated using SIF and EVI (here denoted asΔEOS) mainly distributed between −30 and −10 days (Figure 3b). This indicates that the difference in EOS generated using SIF and EVI is generally larger than that generated using SOS. For both SOS and EOS, the largest difference of phenology from SIF and EVI (i.e., ΔSOS > 30 days or ΔEOS < −30 days) occurred in temperature-limited areas and water-limited areas, while the difference of phenology from SIF and EVI at no climatic limitations was the smallest.

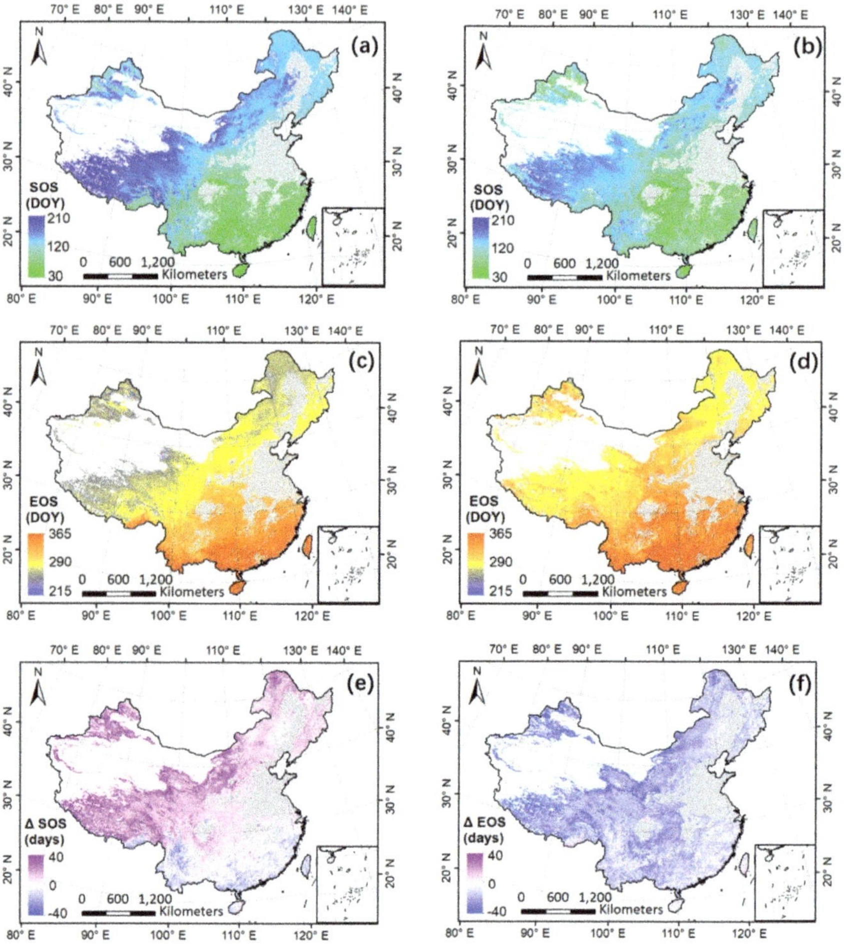

Figure 2. The spatial pattern of multi-year average phenology of natural vegetation from SIF (**a**,**c**) and EVI (**b**,**d**) and their differences, i.e., SOS/EOS derived from SIF subtracted by that derived from EVI (**e**,**f**).

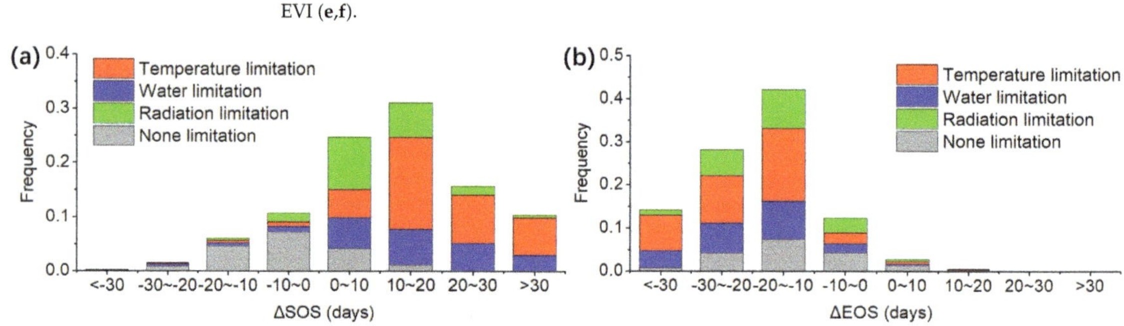

Figure 3. The differences in SOS (**a**) or EOS (**b**) derived from SIF and EVI among different climatic limitation areas.

As shown in Figure 4, we found that increasing the climatic limitation index could result in larger differences in SOS and EOS from SIF and EVI for each dominant climatic limitation area. This finding could explain why the difference in phenology derived from SIF and EVI with no climatic limitations is the smallest in Figure 3. In terms of different climatic limitations, we found that the slope of linear regression of ΔSOS or ΔEOS to water-limitation index was the highest, followed by temperature-limitation index, and then radiation-limitation index, which might be a consequence of divergent responses of SIF and EVI to different climatic or environmental constraints. In addition, the climatic limitation indices were more correlated with ΔSOS than with ΔEOS. A likely cause is that autumn phenology is more complex than spring phenology, as it may be affected by multiple climatic factors, thus weakening the relationship between one dominant climatic limitation index and ΔEOS.

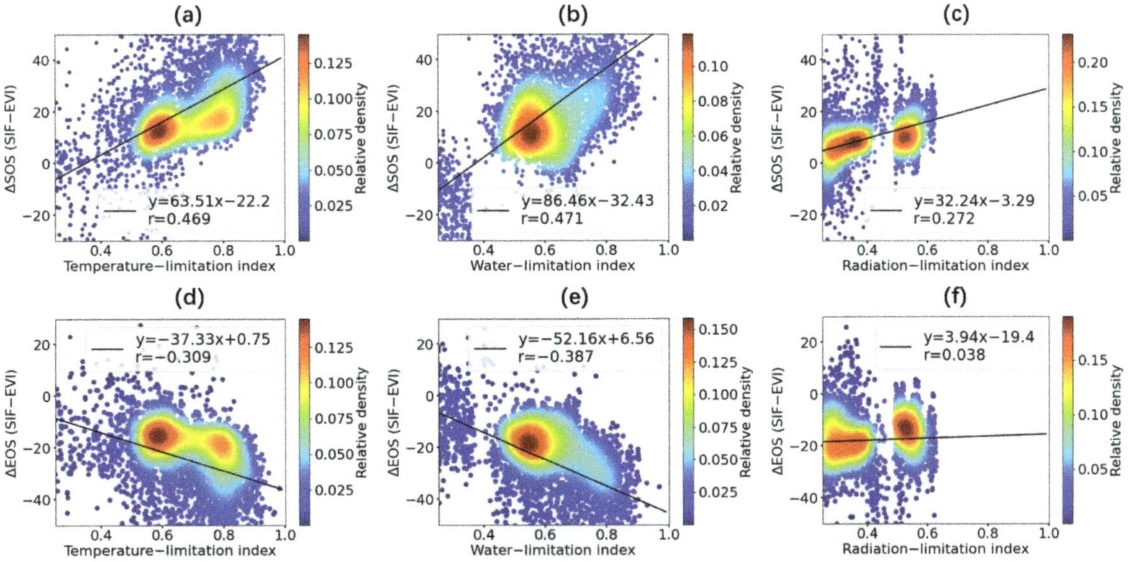

Figure 4. Scatterplot of differences of SOS (**first row**) and EOS (**second row**) derived from SIF and EVI versus climatic limitation index at temperature-dominant area (**a**,**d**), water-dominant area (**b**,**e**) and radiation-dominant area (**c**,**f**). The black line shows the linear regression between phenology from SIF or EVI and climatic limitation indices; r is the correlation coefficient of the linear regression.

Table 1 quantified the relative contributions of phenology from SIF and EVI to ΔSOS and ΔEOS under the dominant climatic limitation areas. We found that the phenology extracted using SIF was more correlated with temperature, water and radiation limiting factors than that using EVI, making it the main cause of the difference of phenology from SIF and EVI. Especially in water-limited areas, the contributions of phenology from SIF to ΔSOS or ΔEOS is much larger than those from EVI (SOS: 90.00% vs. 10.00%, EOS: 80.00% vs. 20.00%), as phenology derived from EVI had a low correlation with the water limitation index. These different responses of SIF and EVI to the water limitation index attributed to the differences in phenology from SIF and EVI in the water-limited area, which was also shown in Figure 4.

Table 1. The relationships between phenology from SIF or EVI and climatic limitation indices. s: slope of the linear regression between phenology from SIF or EVI and climatic limitation indices. r: correlation coefficient of the linear regression. Cr: contribution ratio of phenology from SIF or EVI to the difference of phenology derived from SIF and EVI under climatic limitations.

Parameters	SOS						EOS					
	SIF			EVI			SIF			EVI		
	s	r	Cr	s	r	Cr	s	r	Cr	s	r	Cr
Temperature-limitation	150.23	0.73	62.00%	92.55	0.62	38.00%	−67.31	−0.57	68.50%	−31.07	−0.26	31.50%
Water-limitation	97.10	0.55	90.00%	10.62	0.07	10.00%	−70.82	−0.54	80.00%	−17.52	0.14	20.00%
Radiation-limitation	197.98	0.87	53.65%	166.89	0.80	46.35%	−183.13	−0.90	49.35%	−191.74	−0.88	50.65%

4. Discussion

In this study, we analyzed the characteristics of phenology derived from SIF and EVI for natural vegetated areas in China and found substantial differences between SOS/EOS generated using SIF and EVI. Specifically, the SOS derived from SIF was generally later than that derived from EVI, which was the case in 70% of the total natural vegetated area in China. We found this occurred in climatic limiting areas, where deciduous forests, mixed forests and grasslands were mainly distributed. Those vegetation types initiate photosynthesis after green leaves emerge in spring [17,20]; thus, photosynthesis phenology tends to be later than greenness phenology for SOS, which explains our results. In those areas covered by evergreen forests in the south with no distinct climatic limitations, the SOS derived from SIF was slightly earlier than that from EVI. A higher PAR supply in humid areas would stimulate photosynthesis more quickly, leading to photosynthesis starting earlier than greenness in spring for evergreen forests [15,19,36]. The EOS from SIF was generally earlier than that from EVI, which is consistent with previous studies [15,17,19], implying seasonal hysteresis of EVI in response to photoperiod changes in the period of senescence [15,17].

Furthermore, we revealed that the differences between phenology generated using SIF and EVI were diverse in SOS and EOS. We found that the difference in EOS generated using SIF and EVI was generally larger than that generated using SOS. Possible reasons include the following: (a) The autumn phenology extracted from satellite VIs had higher uncertainty (and perhaps bias) relative to spring phenology [37]. For example, Lu et al. [20] presented that EVI could hardly predict the autumn phenology of deciduous forests accurately with an overall R^2 less than 0.3, while the R^2 of spring phenology was generally higher than 0.7. (b) Seasonal decoupling of physiological status and greenness information occurred in autumn. Specifically, SOS derived from SIF and EVI occur relatively synchronously, but they become increasingly asynchronous as the growing season progresses [38], leading to larger differences in EOS generated using SIF and EVI than that in SOS.

We further inferred that the differences between SIF-based phenology and EVI-based phenology in space have a close relationship with their different responses to climatic limitations. In contrast to information about green biomass proxied by EVI, SIF contains information on the absorbed photosynthetically active radiation by vegetation (APAR) and environmental stresses (especially water stress) related to photosynthetic light-use efficiency (LUE) [10]. Therefore, SIF is more sensitive to climate variability than EVI [39,40]. This is consistent with our finding that phenology from SIF was more correlated with climatic limitations than that from EVI, making it the main cause of the difference between phenology generated using SIF and EVI. Under these divergent responses to climatic limitations, the differences in SOS and EOS from SIF and EVI become larger, along with a higher climatic limitation index. However, in the radiation-limited area, a higher radiation limitation index did not contribute to a larger difference in EOS derived from SIF and EVI. This happened as EOS derived from SIF and EVI had similar regression slopes with the radiation limitation index, suggesting that autumn phenology is more radiation-limited than spring phenology from both greenness and photosynthesis perspectives [41]. In addition, the radiation-limiting area in this study was distributed in northern China, where snow cover existed in autumn and winter, which may introduce the undesired errors of EOS extracting from reflectance-based EVI [42].

Although the GOSIF product was generated using remote sensing data from the MODIS and meteorological reanalysis data as inputs to the predictive SIF model, which may increase the correlation of SIF and climatic factors in a time series, this correlation from data sources will be offset in the spatial statistics adopted in this study. In addition, we employed SIF and EVI to extract phenology from photosynthesis and greenness perspectives, respectively. Other proxies, such as Chlorophyll/Carotenoid Index [43], Normalized Difference Vegetation Index, can be further analyzed in future studies to investigate the unique characteristics of each proxy on remote sensing derived phenology. The relationship between vegetation phenology and multiple climatic factors instead of one dominant climatic limitation index needs to be analyzed further to explore whether and how the impacts of climatic interactions on vegetation dynamics. Moreover, we focused on natural vegetated areas in China as a target, as it provides a natural laboratory with a wide variation of ecosystems and climate types. Further research could be expanded to the hemisphere or global scale to evaluate our findings.

5. Conclusions

This study revealed a substantial difference between phenology extracted using satellite-derived SIF and EVI data across areas limited by different climatic factors (temperature, radiance, water). We inferred that the differences between SIF-based phenology and EVI-based phenology have a close relationship with their different responses to climate limitations. The higher climatic limitation index could result in larger differences in phenology derived from SIF and EVI for each dominant climatic limitation area. The phenology extracted using SIF was more correlated with climatic limitations than that using EVI, especially in water-limited areas, making it the main cause of the difference between phenology extracted using SIF and EVI. These findings improved our understanding of land surface phenology from greenness and photosynthesis perspectives and provided insight into the mechanistic differences between SIF and EVI in characterizing land surface phenology.

Author Contributions: Conceptualization, Q.H.; methodology, C.W. and S.L.; validation, Y.W. and Q.X.; resources, J.H. and Y.C.; writing—original draft preparation, C.W.; writing—review and editing, Q.H., Y.W., S.L., Q.X., J.H. and Y.C. All authors have read and agreed to the published version of the manuscript.

Funding: This research was supported by the National Natural Science Foundation of China (42101391, 41901380), the Young Elite Scientists Sponsorship Program by CAST (2020QNRC001), the "Soil and Water Conservation Monitoring Research in Hubei Province" project (2017052601) funded by Hubei Anyuan Safety and Environmental Protection Technology Co., Ltd. (Wuhan, China), the Fundamental Research Funds for the Central Universities (CCNU21XJ028, CCNU20QN032), and the fellowship of the China Postdoctoral Science Foundation (2021M701361). Qiaoyun Xie acknowledges support from the University of Technology Sydney Chancellor's Postdoctoral Research Fellowship.

Institutional Review Board Statement: Not applicable.

Informed Consent Statement: Not applicable.

Data Availability Statement: The data used to support the findings of this study are included within the article.

Conflicts of Interest: The authors declare no conflict of interest.

References

1. Helmut, L. Phenology and Seasonality Modeling. *Soil Sci.* **1975**, *120*, 461.
2. White, M.A.; Thornton, P.E.; Running, S.W. A continental phenology model for monitoring vegetation responses to interannual climatic variability. *Glob. Biogeochem. Cycle* **1997**, *11*, 217–234. [CrossRef]

3. Xie, Q.; Huete, A.; Hall, C.C.; Medlyn, B.E.; Power, S.A.; Davies, J.M.; Medek, D.E.; Beggs, P.J. Satellite-observed shifts in C3/C4 abundance in Australian grasslands are associated with rainfall patterns. *Remote Sens. Environ.* **2022**, *273*, 112983. [CrossRef]
4. Richardson, A.D.; Keenan, T.F.; Migliavacca, M.; Ryu, Y.; Sonnentag, O.; Toomey, M. Climate change, phenology, and phenological control of vegetation feedbacks to the climate system. *Agric. For. Meteorol.* **2013**, *169*, 156–173. [CrossRef]
5. Melaas, E.K.; Richardson, A.D.; Friedl, M.A.; Dragoni, D.; Gough, C.M.; Herbst, M.; Montagnani, L.; Moors, E. Using FLUXNET data to improve models of springtime vegetation activity onset in forest ecosystems. *Agric. For. Meteorol.* **2013**, *171*, 46–56. [CrossRef]
6. Zhang, X.; Friedl, M.A.; Schaaf, C.B. Global vegetation phenology from Moderate Resolution Imaging Spectroradiometer (MODIS): Evaluation of global patterns and comparison with in situ measurements. *J. Geophys. Res. Biogeosci.* **2006**, *111*, 367–375. [CrossRef]
7. De Jong, R.; de Bruin, S.; de Wit, A.; Schaepman, M.E.; Dent, D.L. Analysis of monotonic greening and browning trends from global NDVI time-series. *Remote Sens. Environ.* **2011**, *115*, 692–702. [CrossRef]
8. Zhang, X.; Friedl, M.A.; Schaaf, C.B.; Strahler, A.H.; Hodges, J.C.; Gao, F.; Reed, B.C.; Huete, A. Monitoring vegetation phenology using MODIS. *Remote Sens. Environ.* **2003**, *84*, 471–475. [CrossRef]
9. Yin, G.; Verger, A.; Filella, I.; Descals, A.; Peñuelas, J. Divergent estimates of forest photosynthetic phenology using structural and physiological vegetation indices. *Geophys. Res. Lett.* **2020**, *47*, e2020GL089167. [CrossRef]
10. Porcar-Castell, A.; Tyystjärvi, E.; Atherton, J.; Van der Tol, C.; Flexas, J.; Pfündel, E.E.; Moreno, J.; Frankenberg, C.; Berry, J.A. Linking chlorophyll a fluorescence to photosynthesis for remote sensing applications: Mechanisms and challenges. *J. Exp. Bot.* **2014**, *65*, 4065–4095. [CrossRef]
11. Smith, W.; Biederman, J.; Scott, R.L.; Moore, D.; He, M.; Kimball, J.; Yan, D.; Hudson, A.; Barnes, M.; MacBean, N. Chlorophyll fluorescence better captures seasonal and interannual gross primary productivity dynamics across dryland ecosystems of southwestern North America. *Geophys. Res. Lett.* **2018**, *45*, 748–757. [CrossRef]
12. Sun, Y.; Frankenberg, C.; Wood, J.D.; Schimel, D.; Jung, M.; Guanter, L.; Drewry, D.; Verma, M.; Porcar-Castell, A.; Griffis, T.J. OCO-2 advances photosynthesis observation from space via solar-induced chlorophyll fluorescence. *Science* **2017**, *358*, eaam5747. [CrossRef] [PubMed]
13. Guanter, L.; Frankenberg, C.; Dudhia, A.; Lewis, P.E.; Gómez-Dans, J.; Kuze, A.; Suto, H.; Grainger, R.G. Retrieval and global assessment of terrestrial chlorophyll fluorescence from GOSAT space measurements. *Remote Sens. Environ.* **2012**, *121*, 236–251. [CrossRef]
14. Wang, C.; Beringer, J.; Hutley, L.B.; Cleverly, J.; Li, J.; Liu, Q.; Sun, Y. Phenology dynamics of dryland ecosystems along the North Australian tropical transect revealed by satellite solar-induced chlorophyll fluorescence. *Geophys. Res. Lett.* **2019**, *46*, 5294–5302. [CrossRef]
15. Walther, S.; Voigt, M.; Thum, T.; Gonsamo, A.; Zhang, Y.; Köhler, P.; Jung, M.; Varlagin, A.; Guanter, L. Satellite chlorophyll fluorescence measurements reveal large-scale decoupling of photosynthesis and greenness dynamics in boreal evergreen forests. *Glob. Chang. Biol.* **2016**, *22*, 2979–2996. [CrossRef]
16. Joiner, J.; Yoshida, Y.; Vasilkov, A.; Schaefer, K.; Jung, M.; Guanter, L.; Zhang, Y.; Garrity, S.; Middleton, E.; Huemmrich, K. The seasonal cycle of satellite chlorophyll fluorescence observations and its relationship to vegetation phenology and ecosystem atmosphere carbon exchange. *Remote Sens. Environ.* **2014**, *152*, 375–391. [CrossRef]
17. Jeong, S.-J.; Schimel, D.; Frankenberg, C.; Drewry, D.T.; Fisher, J.B.; Verma, M.; Berry, J.A.; Lee, J.-E.; Joiner, J. Application of satellite solar-induced chlorophyll fluorescence to understanding large-scale variations in vegetation phenology and function over northern high latitude forests. *Remote Sens. Environ.* **2017**, *190*, 178–187. [CrossRef]
18. Li, X.; Xiao, J.; He, B.; Altaf Arain, M.; Beringer, J.; Desai, A.R.; Emmel, C.; Hollinger, D.Y.; Krasnova, A.; Mammarella, I. Solar-induced chlorophyll fluorescence is strongly correlated with terrestrial photosynthesis for a wide variety of biomes: First global analysis based on OCO-2 and flux tower observations. *Glob. Chang. Biol.* **2018**, *24*, 3990–4008. [CrossRef]
19. Chang, Q.; Xiao, X.; Jiao, W.; Wu, X.; Doughty, R.; Wang, J.; Du, L.; Zou, Z.; Qin, Y. Assessing consistency of spring phenology of snow-covered forests as estimated by vegetation indices, gross primary production, and solar-induced chlorophyll fluorescence. *Agric. For. Meteorol.* **2019**, *275*, 305–316. [CrossRef]
20. Lu, X.; Liu, Z.; Zhou, Y.; Liu, Y.; An, S.; Tang, J. Comparison of phenology estimated from reflectance-based indices and solar-induced chlorophyll fluorescence (SIF) observations in a temperate forest using GPP-based phenology as the standard. *Remote Sens.* **2018**, *10*, 932. [CrossRef]
21. Urban, D.; Guan, K.; Jain, M. Estimating sowing dates from satellite data over the US Midwest: A comparison of multiple sensors and metrics. *Remote Sens. Environ.* **2018**, *211*, 400–412. [CrossRef]
22. Zu, J.; Zhang, Y.; Huang, K.; Liu, Y.; Chen, N.; Cong, N. Biological and climate factors co-regulated spatial-temporal dynamics of vegetation autumn phenology on the Tibetan Plateau. *Int. J. Appl. Earth Obs. Geoinf.* **2018**, *69*, 198–205. [CrossRef]
23. Bonan, G.B. Forests and climate change: Forcings, feedbacks, and the climate benefits of forests. *Science.* **2008**, *320*, 1444–1449. [CrossRef] [PubMed]
24. Nemani, R.R.; Keeling, C.D.; Hashimoto, H.; Jolly, W.M.; Piper, S.C.; Tucker, C.J.; Myneni, R.B.; Running, S.W. Climate-driven increases in global terrestrial net primary production from 1982 to 1999. *Science* **2003**, *300*, 1560–1563. [CrossRef] [PubMed]

25. Ma, X.; Huete, A.; Yu, Q.; Coupe, N.R.; Davies, K.; Broich, M.; Ratana, P.; Beringer, J.; Hutley, L.B.; Cleverly, J. Spatial patterns and temporal dynamics in savanna vegetation phenology across the North Australian Tropical Transect. *Remote Sens. Environ.* **2013**, *139*, 97–115. [CrossRef]
26. Sanders, A.F.; Verstraeten, W.W.; Kooreman, M.L.; Van Leth, T.C.; Beringer, J.; Joiner, J. Spaceborne sun-induced vegetation fluorescence time series from 2007 to 2015 evaluated with Australian flux tower measurements. *Remote Sens.* **2016**, *8*, 895. [CrossRef]
27. Yoshida, Y.; Joiner, J.; Tucker, C.; Berry, J.; Lee, J.-E.; Walker, G.; Reichle, R.; Koster, R.; Lyapustin, A.; Wang, Y. The 2010 Russian drought impact on satellite measurements of solar-induced chlorophyll fluorescence: Insights from modeling and comparisons with parameters derived from satellite reflectances. *Remote Sens. Environ.* **2015**, *166*, 163–177. [CrossRef]
28. Li, X.; Xiao, J. A global, 0.05-degree product of solar-induced chlorophyll fluorescence derived from OCO-2, MODIS, and reanalysis data. *Remote Sens.* **2019**, *11*, 517. [CrossRef]
29. Zhao, J.; Yu, L.; Liu, H.; Huang, H.; Wang, J.; Gong, P. Towards an open and synergistic framework for mapping global land cover. *PeerJ* **2021**, *9*, e11877. [CrossRef]
30. Beck, P.S.; Atzberger, C.; Høgda, K.A.; Johansen, B.; Skidmore, A.K. Improved monitoring of vegetation dynamics at very high latitudes: A new method using MODIS NDVI. *Remote Sens. Environ.* **2006**, *100*, 321–334. [CrossRef]
31. Fisher, J.I.; Mustard, J.F.; Vadeboncoeur, M.A. Green leaf phenology at Landsat resolution: Scaling from the field to the satellite. *Remote Sens. Environ.* **2006**, *100*, 265–279. [CrossRef]
32. Ahl, D.E.; Gower, S.T.; Burrows, S.N.; Shabanov, N.V.; Myneni, R.B.; Knyazikhin, Y. Monitoring spring canopy phenology of a deciduous broadleaf forest using MODIS. *Remote Sens. Environ.* **2006**, *104*, 88–95. [CrossRef]
33. Garonna, I.; De Jong, R.; De Wit, A.J.; Mücher, C.A.; Schmid, B.; Schaepman, M.E. Strong contribution of autumn phenology to changes in satellite-derived growing season length estimates across Europe (1982–2011). *Glob. Chang. Biol.* **2014**, *20*, 3457–3470. [CrossRef] [PubMed]
34. Sun, Q.; Li, B.; Zhou, G.; Jiang, Y.; Yuan, Y. Delayed autumn leaf senescence date prolongs the growing season length of herbaceous plants on the Qinghai–Tibetan Plateau. *Agric. For. Meteorol.* **2020**, *284*, 107896. [CrossRef]
35. Wang, C.; Li, J.; Liu, Q.; Zhong, B.; Wu, S.; Xia, C. Analysis of differences in phenology extracted from the enhanced vegetation index and the leaf area index. *Sensors* **2017**, *17*, 1982. [CrossRef]
36. Zhou, L.; Chi, Y.; Liu, X.; Dai, X.; Yang, F. Land surface phenology tracked by remotely sensed sun-induced chlorophyll fluorescence in subtropical evergreen coniferous forests. *Acta Ecol. Sin.* **2020**, *40*, 4114–4125.
37. Ganguly, S.; Friedl, M.A.; Tan, B.; Zhang, X.; Verma, M. Land surface phenology from MODIS: Characterization of the Collection 5 global land cover dynamics product. *Remote Sens. Environ.* **2010**, *114*, 1805–1816. [CrossRef]
38. Wang, X.; Dannenberg, M.P.; Yan, D.; Jones, M.O.; Kimball, J.S.; Moore, D.J.; van Leeuwen, W.J.; Didan, K.; Smith, W.K. Globally consistent patterns of asynchrony in vegetation phenology derived from optical, microwave, and fluorescence satellite data. *J. Geophys. Res. Biogeosci.* **2020**, *125*, e2020JG005732. [CrossRef]
39. Li, X.; Xiao, J. Global climatic controls on interannual variability of ecosystem productivity: Similarities and differences inferred from solar-induced chlorophyll fluorescence and enhanced vegetation index. *Agric. For. Meteorol.* **2020**, *288*, 108018. [CrossRef]
40. Yang, Y.; Chen, R.; Yin, G.; Wang, C.; Liu, G.; Verger, A.; Descals, A.; Filella, I.; Peñuelas, J. Divergent performances of vegetation indices in extracting photosynthetic phenology for northern deciduous broadleaf forests. *IEEE Geosci. Remote Sens. Lett.* **2022**, *19*, 1–5. [CrossRef]
41. Zhang, Y.; Commane, R.; Zhou, S.; Williams, A.P.; Gentine, P. Light limitation regulates the response of autumn terrestrial carbon uptake to warming. *Nat. Clim. Chang.* **2020**, *10*, 739–743. [CrossRef]
42. Qiu, B.; Li, W.; Wang, X.; Shang, L.; Song, C.; Guo, W.; Zhang, Y. Satellite-observed solar-induced chlorophyll fluorescence reveals higher sensitivity of alpine ecosystems to snow cover on the Tibetan Plateau. *Agric. For. Meteorol.* **2019**, *271*, 126–134. [CrossRef]
43. Gamon, J.A.; Huemmrich, K.F.; Wong, C.Y.; Ensminger, I.; Garrity, S.; Hollinger, D.Y.; Noormets, A.; Peñuelas, J. A remotely sensed pigment index reveals photosynthetic phenology in evergreen conifers. *Proc. Natl. Acad. Sci. USA* **2016**, *113*, 13087–13092. [CrossRef] [PubMed]

remote sensing

Article

WUE and CO_2 Estimations by Eddy Covariance and Remote Sensing in Different Tropical Biomes

Gabriel B. Costa [1,2,3,*], Cláudio M. Santos e Silva [2,4], Keila R. Mendes [4], José G. M. dos Santos [5], Theomar T. A. T. Neves [6], Alex S. Silva [6], Thiago R. Rodrigues [7], Jonh B. Silva [8], Higo J. Dalmagro [9], Pedro R. Mutti [2,4], Hildo G. G. C. Nunes [10], Lucas V. Peres [6], Raoni A. S. Santana [6], Losany B. Viana [6], Gabriele V. Almeida [6], Bergson G. Bezerra [2,4], Thiago V. Marques [11], Rosaria R. Ferreira [4], Cristiano P. Oliveira [2,4], Weber A. Gonçalves [2,4], Suany Campos [2] and Maria U. G. Andrade [2]

1 Biosciences Post-Graduate Program (PPG-BIO), Federal University of Western Pará (UFOPA), Santarém 68035-110, Brazil

2 Climate Sciences Post-Graduate Program (PPGCC), Federal University of Rio Grande do Norte (UFRN), Avenue Senador Salgado Filho, 3000, Lagoa Nova, Natal 59078-970, Brazil; claudio.silva@ufrn.br (C.M.S.e.S.); pedro.mutti@ufrn.br (P.R.M.); bergson.bezerra@ufrn.br (B.G.B.); cristiano.prestrelo@ufrn.br (C.P.O.); weber.goncalves@ufrn.br (W.A.G.); suany.csilva@escola.seduc.pa.gov.br (S.C.); uilhiana.andrade.016@ufrn.edu.br (M.U.G.A.)

3 Anthropic Studies in the Amazon Post-Graduate Program (PPGEAA), Federal University of Pará, Castanhal 68740-222, Brazil

4 Departamento de Ciências Atmosféricas e Climáticas (DCAC), Universidade Federal do Rio Grande do Norte, Natal 59078-970, Brazil; keila.mendes@icmbio.gov.br (K.R.M.); rosaria@ufrn.edu.br (R.R.F.)

5 Wildfire Monitoring Program, National Institute for Space Research, São José dos Campos 12227-010, Brazil; guilherme.martins@inpe.br

6 Institute of Engineering and Geosciences, Federal University of Western Pará (UFOPA), Santarém 68035-110, Brazil; theomar.neves@ufopa.edu.br (T.T.A.T.N.); alex.ss@ufopa.edu.br (A.S.S.); lucas.peres@ufopa.edu.br (L.V.P.); raoni.santana@ufopa.edu.br (R.A.S.S.); losany.viana@discente.ufopa.edu.br (L.B.V.); gabriele.almeida@discente.ufopa.edu.br (G.V.A.)

7 Laboratório de Ciências Atmosféricas, Universidade Federal de Mato Grosso Do Sul, Campo Grande 79070-900, Brazil; thiago.r.rodrigues@ufms.br

8 Programa de Pós-Graduação em Física Ambiental, Instituto de Física, Universidade Federal de Mato Grosso, Cuiabá 78060-900, Brazil; jonh_billy@fisica.ufmt.br

9 Programa de Pós-Graduação em Ciências Ambientais, Universidade de Cuiabá (UNIC), Cuiabá 78060-900, Brazil; higo.dalmagro@educadores.net.br

10 Socio-Environmental and Water Resources, Federal Rural University of the Amazon, Belém 66077-830, Brazil; garibalde13@gmail.com

11 Federal Institute of Education, Science and Technology of Rio Grande do Norte, Natal 59628-330, Brazil; thiago.valentim@ifrn.edu.br

* Correspondence: gabriel.costa@ufopa.edu.br

Citation: Costa, G.B.; Santos e Silva, C.M.; Mendes, K.R.; dos Santos, J.G.M.; Neves, T.T.A.T.; Silva, A.S.; Rodrigues, T.R.; Silva, J.B.; Dalmagro, H.J.; Mutti, P.R.; et al. WUE and CO_2 Estimations by Eddy Covariance and Remote Sensing in Different Tropical Biomes. *Remote Sens.* **2022**, *14*, 3241. https://doi.org/10.3390/rs14143241

Academic Editors: Xiaolin Zhu, Xuanlong Ma, Jiaxin Jin, Yuke Zhou and Qiaoyun Xie

Received: 30 May 2022
Accepted: 23 June 2022
Published: 6 July 2022

Abstract: The analysis of gross primary production (GPP) is crucial to better understand CO_2 exchanges between terrestrial ecosystems and the atmosphere, while the quantification of water-use efficiency (WUE) allows for the estimation of the compensation between carbon gained and water lost by the ecosystem. Understanding these dynamics is essential to better comprehend the responses of environments to ongoing climatic changes. The objective of the present study was to analyze, through AMERIFLUX and LBA network measurements, the variability of GPP and WUE in four distinct tropical biomes in Brazil: Pantanal, Amazonia, Caatinga and Cerrado (savanna). Furthermore, data measured by eddy covariance systems were used to assess remotely sensed GPP products (MOD17). We found a distinct seasonality of meteorological variables and energy fluxes with different latent heat controls regarding available energy in each site. Remotely sensed GPP was satisfactorily related with observed data, despite weak correlations in interannual estimates and consistent overestimations and underestimations during certain months. WUE was strongly dependent on water availability, with values of 0.95 gC kg^{-1} H_2O (5.79 gC kg^{-1} H_2O) in the wetter (drier) sites. These values reveal new thresholds that had not been previously reported in the literature. Our findings have crucial implications for ecosystem management and the design of climate policies regarding the conservation of tropical biomes, since WUE is expected to change in the ongoing climate change scenario th. indicates an increase in frequency and severity of dry periods.

Keywords: gross primary production; evapotranspiration; water use efficiency

1. Introduction

Discussions on climate change have become increasingly more relevant in the general scientific community, particularly since the creation of the Intergovernmental Panel on Climate Change (IPCC), which is composed of a diverse group of worldwide researchers, focusing on climate change studies and its impacts on society. Since the industrial revolution, the concentration of carbon dioxide (CO_2) in the atmosphere has increased with the use of fossil fuels, deforestation, the use of nitrogen in agriculture and livestock farming, which are reported to be the main uses responsible for the anthropic greenhouse effect. Studies have already shown the relevance of biosphere–atmosphere interactions in Brazilian biomes regarding planetary climate regulation due, for example, to water, energy, and carbon exchanges with the atmosphere [1–6]. However, there are still uncertainties regarding these processes due to the remarkable diversity of physiognomies, landscapes and other biophysical aspects that might play a role in differentiating atmospheric patterns from one place to another within each biome.

To reduce these uncertainties, in situ measurements are needed to better understand the particularities of each environment. Furthermore, these observed data can also be used to assess soil–vegetation–atmosphere models [7–10] and to analyze satellite-derived estimates of water and/or CO_2 balance components [11–16]. Both these models and remote sensing data are extremely important in providing reliable information on CO_2 exchanges over tropical forests where flux tower coverage is scarce or non-existent, such as in many parts of Brazil.

Additionally, certain forest physiognomies are not endemic to Brazil, but occur in several other regions of the globe, and therefore, their particularities regarding biophysical patterns need to be understood in detail. Wetlands in tropical rainforests such as the Amazon are environments where organic production rates are high and anoxic conditions are frequent, and therefore, they represent crucial zones for the global balance of greenhouse gases in the atmosphere. Cerrado ecosystems (savannas), however, are located at the tropics and subtropics and are characterized by marked wet and dry seasons. They cover approximately 60%, 50% and 45% of the total area of Africa, Australia and South America, respectively [17]. These ecosystems play an important role in the cycle of several greenhouse gasses, as reported by the studies [18–20] among others, and on energy fluxes (latent and sensible heat) in the context of climate change, since they cover approximately 20% (2.7 billion ha) of the global surface [17].

Carbon dynamics and biophysical evapotranspiration controls over tropical Brazilian ecosystems and in other regions of the world have also been studied because of the need to better understand the effects of land use change on regional and global biogeochemical cycles [13,15,21]. Evidently, it also allows us to estimate the contribution of tropical biomes to the regional and global climate controls, also comprising the relationship between precipitation, evapotranspiration, plant productivity and greenhouse gas exchanges within the atmosphere [22]. These changes motivated the development of various research fields within climate sciences focusing on the debate over greenhouse gases cycles and anthropic influence over these cycles. CO_2 concentration in the atmosphere surpassed 405 ppm [23] and continues increasing each decade, as well as that of other greenhouse gases. Therefore, it is necessary to develop methods and techniques to quantify these emissions and the fluxes of these gases, improving the understanding on how different environments function and respond to land-use changes.

The eddy covariance method (EC) directly and non-intrusively estimates the vertical transport of CO_2 and other greenhouse gases. Information measured through EC allows for the better comprehension of biogeochemical cycles, energy fluxes and other atmospheric controls. It also provides clues to other pivotal questions regarding ecosystem (in this study, tropical ecosystems) controls over climate. However, due to the lack of an extensive EC measurement network throughout Brazil, other data sources are required, such as satellite-derived data. Our hypothesis is that both gross primary production (GPP) and water-use efficiency (WUE) data respond differently to varied water availability conditions on the main tropical ecosystems in Brazil. Thus, our objective is to analyze the dynamics of GPP and WUE in the Amazon, Cerrado (savanna), Pantanal and Caatinga biomes through data collected through EC systems and to assess the performance of data derived from the Moderate Resolution Imaging Spectroradiometer (MODIS) sensor for the quantification of CO_2 balance components over these environments.

2. Materials and Methods

2.1. Data Policy and Use License

The Ameriflux platform compiles data monitored in three Brazilian biomes: BR-Sa1 (Amazon), BR-CST (Caatinga) and BR-Npw (Pantanal). Data from the Cerrado site (BR-BI, Bananal Island—Javaés) are available at https://daac.ornl.gov (accessed on 25 April 2022). Data provided by the ORNL DAAC can be accessed for free, without restrictions and in accordance with NASA's Earth Science Program. Ameriflux data are shared under CC-BY-4.0 data usage license (Creative Commons by Attribution 4.0 International). This license states that the use of data is free to share (copy and redistribute the material in any medium or format) and/or adapt (remix, transform, and build upon the material). The scientific literature references describing each of the data sites are: BR-Sa1 [24], BR-CST [25] and BR-Npw [26].

2.2. Description of Study Area

2.2.1. Cerrado Site

Measurements were conducted on an experimental site located on an inundation plain at Cantão State Park, 260 km west of the city of Palmas (Tocantins state, Figure 1) as a part of the Large-Scale Biosphere-Atmosphere Experiment in Amazonia (LBA). A micrometeorological tower equipped with an eddy covariance system and instruments for meteorological variable measurements were used. The tower was located 2 km east of the Araguaia River (9°49′27.9″S, 50°08′92.8″W, 120 m in altitude), 1 km east of the Javaézinho River, at the northern border of the Bananal Island, and south of Cantão State Park. The Araguaia plains, where the Bananal Island is located, is characterized by an exuberant landscape comprising Cerrado and Amazon biomes extending throughout three conservation units: Araguaia National Park, Cantão State Park and the Cantão Environmental Protection Area. The Bananal Island encompasses a total area of 21,000 km^2 (approximately 80 km $\times$ 260 km) and is the largest fluvial island in the world, covered mainly by savannas and pastures with seasonal inundations occurring from February to June [3]. The instrumental fetch area covers different types of physiognomies: cerradão, semi deciduous forests (trees with an average height of 20 m), cerrado s.s. and bare field areas with isolated lagoons. The climate in the region is warm and wet, with mean annual precipitation of approximately 1466 mm $year^{-1}$ and 90% of the total precipitation occurring in the wet season between October and April [3]. Measurements comprise the period from January 2004 to December 2006.

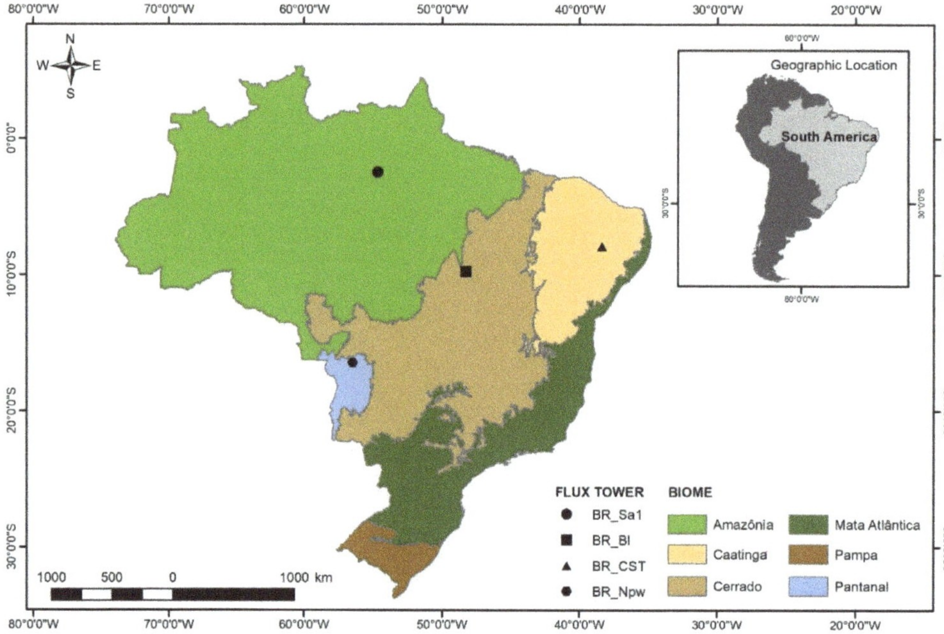

Figure 1. Location of the study sites with the native extension of each biome.

2.2.2. Caatinga Site

Measurements were carried out on a site managed by the Chico Mendes Institute for Biodiversity Conservation (ICMBio), and the micrometeorological tower is part of the NOWCDCB (National Observatory of Water and Carbon Dynamics in the Caatinga Biome, Figure 1) monitoring network. The campaign comprised the period from 1 January 2014 to 31 July 2015. The tower is located in a preserved Caatinga segment (BR-CST) within the Pajeú river watershed (7°58′05.20″S, 38°23′02.62″W, 430 m in altitude) in Serra Talhada city, Pernambuco state, Northeast Brazil. According to Köppen's classification, the climate is semiarid (BSwh), with summer rainfalls occurring between December and May (85% of total precipitation) as reported by [27]. Mean annual precipitation is approximately 640 mm, and average air temperatures range from 23.1 to 26.7 °C [27]. Native species at the site are *Mimosa hostilis*, *Mimosa verrucosa* and *Croton sonderianus*, while *Anadenanthera macrocarpa*, *Spondias tuberosa*, *Caesalpinia pyramidalis*, and *Ziziphus joazeiro* can also be found, with a mean height of 8.0 m [27]. Rainfall was provided by a "Instituto Nacional de Meteorologia" (INMET) station near of site.

2.2.3. Pantanal Site

The study was conducted at the Brazilian Northern Pantanal Wetland (BR-Npw) flux tower (Figure 1) located approximately 35 km SE of Pocone, Mato Grosso, Brazil (16°29′53.71″S: 56°24′45.91″W; 120 m altitude). The site is part of a research station managed by the Federal University of Mato Grosso (UFMT) within a national reserve managed by the Brazilian Social Service of Commerce (SESC Pantanal) [28,29]. According to Köppen classification, the regional climate is Aw, which is defined as a hot and wet climate with rainfall in the summer and dry in the winter [30]. The accumulated precipitation varies from 800 to 1500 mm/year [31,32]. The air temperature ranges between 29 and 32 °C (maximum) and between 17 and 20 °C (minimum) [33,34]. The soil type in the region is classified as Dystric Gleysol [35] with an average concentration of 429 g/kg of sand, 254 g/kg of silt, and 317 g/kg of clay, mean soil organic matter (SOM) (0–0.10 m depth) of 17 g/kg and a soil pH of 4.7 [36]. The vegetation of the site is typical of "scrub" forests in the region, with a mean leaf area of 7.4 m^2 m^{-2} and a mean height of 6 m [37], dominated by *Combretum lanceolatum*, Phol (Combretaceae), a common species found on the riverbanks of the Pantanal region [38]. Vascular or semi-aquatic plants such as *Thalia geniculata* and *Nymphaea* sp. occur in more open areas [39]. These are dense forests that are referred to as "hyperseasonal" because they are subjected to both seasonal flooding and drought [40].

Our data were collected from 1 January 2015 to 31 December 2016. Micrometeorological variables were measured 20 m aboveground, close to the eddy covariance sensors. Air temperature (*Ta*, °C) and relative humidity (*RH*, %) were measured using a thermohygrometer (HMP45AC, Vaisala Inc., Woburn, MA, USA). Precipitation (*Ppt*, mm) was measured 2 m above the ground using a micrometeorological station (WXT520, Vaisala Inc., Helsinki, Finland) installed in an open area to avoid interception by the tower or tree canopy. The flood stage was determined by measuring water levels (*WL*) above the ground at the study site. These inundation levels (±1%) were measured along with water temperature (±0.3 °C) using a CTD-10 (Decagon Devices Inc., Pullman, WA, USA, ± 0.05% full scale at 20 °C) in 2015 and 2016. Due to instrument malfunction in 2014, the data for this year are not available. The start of each flood cycle began with the first reading of standing water at the site and ended when sensors indicated the absence of standing water. These flood cycles were then compared to the stage of the Cuiaba River collected by the RPPN-SESC Pantanal park rangers (pers. comm.) approximately 1 km away.

2.2.4. Amazon Site

Measurements were conducted on a site located at Tapajós National Forest (TNF, 2°51′S, 54°58′W, Figure 1), near the Santarém-Cuiabá highway (BR-163). The TNF is limited by the Tapajós River in the west and by the BR-163 highway in the east, extending 150 km to the south of Santarém city, Pará state. At the eastern side of the BR-163, the landscape is dominated by agriculture. The tower was installed approximately 6 km west of the highway. The canopy has a significant number of large emergent trees (to 55 m height), *Manilkara huberi* (Ducke) Chev., *Hymenaea courbaril* L., *Betholletia excelsa* Humb. and Bonpl., and *Tachigalia* spp., and a closed canopy at ~40 m [41]; this forest can be considered primary, or "oldgrowth" [42]. Analyzed data comprise CO_2 and energy fluxes and meteorological data. Measurements comprise daily and monthly means of hourly data in the period from January 2009 to December 2011. CO_2 fluxes were measured at 58 m in height through a closed-path analyzer (LICOR-6262) while a Campbell CSAT3 anemometer was used for tridimensional wind measurements. The 65 m micrometeorological tower is located at an area emerging from within the primary forest with a dense canopy of approximately 40 m in height, reaching up to 55 m for some emerging trees [43]. Rainfall was provided by a INMET station near of site (Belterra).

Figure 1 shows the location of the four studied sites and their respective biomes.

2.3. Instrumentation and Data Processing

Instrumentation in each of the sites is described in previous publications [3,27,44,45]. Gaps formed due to the exclusion of spurious data during the rigorous data screening process were filled using a marginal distribution sampling method (MDS) described by [46], which accounts for the covariance between fluxes and meteorological variables as well as the autocorrelation of fluxes. In this algorithm, three conditions are identified, and a procedure is adopted accordingly: (1) when flux data are missing, but meteorological data are available (Rg, Ta and VPD), missing data are replaced by the mean value in similar meteorological conditions over a seven-day window; (2) when only radiation data are available, the missing data are replaced by the mean value in similar meteorological conditions over a seven-day window; (3) when there are no meteorological data available, the missing data are replaced by the mean value in the last hour, thus accounting for the daily variability of each variable. If after these steps, data were still not filled, the procedure was repeated with larger window sizes until the gap could be filled. For the gap filling procedure, an automated online tool developed by the Max Planck Institute for Biogeochemistry (http://www.bgc-jena.mpg.de/~MDIwork/eddyproc/, accessed on 2 February 2022) was used.

2.4. Flux Partitioning

Through CO_2 flux (NEE) partitioning, we obtained: gross primary production (GPP) and ecosystem respiration (R_{eco}). For the Cerrado and Pantanal sites, NEE is given as a proxy of the turbulent flow. At the other sites, NEE was composed of turbulent flow and storage. We used the flux partitioning method based on nighttime hours [46]. Since GPP = 0 at the nighttime period, NEE is given as:

$$NEE = R_{eco}, \text{ for nighttime hours} \tag{1}$$

$$NEE = R_{eco} - GPP, \text{ for daytime hours} \tag{2}$$

where R_{eco} (μmol $m^{-2}s^{-2}$) is the sum of autotrophic and heterotrophic respiration. R_{eco} and GPP were calculated using the online tool provided by the Max Planck Institute for Biogeochemistry (http://www.bgc-jena.mpg.de/~MDIwork/eddyproc/, accessed on 2 February 2022).

Diurnal corrections of missing NEE data were modeled based on daytime data using the common rectangular hyperbolic light-response curve model [47,48]:

$$NEE = \frac{\alpha \cdot \beta \cdot R_g}{\alpha \cdot R_g + \beta} + \gamma \tag{3}$$

where α (μmol C J^{-1}) is light-use efficiency and represents the initial slope of the light-response curve; β (μmol C $m^{-2}s^{-1}$) is the maximum CO_2 absorption rate of the canopy at light saturation; γ (μmol C $m^{-2}s^{-1}$) is ecosystem respiration and R_g (W m^{-2}) is global radiation. GPP was calculated as:

$$\text{GPP} = NEE + \text{R}_{eco} \tag{4}$$

To provide a better comparison with literature data (more in situ studies), we calculated Ecosystem WUE, calculated as:

$$\text{WUE} = \text{GPP}/\text{ET} \tag{5}$$

where ET (mm) is evapotranspiration.

2.5. MODIS Data

Remotely sensed GPP data obtained from the MODIS sensor onboard the Terra satellite were used. These data were calculated based on the concept of light use efficiency [47], through the relation between incident photosynthetically active radiation (PAR), the fraction of photosynthetically active radiation absorbed by plants (FPAR) and the actual light use efficiency (ε) of vegetation [49]:

$$\text{GPP} = \varepsilon * \text{APAR} \tag{6}$$

where APAR is the absorbed photosynthetically active radiation, which is calculated as the product between the FPAR—derived from the MOD15A2H product [50]—and the PAR. PAR values are obtained from the Global Modeling Assimilation Office (GMAO) reanalysis [51] set and correspond to 45% of the total incident solar energy in the visible spectrum (0.4 to 0.7 μm). In the MOD17A2 product, the values of ε are derived from the attenuation of its maximum value (εmax) due to two environmental stresses: (1) minimum temperature (Tmin), which can inhibit photosynthesis, reducing enzymatic activity, and (2) vapor pressure deficit (VPD), because high VPD can reduce stomatal conductance [52]. The GPP data for the Caatinga biome were derived from the MOD17A2 version 6.0 product [53] from the MODIS sensor on board the Terra satellite. This algorithm provides 8-day composite data in a 500 m spatial resolution.

Reanalysis data from the GMAO dataset (https://gmao.gsfc.nasa.gov/GMAO_products/reanalysis_products.php, accessed on 25 February 2022) used as input parameters for the MOD17A2 product were Tmin and VPD. Therefore, we carried out an overall assessment to check for inconsistencies in this dataset in relation to the same parameters measured at the Cerrado, Caatinga, Pantanal and Amazon biomes. FPAR and leaf area index (LAI) data from the MOD15A2H product were also verified. Furthermore, the MOD12 product [54] for land cover was also discussed regarding the years 2014 and 2015 in the region.

The MOD12 [54] product for land cover was also discussed regarding the Cerrado and Caatinga site for the period from 1 January 2014 to 31 July 2015, the Pantanal site for the period from 1 January 2015 to 31 December 2016, and the Amazon site from 1 January 2009 to 31 December 2011.

3. Results

3.1. Meteorological Conditions

Precipitation and air temperature data in all study sites (Figure 2), as well as daily observations of solar radiation, air temperature and relative humidity (Figure 3), presented a seasonality that was consistent with the local climatology. Highest radiation incidence was registered in the Caatinga and Pantanal sites, which makes them the warmest sites since air temperature is strongly correlated with solar radiation. The highest monthly temperature was found in the Pantanal site (Figure 2A), with 31.6 °C being registered in October, while the Amazon site presented the lowest temperatures (25.4 °C in April), which is also probably associated with the fact that radiation is lower (Figure 3A). The Caatinga and Pantanal sites are in phase regarding solar radiation (Figure 3A) and air temperature (Figure 3B) patterns, despite VPD values being three-fold higher in the Caatinga (Figure 3C), while VPD in the Pantanal is similar to that observed in the Amazon. VPD in the Amazon practically doubles until September, along with lower observed accumulated rainfall and higher air temperatures.

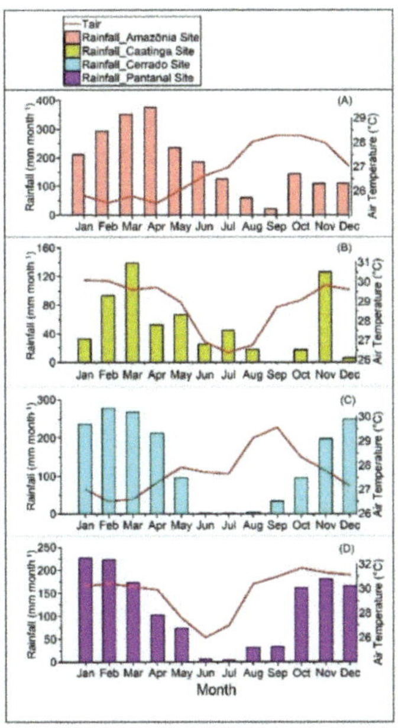

Figure 2. Monthly variation of mean air temperature (lines) and accumulated rainfall (bars) for each site and period: (**A**)—Amazonia: 2009–2011 (rainfall from INMET station); (**B**) Caatinga: January 2014–July 2015 (rainfall from INMET station); (**C**) Cerrado: 2004–2006; (**D**) Pantanal: 2015–2016).

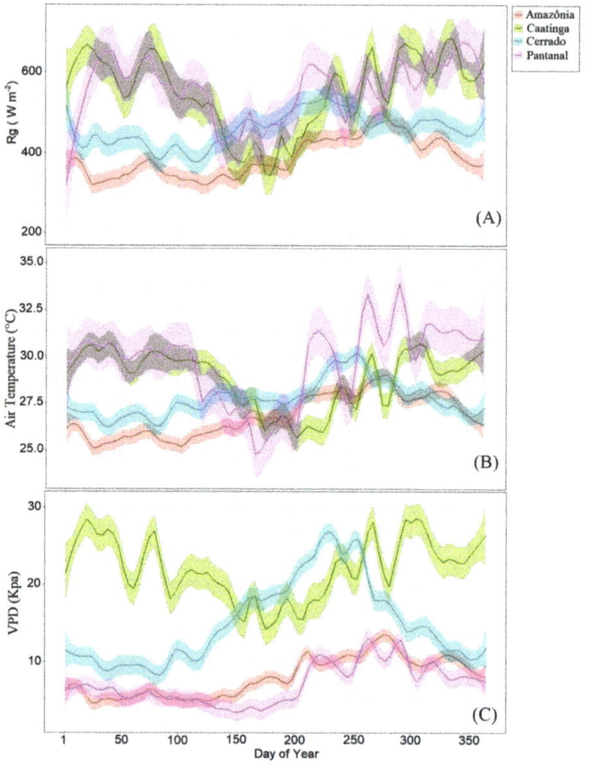

Figure 3. Seasonal variation of the daily mean of meteorological variables: (**A**) global incident radiation (W m^{-2}), (**B**) air temperature (°C) and (**C**) vapor pressure deficit (kPa). The shaded colored areas indicate data standard deviation.

Mean annual precipitation in the Amazon site was approximately 2221 mm, 38% higher than in the Pantanal site (1381 mm) and 25% higher than in the Cerrado site (1668 mm). Caatinga presented an annual accumulated rainfall value of 698.9 mm in 2014 and 376.3 mm until July 2015. The number of months with monthly precipitation <10% of total annual precipitation in each site (Figure 2) varied locally. In the Amazon, seven months met this criterion, while in the Cerrado and Pantanal, six months did as well. In the Caatinga, a total of nine months presented less than 10% of total annual rainfall. Besides the highest precipitation totals, the Amazon also presents the greatest monthly variability of precipitation, particularly in March (Figure 4). September is the month where precipitation in all four sites is the most similar, with the Caatinga site registering null precipitation and the other sites registering precipitation below 50 mm. In March, the difference between the wettest and the driest sites accounts for over 300 mm. Only in the months of June and July did the Caatinga site feature monthly accumulated rainfall higher than the Cerrado and Pantanal sites. Only the Amazon site surpassed 300 mm of precipitation in the months from February to April.

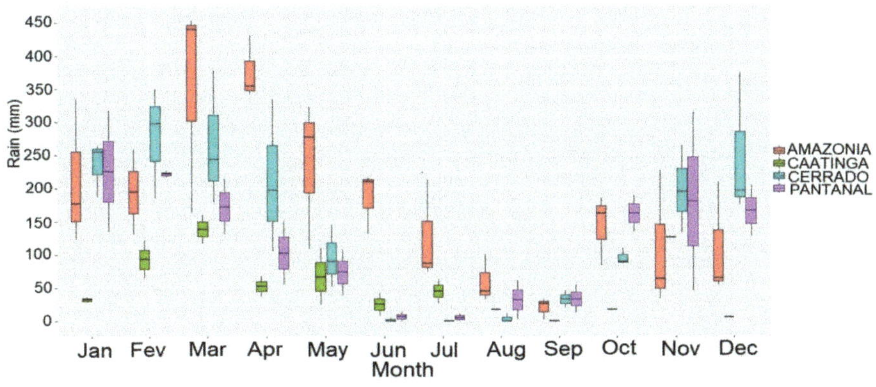

Figure 4. Monthly accumulated rainfall boxplots for each site and studied period (Amazonia: 2009–2011; Caatinga: January 2014–July 2015; Cerrado: 2004–2006; Pantanal: 2015–2016).

3.2. Water and Energy Fluxes

The annual cycle of observed daily ET (Figure 5) showed different patterns in the biomes: (1) Cerrado and Pantanal with maximum water vapor flux in October, decreasing toward March and April as the wet season ends; (2) a well-defined ET cycle in the Amazonia site, peaking in September during the dry season. ET cycles in the Cerrado and Pantanal are similar, with maximum daily ET of approximately 7.0 mm day^{-1}. In July, daily ET values are similar in the Cerrado, Pantanal and Amazon sites, with a similar pattern to what was previously observed for solar radiation and air temperature (Figure 3A,B). The Caatinga site stood out, with values down to three-fold lower if compared to the other sites, despite the similar seasonal pattern (higher daily ET between October and April, lower daily ET between May and September). Maximum ET coincides with higher energy availability (radiation) and higher temperatures in the sites, between October and November.

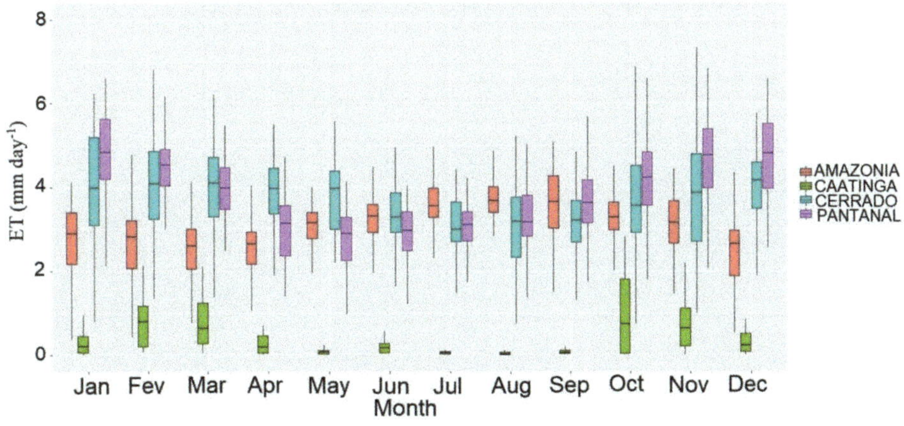

Figure 5. Monthly daily evapotranspiration (mm day^{-1}) boxplots for each site and studied period (Amazonia: 2009–2011; Caatinga: January 2014–July 2015; Cerrado: 2004–2006; Pantanal: 2015–2016).

The monthly seasonal patterns of net radiation and sensible and latent heat fluxes differ strongly from site to site. In the Pantanal, Caatinga and Cerrado, net radiation (Rn) presents values higher than 120 W m^{-2} (Figure 6A) in January and May, with a decreasing pattern the following months. The Amazon site, located further north than the other sites and therefore more influenced by the Intertropical Convergence Zone during these months, presented lower values, of approximately 100 W m^{-2}, with increasing values from May until the peak of the dry season in September. Hourly Rn patterns are similar, with the Amazon site featuring approximately 50 W m^{-2} less than the other sites. However, latent heat flux (LE) (Figure 6B) and sensible heat flux (H) (Figure 6C) have distinct hourly and monthly patterns. Maximum LE in the Pantanal is three-fold higher than in the Caatinga, while the H pattern is the inverse. Rn, H and LE patterns are in phase regarding Pantanal and Amazonia sites, particularly because LE and H sharply increase in the dry season, following the increase in available energy. An opposite relationship is found between Rn and LE in the Cerrado and Caatinga sites, which increase in September and October while LE reduces in the same period.

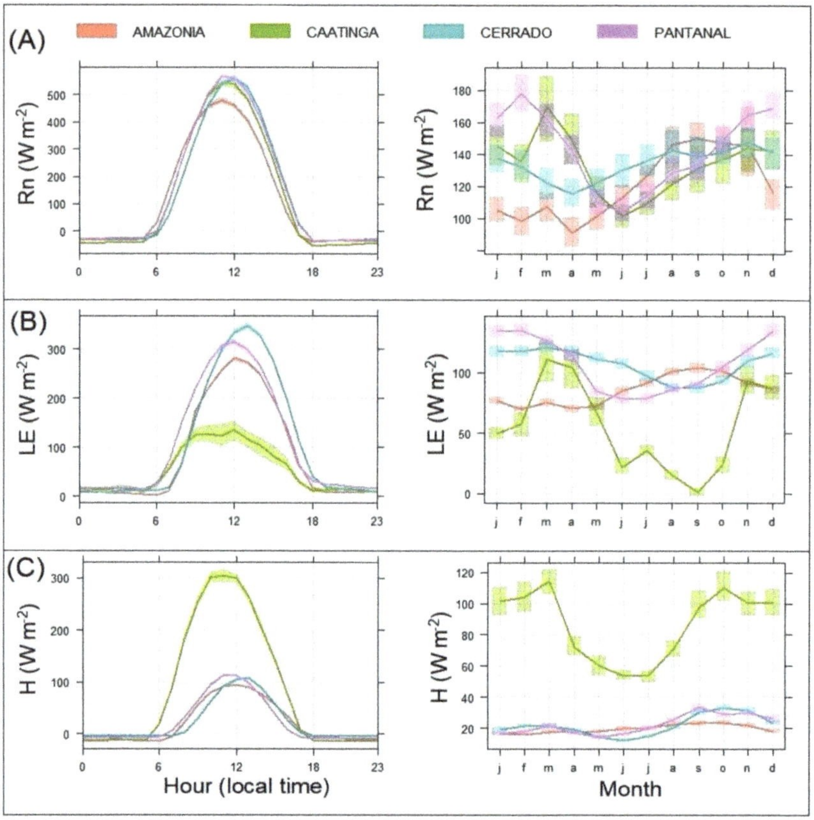

Figure 6. Hourly and monthly variation of energy fluxes: (**A**) net radiation (Rn, W m^{-2}), (**B**) latent heat flux (LE, W m^{-2}) and (**C**) sensible heat flux (H, W m^{-2}).

3.3. Carbon Fluxes and WUE

According to the daily GPP analysis shown in Figure 7, seasonal changes in GPP are more intense in the Caatinga and Pantanal sites if compared to the Amazonia and Cerrado sites. The coefficient of determination (R^2) between observed GPP and evapotranspiration (Figure 8) was also higher in the Caatinga and Pantanal. Maximum values reach up to 9.0 gC m^{-2} d^{-1} in the Cerrado and Amazonia, while the lowest GPP values reach approximately 0.5 gC m^{-2} d^{-1} in the Caatinga. The seasonality and intensity of GPP values in the Amazon and Cerrado are similar, while in the Caatinga and Pantanal, they are similar only between June and July. Regarding maximum values, GPP peaks in April in the Cerrado, coinciding with the period of least radiation availability after the site inundation, while in the Amazon, GPP peaks in October, characterized by high radiation, air temperature and an increase in precipitation after an annual minimum in September.

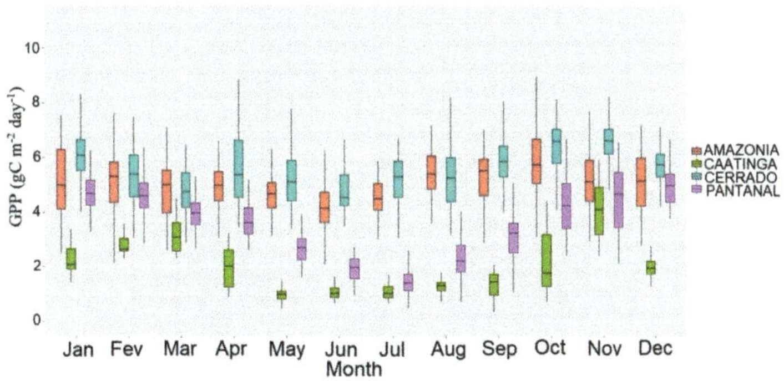

Figure 7. Monthly GPP (gC m^{-2} day^{-1}) boxplot for each site and study period (Amazonia: 2009–2011; Caatinga: January 2014–July 2015; Cerrado: 2004–2006; Pantanal: 2015–2016).

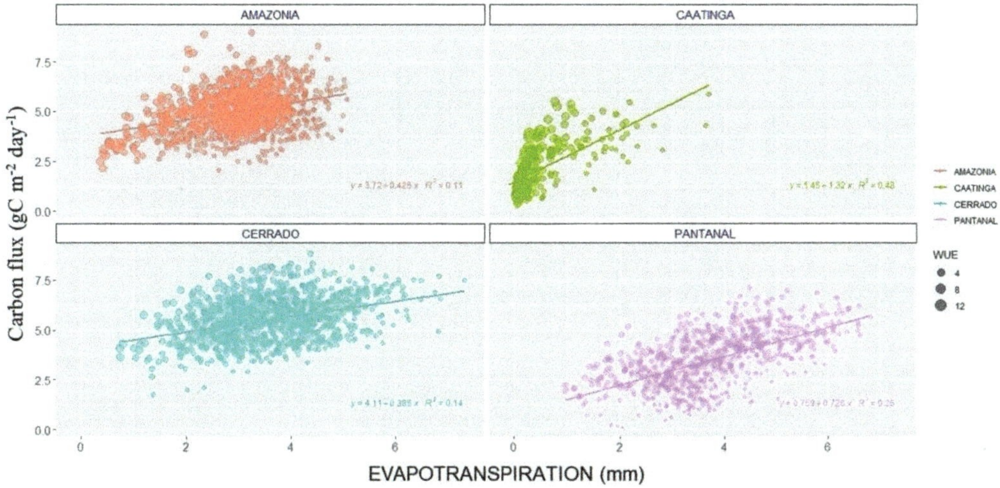

Figure 8. Correlation between daily GPP averages and evapotranspiration for each site. The size of the circles indicates the intensity of water-use efficiency (WUE) estimated at each day (gC kg H_2O day^{-1}).

We also assessed observed GPP data by comparing them with MODIS satellite data on a monthly scale (Figure 9). Results show a high overestimation of satellite data in drier periods, especially in the Amazonia site (Figure 9A). In the Caatinga site (Figure 9B), measurements are more similar from September to November if we consider the median values and monthly variability. In the Cerrado site (Figure 9C), MODIS GPP underestimates observed data in the dry season (August and September), while in the Pantanal (Figure 9D), daily variability of MODIS data is much more prominent than observed data, despite the coherence in representing the seasonal cycle. Given these patterns, the best correlations were found in the Pantanal (R^2 = 0.31) and Caatinga (R^2 = 0.27) sites (Figure 10), with values similar to what is found in the general literature regarding MODIS GPP assessment (Table 1). Due to the importance of investigating specific local water cycles and the effect of drought on the water balance and carbon uptake, we calculated WUE (Figure 11), showing interesting variability aspects depending on the studied biome. WUE is lower in the site with the highest evapotranspiration (Pantanal), while a remarkable variability and the highest WUE values are found in the Caatinga site. WUE in the other sites did not present great seasonal differences, with values reaching approximately 4.0 gC kg H_2O day^{-1} in the driest months (between September and October). WUE surpassed 15 gC kg H_2O day^{-1} in the driest month (September) of the driest site (Caatinga), sharply decreasing in October with the occurrence of rainfall. Mean annual WUE values estimated in this study delineate new thresholds if compared to previously reported values in the literature (Table 2), with the Pantanal site featuring WUE lower than 1.0 gC kg H_2O year^{-1} and the Caatinga site featuring values near 5.8 gC kg H_2O year^{-1}, which is much higher than in other dry forests studied in the literature.

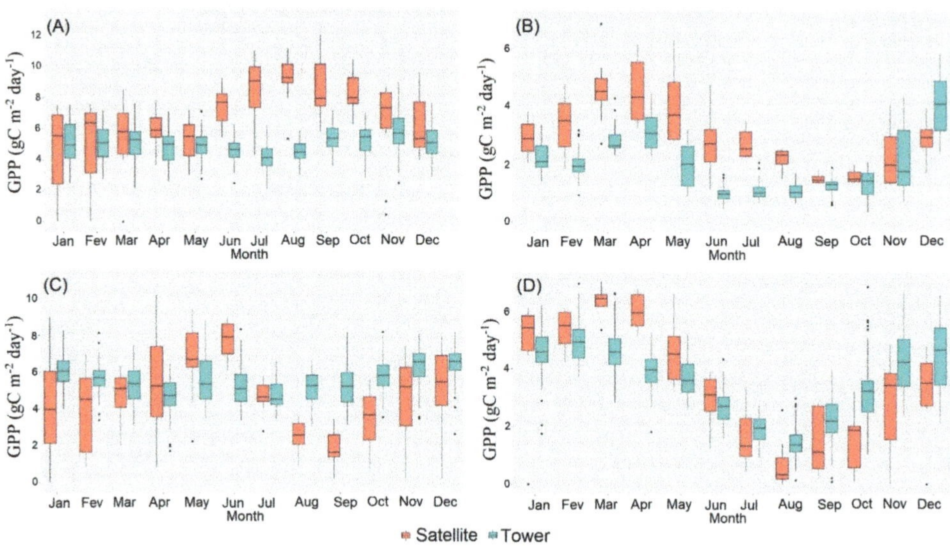

Figure 9. Monthly GPP boxplot (gC m^{-2} day^{-1}) for eddy covariance observed data (Tower) and MODIS—derived data (Satellite) for the: (**A**) Amazonia; (**B**) Caatinga; (**C**) Cerrado; (**D**) Pantanal sites.

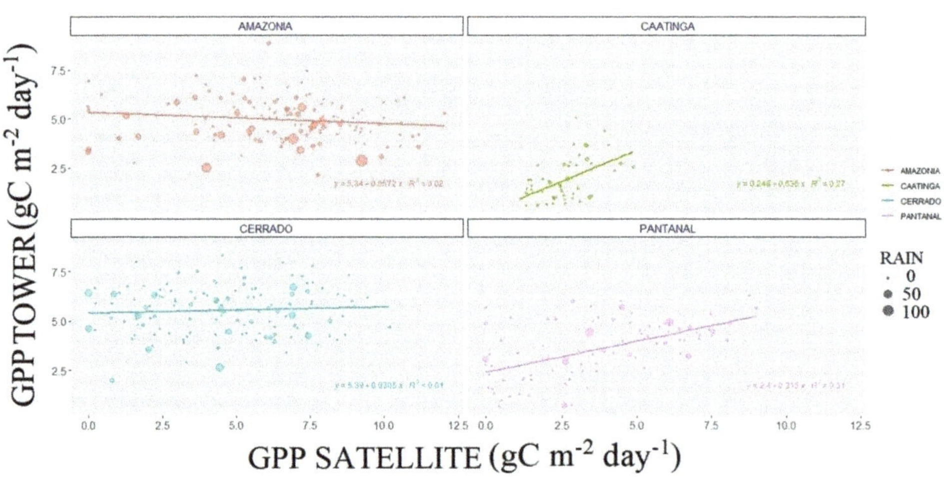

Figure 10. Correlation between daily tower GPP × MODIS GPP for each site. The size of the circles indicates the intensity of daily accumulated rainfall (mm).

Table 1. Statistical summary of studies comparing MODIS GPP and eddy covariance estimations for different land covers using linear regression models.

Forest Type	Slope	R^2	Reference
Alpine grassland	1.30	0.50	Zhu et al. [55]
Alpine grassland	0.58	0.17	Zhu et al.
Dry tropical forest	0.24	0.27	**This study**
Floodplain forest	5.39	0.01	**This study**
Primary forest	5.34	0.02	**This study**
Semi-deciduous forest	0.49	0.32	Danelichen et al. [56]
Temperate grassland	1.59	0.70	Zhu et al., 2018
Temperate grassland	0.50	0.40	Zhu et al., 2018
Tropical grassland	0.89	0.53	Zhu et al., 2018
Tropical peatland	0.23	0.16	Wang et al. [57]
Tropical grassland	0.91	0.63	Zhu et al., 2018
Tropical grassland	0.89	0.53	Zhu et al., 2018
Wetland	2.40	0.31	**This study**

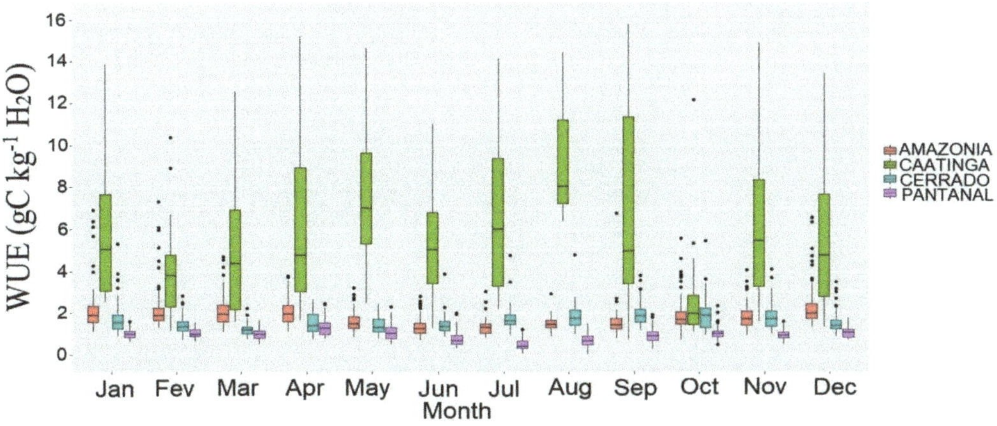

Figure 11. Monthly WUE (gC kg H_2O day^{-1}) boxplot for each site and studied period: (Amazonia: 2009–2011; Caatinga: 01/2014–07/2015; Cerrado: 2004–2006; Pantanal: 2015–2016).

Table 2. Comparison between mean annual WUE retrieved in the present study with values for different types of forests found in the literature.

Forest Types	WUE (g C kg^{-1}H_2O)	References
Wetland	0.95	**This study**
Boreal treeless wetland	1.2	Kuglitsch et al. [58]
Floodplain forest	1.61	**This study**
Maritime pine	1.69	Berbigier et al. [59]
Primary forest	1.82	**This study**
Deciduous broadleaf forest	1.87	Wang et al.
Old-growth forest	1.83	Liu et al. [60]
Evergreen broadleaf forest	2.35	Tang et al. [61]
Conifer plantation forest	2.53	Yu et al. [62]
Deciduous broadleaf forest	2.57	Yu et al.
Eucalypt plantation	2.87	Rodrigues et al. [63]
Ponderosa pine	2.97	Law et al. [64]
Evergreen broadleaf forest	3.13	Tang et al.
Boreal aspen	3.70	Krishnan et al. [65]

Table 2. *Cont.*

Forest Types	WUE (g C kg^{-1}H$_2$O)	References
Temperate broad-leaved deciduous	5.0	Kuglitsch et al.
Douglas-fir	5.40	Ponton et al. [66]
Dry tropical forest	5.79	**This study**

4. Discussion

Evapotranspiration showed well-defined seasonality in all sites, varying at the annual scale with local precipitation, radiation availability and increasing temperatures. Highest evapotranspiration values were registered in the warmer months and/or in months with more radiation or precipitation. Evapotranspiration reached maximum values in the Cerrado and Pantanal sites in wetter months, reaching up to 7 mm day^{-1} in November, with a mean value slightly above 3 mm day^{-1}. In the Caatinga site, even in the month with the highest evapotranspiration rates (October), overall values did not reach the same intensity of the other sites, showing the particularity of this site regarding its arid climate (BSh) according to Köppen's classification (Liang et al., 2020) [67]. Crucial studies have already been conducted to estimate the magnitude, seasonality and controls of ET at the local scale using eddy covariance measurements in Brazil (Da Rocha et al., 2009; Costa et al., 2010) [6,68]. These studies usually show that dry season evapotranspiration is higher than in the wet season, and Rn is the main ET control in tropical rainforests (such as the Amazon), while it is not true for the Caatinga, Cerrado and Pantanal sites. Our results showed that in these sites, evapotranspiration decreases throughout the dry season, reaching the lowest rates between August and September, when the Amazon features its highest values. Even during the dry season, high evapotranspiration rates can be explained by the predominance of pioneer tree species (*Combretum lanceolatum* and *Vochysia divergens*) with high photosynthesis rates and stomatal conductance (gs) (Dalmagro et al., 2013; 2016) [69,70], associated with the ability of these species to extract water from deep storages containing similar water content to that of inundation areas and wet periods (Sanches et al., 2011; Vourlitis et al., 2011; Dalmagro et al., 2013; da Silva et al., 2021) [29,69,71,72].

In the dry season of the Cerrado site, an inverse relationship was found between LE and Rn, which is typical of savanna ecosystems, where the root system does not reach deep water storages, but part of the vegetation is adapted and relies on senescence mechanisms of tree leaves and dormancy of grasses (3). A pulse in productivity (increased GPP) that is directly related to the increase in water availability (rainfall) can be observed, particularly from October on, which marks the beginning of the transition from the dry to the wet season. As expected, all sites reduced productivity in response to increased rainfall variability, with the most productive ecosystems being those with the highest precipitation rates (Amazon and Cerrado). The low rainfall rates observed in the Caatinga in October were sufficient to drive a sharp increase in productivity, which indicates that more frequent rainfall events in this environment could lead to more nutrient availability and the mitigation of water stress through leaf absorption mechanisms. A further indicator of these aspects can be observed through the remarkable variability in evapotranspiration during this month in the Caatinga. Furthermore, this biome presented the best correlations between GPP and evapotranspiration. The comparison between observed and remotely sensed data showed that both datasets represent the same seasonality, despite their weak correlation. Satellite data accurately represents the strong response of the Pantanal and Caatinga sites to local precipitation, while the results are not satisfactory regarding the dry period in the Amazon and Cerrado. The weak correlation between GPP data in tropical forests is well documented (Zhu et al., 2018) [55] and is generally attributed to light-use efficiency, which is underestimated by the MODIS GPP algorithm (Sjöström et al., 2013) [73]. Additionally, potential uncertainties regarding the FPAR might also affect MODIS GPP estimates, which have also been discussed in the literature (Liu et al., 2017) [60]. Despite its wide use, our results corroborate previous studies reporting the need to use MODIS GPP data with caution for interannual and intra-annual studies, as suggested by Zhu et al. [55]. Our study revealed that humid biomes present lower WUE, and drier biomes present higher WUE, with decreasing values with the onset of the wet season and maximum values in the dry season. This result is consistent with previous studies in other environments (Liu et al., 2017; Singh et al., 2014; Yu et al., 2008) [60,62,74]. Ito and Inatomi [75] claim that forest ecosystems presented WUE values comparable or superior to arid (such as caatinga) and shrub biomes (such as Cerrado and Pantanal), with an intermediate value of 0.6–1.2 g C kg^{-1} H_2O for humid forest biomes. The average WUE for the Pantanal is within this range.

WUE responds differently to seasonal water availability in each biome. Overall, more productive biomes present high WUE (Xue et al., 2015) [76]. In sparsely vegetated areas, such as the Cerrado (savannas) and the Pantanal wetlands, WUE varies greatly throughout the year, despite presenting lower values than the Caatinga. From January to April, when precipitation is high in all sites (except Caatinga), the Amazon site featured an increase in WUE daily values, while the Cerrado WUE decreased, and the Pantanal WUE remained unaltered. From May, rainfall triggers a sharp decrease in WUE over these regions until September. Our results reveal important implications for the understanding of climate change effects on carbon and water exchange processes in tropical biomes, because the projected reduction in water availability over these sites due to the increasing number of dry days [77] may lead to an increase in WUE at the ecosystem level. However, increasing temperatures may further increase or reduce ecosystem WUE at the monthly scale. Consequently, changes in ecosystem WUE due to climate change will depend on the relative impact of these changes in precipitation and temperature. Caatinga WUE is highly dependent on water availability, with lower values and variability in the wetter months and higher values and variability in the drier months. The results found here corroborate other studies in other parts of the Caatinga biome [78–82], and it is important to report that owing to measurement difficulties, few studies have systematically compared global patterns of WUE of terrestrial ecosystems across different biomes or have analyzed the seasonal variability of WUE in relation to weather conditions, because ecosystem WUE is slightly different from plant WUE [61]. Plant physiologists consider WUE at leaf or stand scales and are mainly interested in relations between total or above-ground biomass, stem biomass or net CO_2 uptake to transpiration or evapotranspiration (ET), and although uncertainties associated with site-to-site variation in site quality criteria, flux measurement methods, calculations and data quality control still exist, ongoing standardization and quality assurance efforts enable global integration with other tools [61].

5. Conclusions

In this study, we presented data on the seasonal variations of energy fluxes, climatic variables, GPP, ET and WUE for different tropical biomes in Brazil. Furthermore, correlations between observed carbon exchange data and remotely sensed data were also assessed. Results showed that GPP and ET responses to meteorological variables (solar radiation, air temperature, precipitation and VPD) are in phase, suggesting that this meteorological variability controls photosynthesis and ET in a similar fashion on a monthly scale, despite both direct and inverse relationships having been found depending on the type of environment. Based on our results, our study concludes that inconsistent MODIS GPP estimates for some months and sites indicate that the parametrizations used in the MOD17A2H GPP algorithm (such as FPAR) may need to be enhanced over certain land covers in order to improve estimates. WUE in the studied sites varied annually from 0.95 to 5.79 gC kg^{-1} H_2O, with minimum and maximum values differing from usually found values for other environments worldwide. This study will aid future studies regarding the influence of global warming and water stress on carbon and water fluxes in similar tropical forests.

Author Contributions: In the present research article the individual contributions was followed by: Conceptualization, G.B.C., K.R.M., C.M.S.e.S. and B.G.B.; methodology, G.B.C., K.R.M., L.B.V., G.V.A., P.R.M., T.V.M. and R.R.F.; software, G.B.C., K.R.M., L.B.V., G.V.A., T.V.M. and R.R.F.; validation, J.G.M.d.S., T.T.A.T.N., P.R.M., H.G.G.C.N., L.V.P. and R.A.S.S.; formal analysis, G.B.C., K.R.M., C.M.S.e.S., J.G.M.d.S., B.G.B., A.S.S., C.P.O., T.R.R., J.B.S., W.A.G., and S.C.; investigation, G.B.C., K.R.M., T.R.R., H.J.D., C.M.S.e.S. and B.G.B.; resources, G.B.C., B.G.B. and H.J.D.; data curation, G.B.C., K.R.M., T.R.R., J.B.S., L.B.V. and G.V.A.; writing—original draft preparation, G.B.C., K.R.M., J.G.M.d.S., C.M.S.e.S. and B.G.B.; writing—review and editing, T.V.M., T.T.A.T.N., P.R.M., C.P.O., A.S.S.,T.R.R., J.B.S., C.P.O., W.A.G., S.C. and M.U.G.A.; visualization, G.B.C., K.R.M. and C.M.S.e.S.; supervision, G.B.C., K.R.M., C.M.S.e.S. and T.R.R.; project administration, H.J.D., C.M.S.e.S. and B.G.B.; funding acquisition, G.B.C., H.J.D., C.M.S.e.S. and B.G.B. All authors have read and agreed to the published version of the manuscript.

Funding: The authors are also thankful to the Coordination for the Improvement of Higher Education Personnel (CAPES) for the postdoctoral funding granted to KRM and to the National Council for Scientifc and Technological Development (CNPq) for the research productivity grant of C.M.S.e.S. (Process n° 303802/2017-0), the financial support of CNPq, through undergraduate research project (PIBIC-UFOPA for L.B.V. and G.V.A.) and the project NOWCDCB: National Observatory of Water and Carbon Dynamics in the Caatinga Biome (INCT-MCTI/CNPq/CAPES/FAPs 16/2014, grant: 465764/2014-2) and (MCTI/CNPq N° 28/2018, grant 420854/2018-5). The APC was funded by PAPCIQ–UFOPA.

Data Availability Statement: Not applicable.

Acknowledgments: The authors are thankful to AMERIFLUX and LBA project for provide free access to the EC data used in this study.

Conflicts of Interest: The authors declare no conflict of interest and the funders had no role in the design of the study; in the collection, analyses, or interpretation of data; in the writing of the manuscript, or in the decision to publish the results.

References

1. Andreae, M.O.; Artaxo, P.; Brandão, C.; Carswell, F.E.; Ciccioli, P.; da Costa, A.L.; Culf, A.D.; Esteves, J.L.; Ash, J.H.C.; Grace, J.; et al. Biogeochemical cycling of carbon, water, energy, trace gases, and aerosols in Amazonia: The LBA-EUSTACH experiments. *J. Geophys. Res.* **2002**, *107*, 25. [CrossRef]
2. Araújo, A.C.; Nobre, A.D.; Kruijt, B.; Elbers, J.A.; Dallarosa, R.; Stefani, P.; Randow, C.; von Manzi, A.O.; Culf, A.D.; Gash, J.H.C.; et al. Comparative measurements of carbon dioxide fluxes from two nearby towers in a central Amazonian rainforest: The Manaus LBA site. *J. Geophys Res.* **2002**, *107*, 8090. [CrossRef]
3. Borma, L.D.S.; da Rocha, H.R.; Cabral, O.M.; von Randow, C.; Collicchio, E.; Kurzatkowski, D.; Brugger, P.J.; Freitas, H.; Tannus, R.; Oliveira, L.; et al. Atmosphere and hydrological controls of the evapotranspiration over aflood plain forestin the Bananal Island region, Amazonia. *J. Geophys. Res.-Biogeo.* **2009**, *114*, G01003. [CrossRef]
4. Carswell, F.E.; Costa, A.L.; Palheta, M.; Malhi, Y.; Meir, P.; Costa, J.d.P.R.; Ruivo, M.d.L.; Leal, L.d.S.M.; Costa, J.M.N.; Clement, R.J.; et al. Seasonality in CO2 and H2O flux at an eastern Amazonian rain forest. *J. Geophys. Res.* **2002**, *107*, 16.
5. Rocha, H.R.; Goulden, M.L.; Miller, S.D.; Menton, M.C.; Pinto, L.D.V.O.; De Freitas, H.C.; E Silva Figueira, A.M. Seasonality of water and heat fluxes over a tropical forest in Eastern Amazonia. *Ecol. Appl.* **2004**, *14*, 22–32. [CrossRef]
6. Rocha, H.R.; Manzi, A.O.; Cabral, O.M.; Miller, S.D.; Goulden, M.L.; Saleska, S.R.; Restrepo-Coupe, N.; Wofsy, S.C.; Borma, L.S.; Artaxo, P.; et al. Patterns of water and heat flux across a biome gradient from tropical forest to savanna in Brazil. *J. Geophys. Res.* **2009**, *114*, 8. [CrossRef]
7. Sellers, P.J.; Shuttleworth, W.J.; Dorman, J.L.; Dalcher, A.; Roberts, J.M. Calibrating the Simple Biosphere Model for Amazonian tropical forest using field and remote sensing data. Part I: Average calibration with field data. *J. Appl. Meteorol.* **1989**, *28*, 727–759.
8. Saad, S.I.; da Rocha, H.R.; Silva Dias, M.A.F.; Rosolem, R. Can the deforestation breeze change the rainfall in Amazonia? A case study for the BR-163 highway region. *Earth Interact.* **2010**, *14*, 1–25. [CrossRef]
9. Bai, Y.; Li, X.; Zhou, S.; Yang, X.; Yu, K.; Wang, M.; Liu, S.; Wang, P.; Wu, X.; Wang, X.; et al. Quantifying plant transpiration and canopy conductance using eddy flux data: An underlying water use efficiency method. *Agric. For. Meteorol.* **2019**, *271*, 375–384. [CrossRef]
10. Mendes, K.R.; Campos, S.; Mutti, P.R.; Ferreira, R.R.; Ramos, T.M.; Marques, T.V.; dos Reis, J.S.; de Lima Vieira, M.M.; Silva, A.C.N.; Marques, A.M.S.; et al. Assessment of SITE for CO2 and Energy Fluxes Simulations in a Seasonally Dry Tropical Forest (Caatinga Ecosystem). *Forests* **2021**, *12*, 86. [CrossRef]
11. Ruhoff, A.L.; Paz, A.R.; Collischonn, W.; Aragao, L.E.O.C.; Rocha, H.R.; Malhi, Y.S. A MODIS-Based Energy Balance to Estimate Evapotranspiration for Clear-Sky Days in Brazilian Tropical Savannas. *Remote Sens.* **2012**, *4*, 703–725. [CrossRef]

12. Nascimento, G.S.; Ruhoff, A.; Cavalcanti, J.R.; Marques, D.M.; Roberti, D.R.; da Rocha, H.R.; Munar, A.M.; Fragoso, C.R.; de Oliveira, M.B.L. Assessing CERES Surface Radiation Components for Tropical and Subtropical Biomes. *IEEE J. Sel. Top. Appl. Earth Obs. Remote Sens.* **2019**, *1*, 3826–3840. [CrossRef]
13. Fonseca, L.D.M.; Dalagnol, R.; Malhi, Y.; Rifai, S.W.; Costa, G.B.; Silva, T.S.F.; da Rocha, H.R.; Tavares, I.B.; Borma, L.S. Phenology and Seasonal Ecosystem Productivity in an Amazonian Floodlain Forest. *Remote Sens.* **2019**, *11*, 1530. [CrossRef]
14. Moreira, A.A.; Ruhoff, A.L.; Roberti, D.R.; de Arruda Souza, V.; da Rocha, H.R.; Paiva, R.C.D. Assessment of terrestrial water balance using remote sensing data in South America. *J. Hydrol.* **2019**, *575*, 131–147. [CrossRef]
15. Myneni, R.; Knyazikhin, Y.; Park, T. MCD15A3H MODIS/Terra+Aqua Leaf Area Index/FPAR 4-Day L4 Global 500m SIN Grid V006. NASA EOSDIS Land Processes DAAC. Available online: https://lpdaac.usgs.gov/products/mcd15a3hv006/ (accessed on 26 April 2022).
16. Laipelt, L.; Ruhoff, A.L.; Fleischmann, A.S.; Kayser, R.H.B.; Kich, E.M.; da Rocha, H.R.; Neale, C.M.U. Assessment of an Automated Calibration of the SEBAL Algorithm to Estimate Dry-Season Surface-Energy Partitioning in a Forest–Savanna Transition in Brazil. *Remote Sens.* **2020**, *12*, 1108. [CrossRef]
17. Ferreira, R.R.; Mutti, P.R.; Mendes, K.R.; Campos, S.; Marques, T.V.; Oliveira, C.P.; Gonçalves, W.; Mota, J.; Difante, G.; Urbano, S.A.; et al. An assessment of the MOD17A2 gross primary production product in the Caatinga biome, Brazil. *Int. J. Remote Sens.* **2021**, *42*, 1275–1291. [CrossRef]
18. Fei, X.H.; Jin, Y.Q.; Zhang, Y.P.; Sha, L.Q.; Liu, Y.T.; Song, Q.H.; Zhou, W.; Liang, N.; Yu, G.; Zhang, L.; et al. Eddy covariance and biometric measurements show that a savanna ecosystem in Southwest China is a carbon sink. *Sci. Rep.* **2017**, *7*, 41025. [CrossRef]
19. Kanniah, K.D.; Beringer, J.; Hutley, L.B. Environmental controls on the spatial variability of savanna productivity in the Northern Territory, Australia. *Agric. For. Meteorol.* **2011**, *151*, 1429–1439. [CrossRef]
20. Quansah, E.; Mauder, M.; Balogun, A.A.; Amekudzi, L.K.; Hingerl, L.; Bliefernicht, J.; Kunstmann, H. Carbon dioxide fluxes from contrasting ecosystems in the Sudanian Savanna in West Africa. *Carbon Balance Manag.* **2015**, *10*, 1. [CrossRef]
21. Scheiter, S.; Higgins, S.I.; Beringer, J.; Hutley, L.B. Climate change and long-term fire management impacts on Australian savannas. *New Phytol.* **2015**, *205*, 1211–1226. [CrossRef]
22. Scott, R.L.; Hamerlynck, E.P.; Jenerette, G.D.; Moran, M.S.; Huxman, T.E.; Barron-Gafford, G.A. Carbon dioxide exchange in a semidesert grassland through drought-induced vegetation change. *J. Geophys. Res.* **2010**, *115*, G03026. [CrossRef]
23. Davidson, E.A.; de Araújo, A.C.; Artaxo, P.; Balch, J.K.; Brown, I.F.; Bustamante, M.M.; Coe, M.T.; DeFries, R.S.; Keller, M.; Longo, M.; et al. The Amazon basin in transition. *Nature* **2012**, *481*, 321–328. [CrossRef] [PubMed]
24. WMO. WMO Greenhouse Gas Bulletin (GHG Bulletin)—No.15: The State of Greenhouse Gases in the Atmosphere Based on Global Observations through 2018. 2019. Available online: https://library.wmo.int/doc_num.php?explnum_id=10437 (accessed on 27 March 2022).
25. Flux Tower BR-Sa1. Available online: https://ameriflux.lbl.gov/sites/siteinfo/BR-Sa1 (accessed on 1 February 2022).
26. Flux Tower BR-CST. Available online: https://ameriflux.lbl.gov/sites/siteinfo/BR-CST (accessed on 1 February 2022).
27. Flux Tower BR-Npw. Available online: https://ameriflux.lbl.gov/sites/siteinfo/BR-Npw (accessed on 1 February 2022).
28. Alcântara, L.R.P.; Coutinho, A.P.; dos Santos Neto, S.M.; Carvalho de Gusmão da Cunha Rabelo, A.E.; Antonino, A.C.D. Modeling of the Hydrological Processes in Caatinga and Pasture Areas in the Brazilian Semi-Arid. *Water* **2021**, *13*, 1877. [CrossRef]
29. Dalmagro, H.J.; Zanella de Arruda, P.H.; Vourlitis, G.L.; Lathuilliere, M.J.; de Nogueira, S.J.; Couto, E.G.; Johnson, M.S. Radiative forcing of methane fluxes offsets net carbon dioxide uptake for a tropical flooded forest. *Glob. Change Biol.* **2019**, *25*, 1967–1981. [CrossRef]
30. Da Silva, J.B.; Valle Junior, L.C.G.; Faria, T.O.; Marques, J.B.; Dalmagro, H.J.; Nogueira, J.S.; Vourlitis, G.L.; Rodrigues, T.R. Temporal Variability in Evapotranspiration and Energy Partitioning over a Seasonally Flooded Scrub Forest of the Brazilian Pantanal. *Agric. For. Meteorol.* **2021**, *308*, 108559. [CrossRef]
31. Alvares, C.A.; Stape, J.L.; Sentelhas, P.C.; de Moraes Gonçalves, J.L.; Sparovek, G. Köppen's climate classification map for Brazil. *Meteorol. Z.* **2013**, *22*, 711–728. [CrossRef]
32. Fantin-Cruz, I.; Girard, P.; Zeilhofer, P.; Collischonn, W.; da Cunha, C.N. Unidades fitofisionômicas em mesoescala no Pantanal Norte e suas relações com a geomorfologia. *Biota Neotropica* **2010**, *10*, 31–38. [CrossRef]
33. Johnson, M.S.; Couto, E.G.; Pinto, O.B.; Milesi, J.; Santos Amorim, R.S.; Messias, I.A.M.; Biudes, M.S. Soil CO2 Dynamics in a Tree Island Soil of the Pantanal: The Role of Soil Water Potential. *PLoS ONE* **2013**, *8*, e64874. [CrossRef]
34. Rodrigues, T.R.; Curado, L.F.A.; Novais, J.W.Z.; de Oliveira, A.G.; de Paulo, S.R.; Nogueira, D.S. Distribuição sazonal dos componentes do balanço de energia no norte do Pantanal. *Rev. Ciências Agro-Ambient.* **2011**, *9*, 165–175.
35. Biudes, M.S.; Machado, N.G.; Danelichen, V.H.d.M.; Souza, M.C.; Vourlitis, G.L.; Nogueira, J.d.S. Ground and remote sensing-based measurements of leaf area index in a transitional forest and seasonal flooded forest in Brazil. *Int. J. Biometeorol.* **2014**, *58*, 1181–1193. [CrossRef]
36. Couto, E.; Klinger, P.; Jacomine, T.; Nunes Da Cunha, C.; Vechiatto, A. Guide of technique excursion of the XIV RBMCSA. In *XIV Reunião Brasileira de Manejo e Conservação do solo e da água*; UFMT: Cuiabá, Brazil, 2016; 68p.
37. Vourlitis, G.L.; de Almeida Lobo, F.; Lawrence, S.; Holt, K.; Zappia, A.; Pinto, O.B.; de Souza Nogueira, J. Nutrient resorption in tropical savanna forests and woodlands of central Brazil. *Plant Ecol.* **2014**, *215*, 963–975. [CrossRef]
38. Prado, M.J.D. Intercâmbio Gasoso e Relações Hídricas de Duas Espécies de Combretum no Pantanal Mato-Grossense. Master's Thesis, Federal University of Mato Grosso, Campo Grande, Brazil, 2015; p. 56.

39. Santos, S.A.; da Cunha, C.N.; Tomás, W.; de Abreu, U.G.P.; Arieira, J. *Plantas Invasoras no Pantanal: Como Entender o Problema e Soluções de Manejo por Meio de Diagnóstico Participativo*; Boletim de Pesquisa e Desenvolvimento 66; Embrapa Pantanal: Corumbá, Brazil, 2006; p. 45.
40. Pott, V.J.; Pott, A. *Plantas do Pantanal Corumbá: Embrapa—Centro de Pesquisa Agropecuária do Pantanal*; Embrapa Pantanal: Corumbá, Brazil, 1994; 320p.
41. Junk, W.J.; da Cunha, C.N.; Wantzen, K.M.; Petermann, P.; Strüssmann, C.; Marques, M.I.; Adis, J. Biodiversity and its conservation in the Pantanal of Mato Grosso, Brazil. *Aquat. Sci.* **2006**, *68*, 278–309. [CrossRef]
42. Rice, A.H.; Pyle, E.H.; Saleska, S.R.; Hutyra, L.; Palace, M.; Keller, M.; de Camargo, P.B.; Portilho, K.; Marques, D.F.; Wofsy, S.C. Carbon balance and vegetation dynamics in an old-growth Amazonian forest. *Ecol. Appl.* **2004**, *14*, 55–71. [CrossRef]
43. Clark, D.B. Abolishing virginity. *J. Trop. Ecol.* **1996**, *12*, 735–739. [CrossRef]
44. Stark, S.; Leitold, V.; Wu, J.; Hunter, M.; Castilho, C.; Costa, F.; McMahon, S.; Parker, G.; Shimabukuro, M.; Lefsky, M.; et al. Amazon forest carbon dynamics predicted by profiles of canopy leaf area and light environment. *Ecol. Lett.* **2012**, *15*, 1406–1414. [CrossRef]
45. Silva, A.C.; Rêgo Mendes, K.; Santos e Silva, C.M.; Torres Rodrigues, D.; Brito Costa, G.; Thainara Corrêa da Silva, D.; Rodrigues Mutti, P.; Rodrigues Ferreira, R.; Guedes Bezerra, B. Energy Balance, CO2 Balance, and Meteorological Aspects of Desertification Hotspots in Northeast Brazil. *Water* **2021**, *13*, 2962. [CrossRef]
46. Hutyra, L.R.; Munger, J.W.; Saleska, S.R.; Gottlieb, E.; Daube, B.C.; Dunn, A.L.; Amaral, D.F.; de Camargo, P.B.; Wofsy, S.C. Seasonal controls on the exchange of carbon and water in an Amazonian rain forest. *J. Geophys. Res.* **2007**, *112*, 1–16. [CrossRef]
47. Reichstein, M.; Falge, E.; Baldocchi, D.; Papale, D.; Aubinet, M.; Berbigier, P.; Bernhofer, C.; Buchmann, N.; Giulmanov, T.; Granier, A.; et al. On the separation of net ecosystem exchange into assimilation and ecosystem respiration: Review and improved algorithm. *Glob. Change Biol.* **2005**, *11*, 1424–1439. [CrossRef]
48. Falge, E.; Baldocchi, D.; Olson, R.; Anthoni, P.; Aubinet, M.; Bernhofer, C.; Burba, G.; Ceulemans, R.; Clement, R.; Dolman, H.; et al. Gap filling strategies for defensible annual sums of net ecosystem exchange. *Agric. For. Meteorol.* **2001**, *107*, 43–69. [CrossRef]
49. Lasslop, G.; Reichstein, M.; Papale, D.; Richardson, A.D.; Arneth, A.; Barr, A.; Stoy, P.; Wohlfahrt, G. Separation of net ecosystem exchange into assimilation and respiration using a light response curve approach: Critical issues and global evaluation. *Glob. Change Biol.* **2010**, *16*, 187–208. [CrossRef]
50. Monteith, J.L. Solar Radiation and Productivity in Tropical Ecosystems. *J. Appl. Ecol.* **1972**, *9*, 747. [CrossRef]
51. Myneni, R.; Knyazikhin, Y.; Park, T. 15. MOD15A2H MODIS/Terra Leaf Area Index/FPAR 8-Day L4 Global 500 m SIN Grid V006 [Data Set]. NASA EOSDIS Land Processes DAAC. Available online: https://catalog.data.gov/dataset/modis-terraaqua-leaf-area-index-fpar-4-day-l4-global-500m-sin-grid-v006 (accessed on 26 April 2022).
52. Rienecker, M.M.; Suarez, M.J.; Gelaro, R.; Todling, R.; Bacmeister, J.; Liu, E.; Woollen, J. 15. MERRA: NASA's Modern-Era Retrospective Analysis for Research and Applications. *J. Clim.* **2011**, *24*, 3624–3648. [CrossRef]
53. Pei, Y.; Dong, J.; Zhang, Y.; Yang, J.; Zhang, Y.; Jiang, C.; Xiao, X. Performance of Four State-of-the-art GPP Products (VPM, MOD17, BESS and PML) for Grasslands in Drought Years. *Ecol. Inform.* **2020**, *56*, 101052. [CrossRef]
54. Running, S.W.; Nemani, R.R.; Heinsch, F.A.; Zhao, M.; Reeves, M.; Hashimoto, H. A Ontinuous Satellite–derived Measure of Global Terrestrial Primary Production. *Bioscience* **2004**, *54*, 547–560. [CrossRef]
55. Friedl, M.A.; Sulla-Menashe, D.; Tan, B.; Schneider, A.; Ramankutty, N.; Sibley, A.; Huang, X. MODIS Collection 5 global land cover: Algorithm refinements and characterization of new datasets. *Remote Sens. Environ.* **2010**, *114*, 168–182. [CrossRef]
56. Zhu, X.; Pei, Y.; Zheng, Z.; Dong, J.; Zhang, Y.; Wang, J.; Chen, L.; Doughty, R.; Zhang, G.; Xiao, X. Underestimates of Grassland Gross Primary Production in MODIS Standard Products. *Remote Sens.* **2018**, *10*, 1771. [CrossRef]
57. Danelichen, V.H.M.; Biudes, M.S.; Velasque, M.C.S.; Machado, N.G.; Gomes, R.S.R.; Vourlitis, G.L.; Nogueira, J.S. Estimating of gross primary production in an Amazon-Cerrado transitional forest using MODIS and Landsat imagery. *An. Acad. Bras. Cienc.* **2015**, *87*, 1545–1564. [CrossRef]
58. Wang, J.; Dong, J.; Liu, J.; Huang, M.; Li, G.; Running, S.W.; Smith, W.K.; Harris, W.; Saigusa, N.; Kondo, H.; et al. Comparison of gross primary productivity derived from GIMMS NDVI3g, GIMMS, and MODIS in southeast Asia. *Remote Sens.* **2014**, *6*, 2108–2133. [CrossRef]
59. Kuglitsch, F.G.; Reichstein, M.; Beer, C.; Carrara, A.; Ceulemans, R.; Granier, A.; Janssens, I.A.; Köstner, B.; Lindroth, A.; Loustau, D.; et al. Characterisation of ecosystem water-use efficiency of European forests from eddy-covariance measurements. *Biogeosciences Discuss.* **2008**, *5*, 4481–4519. [CrossRef]
60. Berbigier Gier, P.; Bonnefond, J.-M.; Mellmann, P. CO2 and water vapour fluxes for 2 years above Euroflux forest site. *Agric. Forest Meterol.* **2001**, *108*, 183–197. [CrossRef]
61. Liu, X.; Chen, X.; Li, R.; Long, F.; Zhang, L.; Zhang, Q.; Li, J. Water-use efficiency of an old-growth forest in lower subtropical China. *Sci. Rep.* **2017**, *7*, 42761. [CrossRef] [PubMed]
62. Tang, X.; Li, H.; Desai, A.R.; Nagy, Z.; Luo, J.; Kolb, T.E.; Olioso, A.; Xu, X.; Yao, L.; Kutsch, W.; et al. How is water-use efficiency of terrestrial ecosystems distributed and changing on Earth? *Sci. Rep.* **2014**, *4*, 7483. [CrossRef] [PubMed]
63. Yu, G.R.; Zhang, L.M.; Sun, X.M.; Fu, Y.L.; Wen, X.F.; Wang, Q.F.; Li, S.G.; Ren, C.Y.; Song, X.I.A.; Liu, Y.F.; et al. Environmental controls over carbon exchange of three forest ecosystems in eastern China. Glob. *Change Biol.* **2008**, *14*, 2555–2571. [CrossRef]
64. Rodrigues, A.; Pita, G.; Mateus, J.; Kurz-Besson, C.; Casquilho, M.; Cerasoli, S.; Gomes, A.; Pereira, J. Eight years of continuous carbon fluxes measurements in a Portuguese eucalypt stand under two main events: Drought and felling. *Agric. Forest Meterol.* **2011**, *151*, 493–507. [CrossRef]

65. Law, B.E.; Williams, M.; Anthoni, P.M.; Baldocchi, D.D.; Unsworth, M.H. Measuring and modelling seasonal variation of carbon dioxide and water vapour exchange of a Pinus ponderosa forest subject to soil water deficit. *Glob. Change Biol.* **2000**, *6*, 613–630. [CrossRef]
66. Krishnan, P.; Black, T.A.; Grant, N.J.; Barr, A.G.; Hogg, E.T.H.; Jassal, R.S.; Morgenstern, K. Impact of changing soil moisture distribution on net ecosystem productivity of a boreal aspen forest during and following drought. *Agric. Forest Meterol.* **2006**, *139*, 208–223. [CrossRef]
67. Ponton, S.; Flanagan, L.B.; Alstad, K.P.; Johnson, B.G.; Morgenstern, K.A.I.; Kljun, N.; Black, T.A.; Barr, A.G. Comparison of ecosystem water-use efficiency among Douglas-fir forest, aspen forest and grassland using eddy covariance and carbon isotope techniques. *Glob. Change Biol.* **2006**, *12*, 294–310. [CrossRef]
68. Liang, S.; Li, X.; Wang, J. Estimate of vegetation production of terrestrial ecosystem. In *Advanced Remote Sensing: Terrestrial Information Extraction and Applications*; Academic Press: Oxford, UK, 2020; pp. 581–620.
69. Costa, M.H.; Biajoli, M.C.; Sanches, L.; Malhado, A.C.M.; Hutyra, L.R.; da Rocha, H.R.; Aguiar, R.G.; de Araújo, A.C. Atmospheric versus vegetation controls of Amazonian tropical rain forest evapotranspiration: Are the wet and seasonally dry rain forests any different? *J. Geophys. Res.* **2010**, *115*, 9. [CrossRef]
70. Dalmagro, H.J.; de Lobo, F.A.; Vourlitis, G.L.; Dalmolin, Â.C.; Antunes, M.Z.; Ortíz, C.E.R.; de Nogueira, J.S. Photosynthetic parameters of two invasive tree species of the Brazilian Pantanal in response to seasonal flooding. *Photosynthetica* **2013**, *51*, 281–294. [CrossRef]
71. Dalmagro, H.J.; Lathuillière, M.J.; Vourlitis, G.L.; Campos, R.C.; Pinto, O.B.; Johnson, M.S.; Ortíz, C.E.; Lobo, F.D.A.; Couto, E.G. Physiological responses to extreme hydrological events in the Pantanal wetland: Heterogeneity of a plant community containing super-dominant species. *J. Veg. Sci.* **2016**, *27*, 568–577. [CrossRef]
72. Sanches, L.; Vourlitis, G.L.; de Carvalho Alves, M.; Pinto-Júnior, O.B.; de Souza Nogueira, J. Seasonal patterns of evapotranspiration for a Vochysia divergens forest in the Brazilian Pantanal. *Wetlands* **2011**, *31*, 1215–1225. [CrossRef]
73. Vourlitis, G.L.; de Almeida Lobo, F.; Biudes, M.S.; Rodríguez Ortíz, C.E.; de Souza Nogueira, J. Spatial Variations in Soil Chemistry and Organic Matter Content across a Vochysia Divergens Invasion Front in the Brazilian Pantanal. *Soil Sci. Soc. Am. J.* **2011**, *75*, 1554–1561. [CrossRef]
74. Sjöström, M.; Ardö, J.; Arneth, A.; Boulain, N.; Cappelaere, B.; Eklundh, L.; de Grandcourt, A.; Kutsch, W.L.; Merbold, L.; Nouvellon, Y.; et al. Exploring the potential of MODIS EVI for modeling gross primary production across African ecosystems. *Remote Sens. Env.* **2011**, *115*, 1081–1089. [CrossRef]
75. Singh, N.; Patel, N.R.; Bhattacharya, B.K.; Soni, P.; Parida, B.R.; Parihar, J.S. Analyzing the dynamics and inter-linkages of carbon and water fluxes in subtropical pine (Pinus roxburghii) ecosystem. *Agric. Forest Meterol.* **2014**, *197*, 206–218. [CrossRef]
76. Ito, A.; Inatomi, M. Water-use efficiency of the terrestrial biosphere: A model analysis focusing on interactions between the global carbon and water cycles. *J. Hydrometeorol.* **2012**, *13*, 681–694. [CrossRef]
77. Xue, B.L.; Guo, Q.H.; Otto, A.; Xiao, J.F.; Tao, S.L.; Li, L. Global patterns, trends, and drivers of water use efficiency from 2000 to 2013. *Ecosphere* **2015**, *6*, 1–18. [CrossRef]
78. Marengo, J.A.; Torres, R.R.; Alves, L.M. Drought in Northeast Brazil—past, present, and future. *Theor. Appl. Climatol.* **2017**, *129*, 1189–1200. [CrossRef]
79. Mendes, K.R.; Campos, S.; da Silva, L.L.; Mutti, P.R.; Ferreira, R.R.; Medeiros, S.S.; Perez-Marin, A.M.; Marques, T.V.; Ramos, T.M.; de Lima Vieira, M.M.; et al. Seasonal variation in net ecosystem CO2 exchange of a Brazilian seasonally dry tropical forest. *Sci. Rep.* **2020**, *10*, 9454. [CrossRef]
80. Campos, S.; Mendes, K.R.; da Silva, L.L.; Mutti, P.R.; Medeiros, S.S.; Amorim, L.B.; dos Santos, C.A.C.; Perez-Marin, A.M.; Ramos, T.M.; Marques, T.V.; et al. Closure and partitioning of the energy balance in a preserved area of Brazilian seasonally dry tropical forest. *Agric. For. Meteorol.* **2019**, *271*, 398–412. [CrossRef]
81. Marques, T.V.; Mendes, K.; Mutti, P.; Medeiros, S.; Silva, L.; Perez-Marin, A.M.; Campos, S.; Lúcio, P.S.; Lima, K.; Reis, J.; et al. Environmental and biophysical controls of evapotranspiration from Seasonally Dry Tropical Forests (Caatinga) in the Brazilian Semiarid. *Agric. For. Meteorol.* **2020**, *287*, 107957. [CrossRef]
82. Mendes, K.R.; Batista-Silva, W.; Dias-Pereira, J.; Pereira, M.P.; Souza, E.V.; Serrão, J.E.; Granja, J.A.; Pereira, E.C.; Gallacher, D.J.; Mutti, P.R.; et al. Leaf plasticity across wet and dry seasons in Croton blanchetianus (Euphorbiaceae) at a tropical dry forest. *Sci. Rep.* **2022**, *12*, 954. [CrossRef] [PubMed]

remote sensing

Article

Phenological Responses to Snow Seasonality in the Qilian Mountains Is a Function of Both Elevation and Vegetation Types

Yantao Liu [1], Wei Zhou [1,2,3,*], Si Gao [1], Xuanlong Ma [4] and Kai Yan [1]

1 School of Land Science and Technology, China University of Geosciences Beijing, Beijing 100083, China; ytliu@cugb.edu.cn (Y.L.); gaosi_2021@email.cugb.edu.cn (S.G.); kaiyan@cugb.edu.cn (K.Y.)
2 Key Laboratory of Land Consolidation and Rehabilitation, Ministry of Natural Resources, Beijing 100035, China
3 Technology Innovation Center for Ecological Restoration in Mining Areas, Ministry of Natural Resources, Beijing 100083, China
4 College of Earth and Environmental Sciences, Lanzhou University, Lanzhou 730000, China; xlma@lzu.edu.cn
* Correspondence: zhouw@cugb.edu.cn

Abstract: In high-elevation mountains, seasonal snow cover affects land surface phenology and the functioning of the ecosystem. However, studies regarding the long-term effects of snow cover on phenological changes for high mountains are still limited. Our study is based on MODIS data from 2003 to 2021. First, the NDPI was calculated, time series were reconstructed, and an SG filter was used. Land surface phenology metrics were estimated based on the dynamic thresholding method. Then, snow seasonality metrics were also estimated based on snow seasonality extraction rules. Finally, correlation and significance between snow seasonality and land surface phenology metrics were tested. Changes were analyzed across elevation and vegetation types. Results showed that (1) the asymmetry in the significant correlation between the snow seasonality and land surface phenology metrics suggests that a more snow-prone non-growing season (earlier first snow, later snowmelt, longer snow season and more snow cover days) benefits a more flourishing vegetation growing season in the following year (earlier start and later end of growing season, longer growing season). (2) Vegetation phenology metrics above 3500 m is sensitive to the length of the snow season and the number of snow cover days. The effect of first snow day on vegetation phenology shifts around 3300 m. The later snowmelt favors earlier and longer vegetation growing season regardless of the elevation. (3) The sensitivity of land surface phenology metrics to snow seasonality varied among vegetation types. Grass and shrub are sensitive to last snow day, alpine vegetation to snow season length, desert to number of snow cover days, and forest to first snow day. In this study, we used a more reliable NDPI at high elevations and confirmed the past conclusions about the impact of snow seasonality metrics. We also described in detail the curves of snow seasonal metrics effects with elevation change. This study reveals the relationship between land surface phenology and snow seasonality in the Qilian Mountains and has important implications for quantifying the impact of climate change on ecosystems.

Keywords: land surface phenology; NDPI; Qilian Mountains; snow cover; high elevation

Citation: Liu, Y.; Zhou, W.; Gao, S.; Ma, X.; Yan, K. Phenological Responses to Snow Seasonality in the Qilian Mountains Is a Function of Both Elevation and Vegetation Types. *Remote Sens.* **2022**, *14*, 3629. https://doi.org/10.3390/rs14153629

Academic Editor: Xinghua Li

Received: 27 June 2022
Accepted: 25 July 2022
Published: 29 July 2022

1. Introduction

Evidence suggests that global temperatures have continued to rise over the last two decades and will continue to warm over the next three decades [1], which affects many ecosystems [2–4]. Alpine ecosystems are considered to be particularly sensitive to climate change because of harsh natural environments [5–8]. Therefore, accurate assessment of the impacts of climate change in alpine ecosystems is essential.

Land surface phenology (LSP) is defined as the seasonal change pattern of surface vegetation obtained from remote sensing observations, which is usually used to describe the start, end and length of the vegetation growing season [9–11]. Unlike traditional ground-based observations that can record dates of budburst or flushing, LSP is used to describe the full process of regional greening. This may not correspond to a specific vegetation event but can provide a rapid understanding of the key stages of the overall greening of a region [9]. LSP is one of the most sensitive and easily observable nature features when analyzing the response of vegetation to climate change [12], and exploring its changes provides an important avenue for studying severe climate anomalies. Climate change can interfere with vegetation germination. In addition, changes in LSP may have a significant impact on carbon and water cycles [13–15]. An integrated analysis of the impact of climate change on land surface phenology is important for understanding the impact of future climate change.

Driving factors for trends in land surface phenology have frequently been attributed to changes in temperature and precipitation [16–18]. However, as one of the typical features in areas with stable snowpack, changes in snow seasonality can also cause changes in land surface phenology [19,20]. Specifically, Snow accumulates or melts on soil and vegetation, which can directly alter the hydrothermal conditions under which vegetation grows and develops (Figure 1). Snow directly affects near-surface temperatures in several ways. Snow cover in winter insulates the soil from cold air and maintains soil temperature [20]. Soil temperature is higher than air temperature in early spring under snow cover [21]. Due to the insulating nature of snow, temperature no longer has a direct effect on vegetation [22]. The timing of snowmelt is sometimes a more important factor in the growing season than air temperature [23]. These regulatory effects of snow accumulation and snowmelt on surface temperature have important implications for land surface phenology and soil moisture content. Snow is the main source of freshwater for alpine vegetation, as snowmelt provides the necessary moisture for vegetation to sprout in the form of soil water [24–27]. Snow cover protects vegetation and soil from harsh natural hazards such as wind erosion, freezing damage, and intense solar radiation, which often occur in high-elevation mountains and can seriously hinder vegetation growth [28–31]. It has also been demonstrated that winter snow can indirectly affect the carbon sequestration capacity of vegetation by altering community structure and activity of soil microorganisms [32].

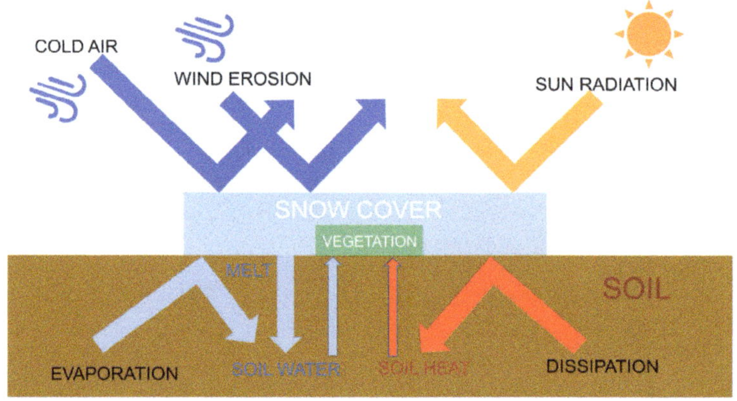

Figure 1. Schematic diagram of snow cover effects.

Although there are many studies focusing on the snow cover effect on land surface phenology, they have inconsistent results [33–38]. In some warm and dry regions, winter snow cover has been shown to favor early spring vegetation germination and prolong the growing season, while the opposite is true in cold, humid regions [35,36], clearly demonstrating that the effect of snow cover varies under different hydrothermal conditions. However, in Inner Mongolia, China [38], and the French Alps [39], two regions with different natural conditions, snow has a similar negative effect on land surface phenology:

late snowmelt delays vegetation emergence. Grass is the predominant vegetation type in these two regions, and the mechanism for the effect of snow on the same species should be consistent. Even so, the impact of snow on a single vegetation type in the same area changes with terrain [33,34]. In short, the influence of snow seasonality on land surface phenology is determined by the coupling of multiple factors such as water and heat conditions, vegetation type and terrain. There remains a lot of views on specific conclusions, especially the effect of snow as a function of elevation gradients and vegetation types, and additional study is required. Additionally, the spring phenology extracted using NDVI is often affected by preseason snow, which may lead to inaccurate conclusions [40,41]. Unraveling the effects of snow on land surface phenology can help identify the mechanisms of change in land surface phenology.

The Qilian Mountains area (QLMA) is an essential part of the Qinghai–Tibet Plateau and is considered as an important ecological barrier in western China. The QLMA has large elevation differences, a variety of vegetation types and significant climate change. In addition, the relatively small area of QLMA could mitigate the impact of spatial differences on the results. It is an ideal laboratory for studying the vegetation response to climate change. Studies on vegetation phenology for the Qilian Mountains region are still limited. We quantified the response of land surface phenology metrics to different snow seasonality metrics from 2002 to 2021. Annual snow seasonality metrics include the first snow day (FSD), last snow day (LSD), snow season length (SSL) and total number of days with snow cover (SCD). Land surface phenology metrics include the start of the growing season (SOS), end of the growing season (EOS) and length of the growing season (LOS) estimated by the normalized difference phenology index (NDPI), which was proven to be an accurate vegetation index for estimating land surface phenology at high elevation [42]. More comprehensive metrics and more reliable vegetation indices enhance the richness and accuracy of our conclusions. In addition, we chose two representative influencing factors, elevation and vegetation type, which may be helpful to explain the complex effects of snow seasonality.

To better understand the relationship between snow and land surface phenology, the following three research questions are proposed:

1. What are the distribution characteristics of snow seasonality and land surface phenology in the Qilian Mountains area?
2. What is the impact of snow on land surface phenology in the study area?
3. How does the phenological response change with elevation and by vegetation types?

2. Materials and Methods

2.1. Study Area

The QLMA lies at the intersection of the Tibetan, Mongolian and Loess plateaus (35.84°–39.97°N, 93.61°–103.90°E), with Qinghai Province in the south and Gansu Province of China in the north (Figure 2). The average elevation of the QLMA is over 3000 m, higher in the center and lower in the surroundings. This area belongs to the mid-latitude high elevation region. Most of the QLMA is in the temperate semiarid zone of the highlands [43], where solar radiation is strong. The average annual precipitation is 300–500 mm, more in the east than in the west. These make QLMA a sensitive area for climate change.

As an important ecological barrier in northwestern China, the QLMA is home to headwaters of rivers fed by snow and glacial meltwater. Grassland vegetation, desert vegetation, shrub and alpine vegetation cover more than 90% of the QLMA. Additional vegetation types include coniferous forest, broadleaved forest and swamp vegetation [44,45]. Because of its rich gene pool of alpine species, the QLMA is considered a key area and a priority area for biodiversity conservation in China [46]. A pilot Qilian Mountain National Park in the QLMA was established in 2017 to protect forests, grasslands, wetlands, surface water resources and glaciers [47]. Collectively, the QLMA is an ideal laboratory for studying climate change and ecosystems in cold and arid regions.

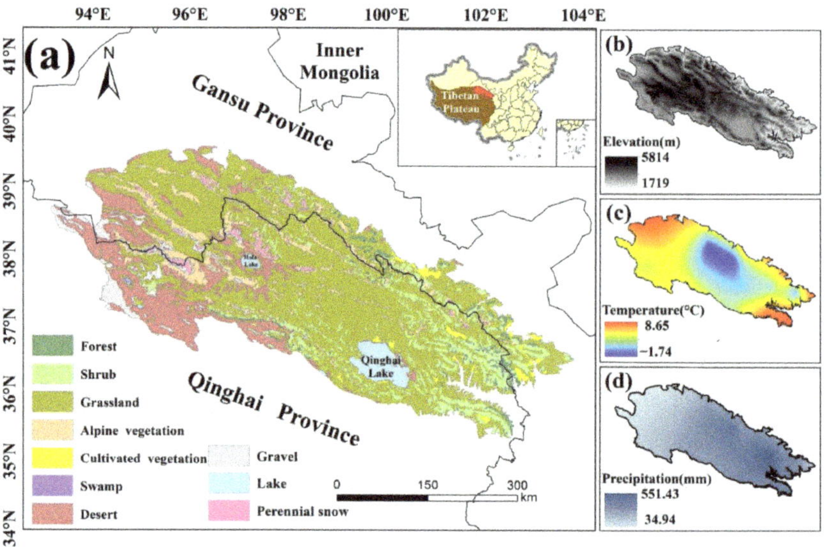

Figure 2. Map of the (**a**) land cover types, (**b**) elevation, (**c**) annual mean temperature, and (**d**) annual mean precipitation of the study area.

2.2. Data Sources and Pre-Processing

The normalized difference snow index (NDSI) was used to create snow pixels with a resolution of 500 m provided by MO(Y)D10A1 V6. The Terra(O) and Aqua(Y) satellites provide a separate daily NDSI from 1 September 2002 to 1 June 2020 [48]. 'NDSI_Snow_Cover_Class' and 'NDSI_Snow_Cover_Algorithm_Flags_QA' are the bands we used to eliminate invalid pixels and 'NDSI_Snow_Cover' is the band to extract NDSI. Data were from NASA's NSIDC DAAC at CIRES, with a large percentage of cloud pixels [49–51]. Some preprocessing was required to eliminate cloud obscuration [52,53]. First, pixels with a non-zero value of the 'NDSI_Snow_Cover_Class' and 'NDSI_Snow_Cover_Algorithm_Flags_QA' band were masked out, as these pixels represent invalid values such as missing data and clouds. Second, daily MOD10A1 and MYD10A1 data were combined using daily maximum values to reduce the number of cloud pixels. Finally, the max of the cloud-free pixels within a three-day time window surrounding the cloud pixels was taken as a replacement value. Experiments proved that these operations can effectively reduce the proportion of cloud pixels and improve the accuracy of snow products [51,54]. These preprocesses were completed using the Google Earth Engine.

The MOD09A1 V6 products provide the surface spectral reflectance of Terra MODIS bands 1–7 at 500 m resolution [55] containing seven bands that have been corrected for atmospheric conditions such as gasses and aerosols. It was used to calculate NDPI and thus estimate land surface phenology from 2003 to 2021. Data were provided by NASA LP DAAC at the USGS EROS Center.

The QLMA vegetation distribution dataset (vegetation pattern data (1:1,000,000) in the Qilian Mountains) was obtained from the National Cryosphere Desert Data Center (NCDC, http://www.ncdc.ac.cn/, accessed on 6 September 2021) and the digital elevation model (DEM) from 30 m ASTER GDEM was downloaded from the Geospatial Data Cloud (http://www.gscloud.cn/, accessed on 13 January 2022). We used these data to distinguish the six types of vegetation and every 100 m in elevation of QLMA. Figure 3 shows the technical workflows.

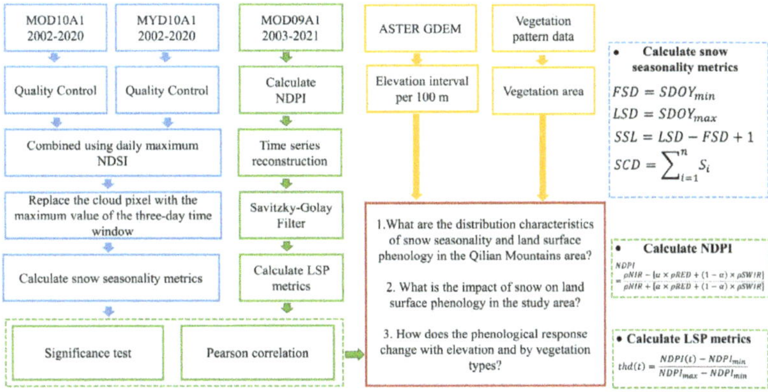

Figure 3. Overview of technical workflows.

2.3. Calculation of Snow Cover Seasonality

Value equal to 40 in the preprocessed 'NDSI_Snow_Cover' layer was chosen as a threshold for identifying snow pixels [56]. Pixels recognized as snow pixels were reclassified as 1, otherwise 0. Previous studies have shown that the snowpack on the Tibetan Plateau increases sharply in September and decreases sharply in the beginning of May [57]. Therefore, 1st September to 1st June of the following year is considered to be the potential snow season to exclude outliers that appear in the summer. A new time series was constructed: 1st September was set to be the first day of year, and 31st May of the following year was set to be the last. The metrics of the snow season were calculated based on methods in [58], where *FSD* and *LSD* are measured in day of year (DOY) and *SSL* and *SCD* are measured in days. The formulas are as follows:

$$FSD = SDOY_{min}, \tag{1}$$

$$LSD = SDOY_{max}, \tag{2}$$

$$SSL = LSD - FSD + 1, \tag{3}$$

$$SCD = \sum_{i=1}^{n} S_i, \tag{4}$$

where *SDOY* represents the distance between the date when a pixel is identified as a snow pixel and previous 1st September, n denotes the total number of days in the potential snow season, and S_i denotes the state of snow cover for any given day within *SDOY*; S_i equals 1 if there is snow and 0 if not.

2.4. Calculation of Land Surface Phenology

Shortwave infrared reflectance was combined with near-infrared and red reflectance to calculate the *NDPI*. It is an index for extracting accurate surface phenology to achieve high contrast between vegetation and background [59]. It was proven to be an accurate vegetation index for estimating land surface phenology in QLMA [42]. The formula is as follows:

$$NDPI = \frac{\rho NIR - [\alpha \times \rho RED + (1-\alpha) \times \rho SWIR]}{\rho NIR + [\alpha \times \rho RED + (1-\alpha) \times \rho SWIR]}, \tag{5}$$

where ρNIR represents the near-infrared band, ρRED represents the red band and $\rho SWIR$ represents the shortwave infrared band. In MOD09A1, ρRED corresponds to band 1, ρNIR to band 2 and $\rho SWIR$ to band 6. α is a constant value for a given sensor and is taken as 0.74 for MODIS products [59].

In this study, we first masked the pixels of cloud shadow, snow and cloud, then calculated *NDPI* values for the filtered images. Then we reconstructed the time series to unify the number of images into 72 in each year by calculating the average value. Finally, we smoothed the curves using the Savitzky–Golay (SG) filter. The window size was set to 5 and the number of polynomials was set to 3. Several methods have been developed to detect land surface phenology based on vegetation indices [60–63]. A simple and effective dynamic threshold method was used to extract land surface phenology from the reconstructed *NDPI* time series (Figure 4) [64]. Pixels with an intra-year *NDPI* change of less than 0.1 were excluded, and these were assumed to be areas without significant seasonality, as the *NDPI* of vegetation with distinct phenological stages generally increases from very small values (close to 0) to above 0.4 in study area. The formula is as follows:

$$thd(t) = \frac{NDPI(t) - NDPI_{min}}{NDPI_{max} - NDPI_{min}}, \quad (6)$$

where $NDPI(t)$ represents the *NDPI* value at the calendar year date sequence *t*; $NDPI_{min}$ represents the maximum value of *NDPI* in a year; $NDPI_{min}$ represents the minimum value of the time series vegetation curve on the left and right part of the curve (SOS corresponds to the left half of the curve and EOS corresponds to the right) in a year bounded by $NDPI_{max}$; and $thd(t)$ represents the percentage corresponding to $NDPI(t)$ after stretching it in time to a range of 0–1. Critical thresholds of 30% and 70% were selected [34].

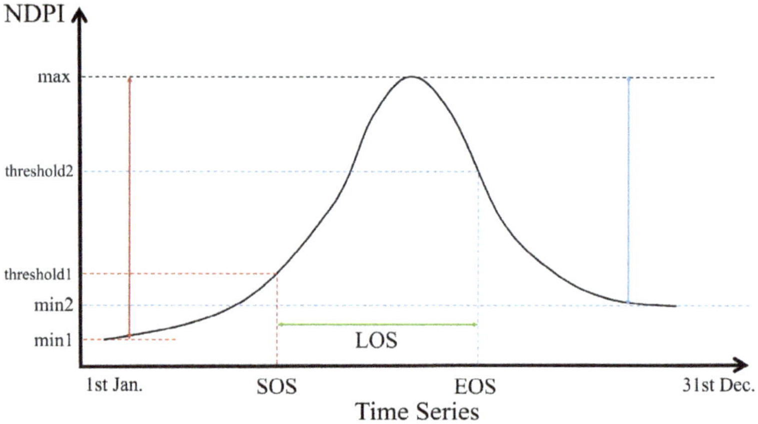

Figure 4. Schematic illustration for extracting LSP metrics using dynamics threshold method. The black curve represents the interannual NDPI for a given pixel. Red lines represent SOS, blue lines represent EOS and green lines represent LOS.

2.5. Correlation Analysis

To assess the relationship between snowpack and land surface phenology, we calculated the Pearson correlation coefficients between four snow season indicators (FSD, LSD, SSL and SCD) and three phenology indicators (SOS, EOS and LOS), and evaluated the significance of the correlation (*p*-value) by performing a *t*-test [36,65]. Based on the vegetation distribution, we focused on the correlation between snow season metrics and phenological metrics in six different vegetation types: forest, shrub, grass, alpine vegetation, cultivated vegetation and desert vegetation. In addition, an elevation gradient was set from 2500 m to 4500 m to study the vertical variation of the correlation between vegetation and snow seasonality. Correlations were calculated for each pixel and then further counted according to different elevations or vegetation types. The spatial patterns of snow seasonality and land surface phenology over the study area were characterized by the mean of the metrics.

3. Results

3.1. Spatial Pattern of Snow Seasonality over the QLMA

The spatial heterogeneity pattern of snow seasonality metrics can be observed in the QLMA (Figure 5). The earliest snowfall occurs in the central area within the QLMA (before October, DOY < 30). This region is also the one with the last snowmelt (after April, DOY > 210), resulting in the longest snow season. For the western and northern regions, the first snowfall occurs weeks later (before December, DOY < 90), and snowmelt occurs late (after April, DOY > 210), the length of the snow season is shorter than in the center. In the northern part of Qinghai Lake and the eastern and southern fringes of the study area, the first snowfall occurs the latest (after December, DOY > 120), the snowmelt occurs earliest (before February, DOY < 150), and the snow season is the shortest. The spatial pattern of SCD is similar to that of SSL, but the spatial differences are not significant because of the widespread presence of intermittent snow. The SCD is less than 30 d in most of the study area, especially in the southeast, but higher in the central and western parts.

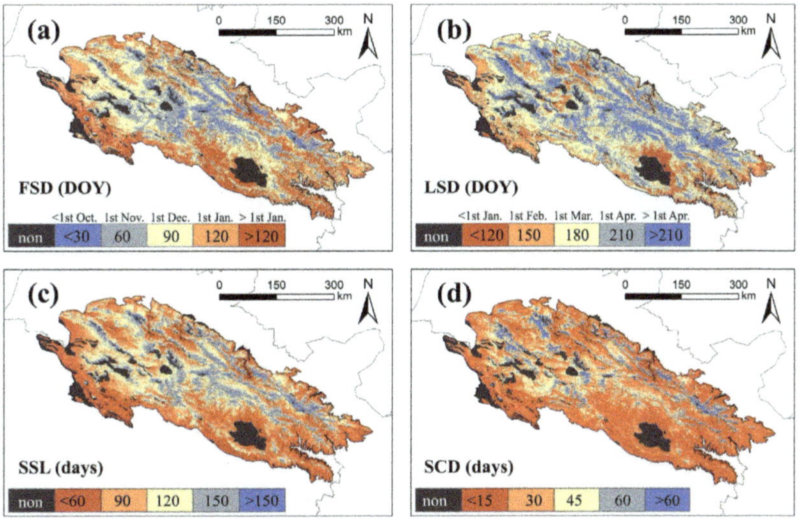

Figure 5. Mean of snow seasonality metrics in QLMA from 2002 to 2020. (**a**) represents the FSD, (**b**) represents the LSD, (**c**) represents the SSL, and (**d**) represents the SCD.

3.2. Land Surface Phenology among Different Vegetation Types

Figure 6 shows the calculated LSP metrics for the different vegetation types. LSP metrics for all vegetation except desert varied with elevation: SOS delays (Figure 6a), EOS advances (Figure 6b) and LOS shortens (Figure 6c) as elevation rises. There are no significant trends in SOS or LOS with elevation for desert, but EOS advances with elevation.

Figure 6d shows the average LSP metrics across elevation. Desert and alpine vegetation have the earliest growing seasons starting in late April. The growing season of forest begins in mid-May, while shrub, grass and cultivated vegetation start growing in late May. The EOS of different vegetation types are relatively close. The EOS of vegetation types other than alpine vegetation generally appear in early September, with desert and forest a few days earlier. The EOS of alpine vegetation is the earliest, occurring around mid-August. Desert has the longest growing season lengths (SSL = 120), SSL in forest (SSL = 105) and alpine vegetation (SSL = 101) are shorter. Shrub, grass and cultivated vegetation have the shortest growing seasons of about 95 d.

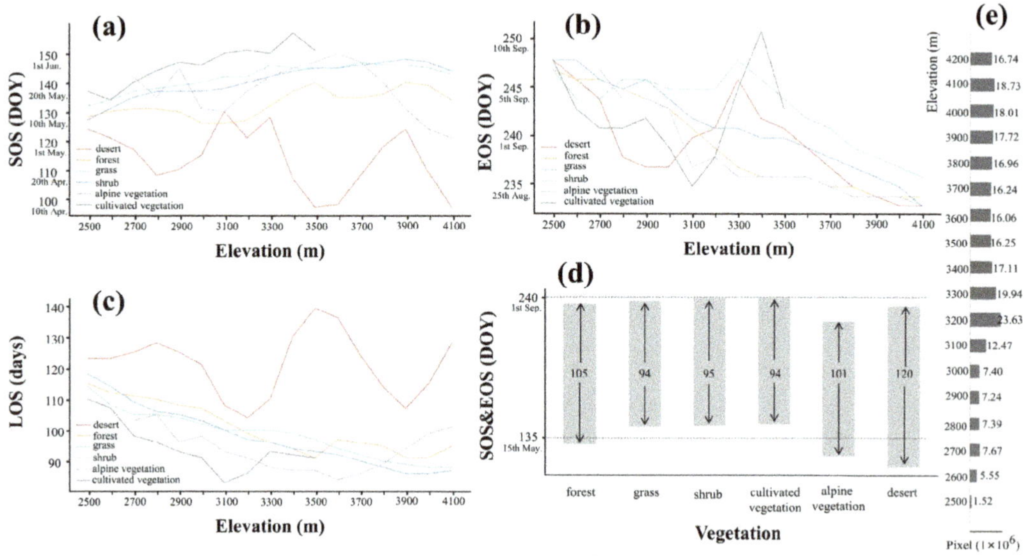

Figure 6. The calculated LSP metrics for the different vegetation types change with elevation. (**a**) represents the SOS, (**b**) represents the EOS and (**c**) represents the LOS. (**d**) shows the average of all elevations. The bar graph represents the average SOS and EOS of six different vegetation types, and the arrows inside the bar indicate their LOS. (**e**) shows the histogram of pixel numbers of different elevation gradients.

3.3. Spatial Pattern of Land Surface Phenology over the QLMA

The spatial pattern of the mean values of land surface phenology metrics in the QLMA area is shown in Figure 7. Many western areas were filtered out because of the lack of seasonal vegetation. The earliest start of the growing season is in the western and northern margins, usually before 1st May, followed by the eastern areas, where the growing season starts within about a month (before 1st June). Vegetation in the central part of the study area has the latest start of the growing season, occurring after 1st June, and in a few areas even later (after 15th June) (Figure 7a). The growing season for most of the vegetation in the study area ends between 15th August and 15th September. Growing seasons in the northern part of Qinghai Lake and the eastern edge of the study area end half a month later. The vegetation with the latest end of the growing season is located in the western region, occurring after 1st October (Figure 7b). The spatial distribution pattern of LOS is similar to that of SOS, with the northern edge and the sporadic areas in west having the longest vegetation growing season, which exceeds 150 d at most. From the eastern area toward the center, the vegetation growing season gradually shortens from up to 120 d to up to 90 d (Figure 7c).

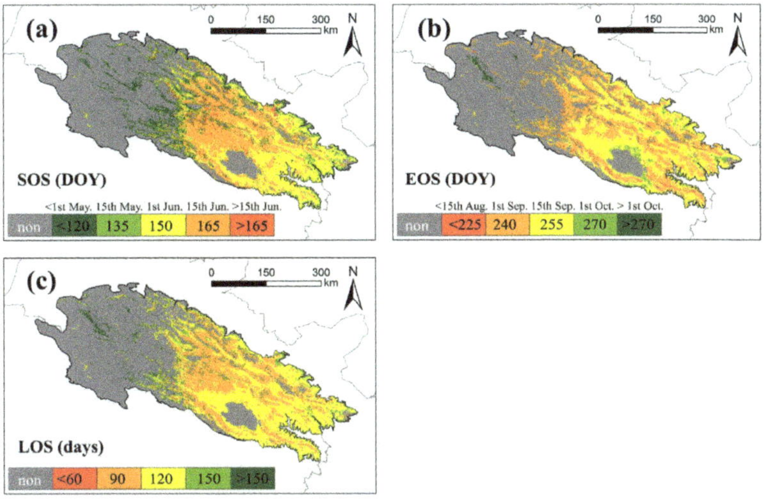

Figure 7. Mean of land surface phenology metrics in QLMA from 2003 to 2021. (**a**) represents the SOS, (**b**) represents the EOS and (**c**) represents the LOS.

3.4. Spatial Pattern of the Correlation between Snow Seasonality and Land Surface Phenology Metrics

Different snow seasonality metrics have different strengths and directions of influence on different LSP metrics, and there is spatial variation in this correlation (Figure 8). Overall, the snow seasonality metrics have a similar impact on EOS and LOS, in contrast to SOS. FSD shows a significant positive correlation with SOS in the southern part of Qinghai Lake and a mostly nonsignificant negative correlation with the central part. FSD shows mainly negative correlation with EOS and LOS in the study area, especially in the central region, where the negative correlation is significant. EOS and LOS in the eastern part of the study area show a nonsignificant positive correlation with FSD. The effect of LSD on LSP is different from that of FSD. LSD shows a significant negative correlation with SOS in the western part of the study area and the southern part of Qinghai Lake. Both EOS and LOS in the central and western parts of the study area show positive correlations with LSD, where LOS in the western part shows a significant positive correlation with LSD. EOS and LOS in the southeast of the study area, however, show a negative but insignificant correlation with LSD. The spatial pattern of SSL and SCD effects on LSP is similar to that of LSD, compared to the more significant correlation of SCD with LSP.

The proportion of significant positive and significant negative correlations indicates the main direction of influence of snow seasonality metrics on LSP metrics (Table 1). All the 12 correlations have a relatively obvious directionality, meaning that there is no positive correlation with the same proportion of negative correlations. Among them, positive correlations dominate in FSD_SOS, LSE_EOS, LSD_LOS, SSL_L0S and SCD_LOS. The proportion of insignificant positive correlations is about 7% more than the proportion of insignificant negative correlations, and the proportion of significant positive correlations exceeds the proportion of significant negative correlations by twice. However, negative correlation is the major of FSD_LOS, SSL_SOS and SCD_SOS. The proportion of insignificant positive correlations is approximately 6% less than the proportion of insignificant negative correlations, and the proportion of significant positive correlations is less than half of the proportion of significant negative correlations.

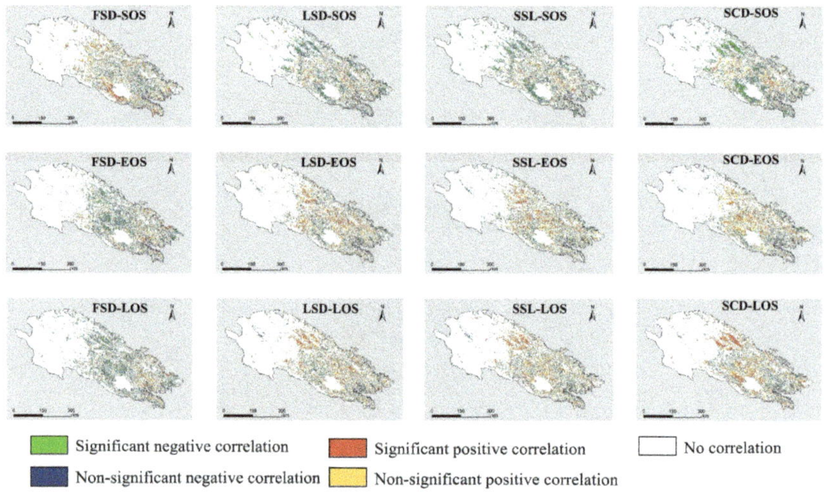

Figure 8. Correlation between different land surface phenology metrics and different snow seasonality metrics. "Significant" means $p < 0.1$.

Table 1. Area of significant correlation between snow season metrics and phenology metrics (%), SN for significant negative ($p < 0.1$) correlation and NN for nonsignificant negative ($p > 0.1$, $r < -0.2$) correlation. SP for significant positive ($p < 0.1$) correlation and NP for nonsignificant positive ($p > 0.1$, $r > 0.2$) correlation.

	SN (%)	NN (%)	NP (%)	SP (%)	SN/SP
FSD_SOS	4.23	11.75	17.79	8.96	0.47
FSD_EOS	8.12	16.68	12.98	5.17	1.57
FSD_LOS	9.04	18.66	11.38	4.34	2.08
LSD_SOS	8.89	16.66	12.54	4.91	1.81
LSD_EOS	4.13	11.40	18.38	8.86	0.47
LSD_LOS	4.04	11.14	18.52	9.55	0.42
SSL_SOS	9.66	17.22	11.47	4.34	2.23
SSL_EOS	4.84	11.35	19.02	8.31	0.58
SSL_LOS	3.87	10.94	18.71	10.00	0.39
SCD_SOS	11.88	17.31	12.90	5.26	2.26
SCD_EOS	4.57	11.30	18.87	7.63	0.60
SCD_LOS	3.94	11.49	18.29	11.13	0.35

3.5. *Elevation-Dependent Correlation between Snow Seasonality and Land Surface* Phenology Metrics

We further investigated the phenological response along the elevation gradient (Figure 9). FSD is significantly negatively correlated with SOS and positively correlated with LOS at low elevation, and the correlation gradually decreases as the elevation increases to 3500 m. The direction of the effect of FSD above 3500 m changes. The correlation with SOS turns from a nonsignificant positive correlation to a significant positive correlation by degrees, and the correlation with LOS changes to a progressively increasing negative correlation. The correlation between FSD and EOS is weak and always fluctuates around 0. There is no shift in the correlation between LSD and LSP metrics with rising elevation. Regardless of the elevation, LSD is always nearly significantly positively correlated with EOS and LOS. LSD is significantly negatively correlated with SOS at low elevation, and the correlation and significance decrease with rising elevation, but there is an increasing trend above 3900 m. The correlation of SSL and SCD with LSP metrics has similar characteristics with elevation. Below 3300 m, they show a nonsignificant weak correlation with LSP metrics. The correlations gradually increase with elevation, and approach stability at around 3500 m. Above 3500 m, SSL and SCD show a significant negative correlation with SOS and a significant positive correlation with EOS and LOS, especially LOS. The effect of SCD is more obvious than SSL.

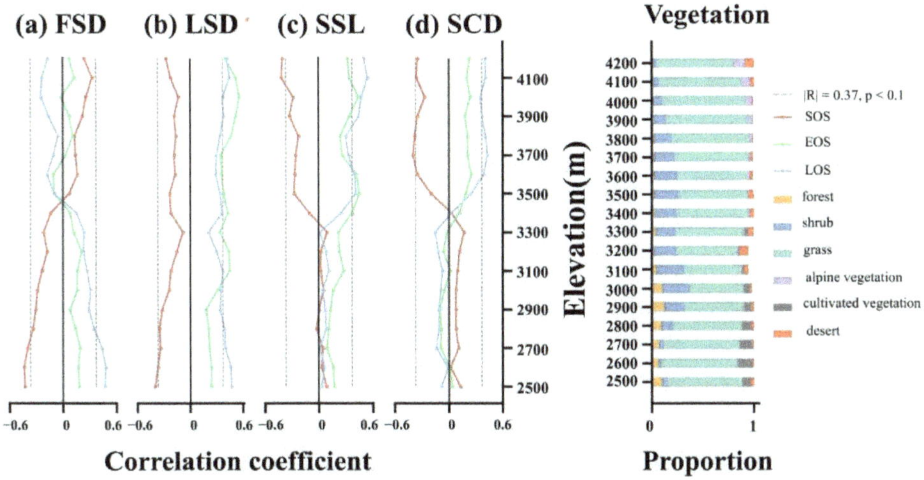

Figure 9. Left panel shows the land surface phenology metrics response to different snow seasonality metrics at different elevation. (**a**) means FSD, (**b**) means LSD, (**c**) means SSL and (**d**) means SCD. Right panel shows the vegetation constitute at different elevations.

Vegetation constitute varies at different elevations. Forest and cultivated vegetation are largely absent above 3000 m. Shrub increases rapidly from 4.95% to a maximum of 27.56% above 2800 m, with most of its distribution concentrated between 2900 m and 3700 m. Grass is the dominant vegetation type in most areas, usually accounting for more than 60% of the vegetation.

3.6. Interspecific Variation in the Response of Land Surface Phenology

To characterize differences in the responses of land surface phenology to snow seasonality, we compared the responses of six vegetation phenological characteristics to snow (Figure 10). The elevation interval with the highest concentration of vegetation distribution is selected: 2600–3500 m for forest and cultivated vegetation, 3100–4000 m for grass, shrub and desert, and 3600–4200 m for alpine vegetation. FSD is significantly negatively correlated with SOS of forest and LOS of alpine vegetation, and significantly negatively correlated with EOS of shrub. EOS of forest, shrub and grass is significantly positively correlated with LSD, as is LOS of shrub and grass. SSL and SCD affect vegetation in almost the same direction. SSL is significantly correlated with EOS for more vegetation types, while SCD is significantly correlated with SOS and LOS of vegetation. The SOS of forest is significantly positively correlated with SCD, while the SOS of desert is significantly negatively correlated with SCD. EOS of shrub and grassland is more sensitive to SSL and is significantly positively correlated. LOS of alpine vegetation is significantly positively correlated with both SSL and SCD, while LOS of desert is significantly positively correlated with SCD only.

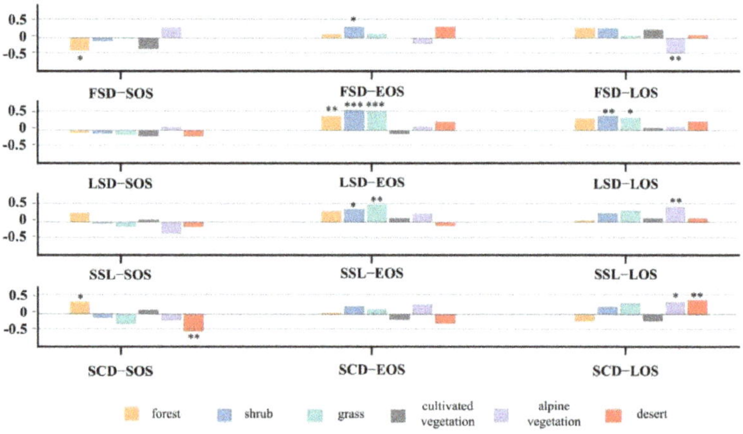

Figure 10. Correlation between vegetation phenological metrics of eight types of vegetation and snow phenological metrics. * $p < 0.1$, ** $p < 0.05$, *** $p < 0.01$.

4. Discussion

4.1. Does Seasonal Snow Seasonality Metrics Affect Land Surface Phenology Metrics?

Our findings demonstrated that land surface phenology at high elevations responds to snow cover seasonality. However, this response relationship is complicated because different snow season metrics affect different phenological metrics (Figure 8, Table 1). The magnitude asymmetry in the significant correlation between the snow seasonality and land surface phenology metrics suggests that a more snow-prone non-growing season (earlier first snow, later snowmelt, longer snow season and more snow cover days) may benefit a more flourishing vegetation growing season the following year (earlier start and later end of the growing season, longer growing season). The area of significant positive correlation between FSD and SOS is more than two times the area of significant negative correlation. In contrast, the area significantly positively correlated with LOS is less than half of the area significantly negatively correlated. These magnitude asymmetries may indicate the direction of the effect of FSD on LSP [33]. That is, FSD is positively correlated with SOS and negatively correlated with EOS and LOS. This suggests that earlier snowfall in the autumn triggers an earlier growing season the following spring and delays the end of the vegetation growing season the following autumn, which also extends the vegetation growing season.

Qiao and Wang [38] found no or negative correlation between FSD and SOS when exploring winter snowpack and spring grassland vegetation phenology in Inner Mongolia, which is not consistent with our findings. This may be attributed to elevation. Significant negative correlations could be observed only in central Inner Mongolia, where the elevation is below 3000 m. However, at higher elevations in the southwest, the correlations are not significant. In QLMA, there are relatively few areas below 3000 m in elevation, resulting in the negative correlation that is not widely observed by us. The findings of Wang et al. [35] in the Tibetan Plateau (TP) are close to ours. They extensively observed a positive correlation between FSD and SOS, especially in the central TP and northwest of the QLMA. The negative correlation between FSD and EOS is consistent with the findings of Qi et al. [34] in QLMA; moreover, we observed a larger area (Figure 8). This finding is equally relevant in other areas with continuous seasonal snowpack [66]. However, the impact of late season transient snowfall events may be limited, which could explain the small impact of FSD in some areas [58,67].

We considered that a later snowmelt may result in a delayed end of the growing season and a longer growing season length. Longer snow seasons and more snow days lead to earlier growing seasons and also prolong growing season length. Our study confirmed the findings of Wang [35] and Qi [34] in the same study area, but differences in elevation and hydrothermal conditions cause this finding to change in other areas. In fact, the effect of different snow seasonality metrics can still be observed even within a region, which may also be due to the elevation and climatic conditions of the different regions [36]. Moreover, different vegetation indices, LSP estimation methods and thresholds for estimating snow seasonality have an impact on the conclusions. However, these effects are usually reflected in intensity and significance. The apparent magnitude asymmetry of correlations within the study area points to the main direction of snow seasonality effects, in which we are able to corroborate each other. Even in different areas, similar effects of snow seasonality can still be observed if elevations and climates are similar, such as at higher elevations in Inner Mongolia and Nepal [38,66].

4.2. Why Do the Effects of Snow Seasonality Metrics Vary with Elevation?

Our study confirmed the high-elevation dependence between snow cover and phenology (Figure 9). The effect of earlier FSD on SOS shifted from delaying to facilitating with increasing elevation. The effect on LOS changes from shortening to lengthening, with the turning point occurring roughly at 3500 m. The correlation between FSD and EOS changes from a nonsignificant positive correlation to almost no correlation. If an earlier and longer growing season is considered to be better, then an earlier FSD is detrimental to vegetation growth at lower elevations and beneficial at higher elevations. This effect of first snow on vegetation phenology with elevation is not unique to QLMA. Paudel and Andersen [66] found no correlation between FSD and EOS at low elevations, but a strong positive correlation at very high elevations. Obviously, the elevation of our study area is far from 'very high', above 5000 m, so we could only observe an insignificant correlation between FSD and EOS. The findings in the Qinghai–Tibetan Plateau and QLMA are more comparable. Qi [34] divided QLMA into four elevation intervals, and the correlation between FSD and EOS is consistent with our findings on the trend of weakening with elevation. Our more detailed division makes the results more obvious. Wang et al. [35] found a positive correlation between FSD and SOS on the TP that gradually increased between 2500–5000 m and then weakened, which is similar to our findings. We further found a change in this correlation not only in intensity but also in direction, that is, a significant negative correlation between FSD and SOS below 3500 m. The study area of Wang et al. [35] is much larger than ours, and the seasonality of snow varies greatly at lower elevations, which may lead to their inability to accurately count correlations below 3500 m. However, that earlier first snowfall at higher elevations favors earlier vegetation phenology is what we all agree on.

We did not observe an effect of LSD on LSP metrics with elevation. Regardless of elevation, later snowmelt is beneficial for earlier vegetation growing season initiation and a longer growing season. Snow melt directly provides the necessary water for vegetation to sprout, and the spring snowpack maintains soil temperatures. No matter what the elevation, accumulated temperature and water are necessary for vegetation to sprout. Paudel and Andersen [66] observed the same conclusion as ours in the low elevation arid zone, and Wang et al. [35] also found a negative correlation between LSD and SOS in the TP, but it was not significant at low elevation. In contrast, Qi et al. [34] found a significant positive correlation between LSD and SOS in the high-elevation interval of QLMA, which may be related to the selection and treatment of the vegetation indices. Compared to estimating SOS based on NDVI, the NDPI we used is shown to better eliminate the effect of pre-season snowpack and avoid misclassification of snow and vegetation pixels [40–42]. In addition, our reconstructed NDPI with a temporal resolution of 4 d also helps to obtain a more accurate SOS.

We also found a similar effect of SSL and SCD on LSP metrics. Vegetation below 3300 m is barely affected by them, while vegetation above 3500 m SOS is significantly negatively correlated with them, and LOS is significantly positively correlated with them. This is consistent with the findings of many other studies, at least at the same elevation [34,35,58]. This may be due to different climatic conditions at different elevations, which could drive differences in correlation [68]. Although the melting of snow will always provide moisture to the soil, different temperature conditions may lead to different effects on the presence of snow. More snow is needed at higher elevations to cover the soil than at relatively warm lower elevations because the snow acts as an insulator [36,69]. Soils are protected from the harsh climatic conditions and severe solar radiation to which they would otherwise be subjected by a snow cover [69–71]. Soil temperatures at high elevations with snow cover are usually higher than in areas without snow [72].

We described for the first time at QLMA the process of reversal of the direction of influence of snow seasonality metrics on LSP metrics with elevation. However, we estimated that the thresholds for LSP (0.3 and 0.7) and the use of seasonal vegetation filter (NDPI > 0.1) may have influenced the strength of the relationship. The vegetation constitute that varies with elevation may be another factor affecting the correlation. Changes in other vegetation proportions can cause fluctuations in correlations (LSD_FSD below 3300 m), but grass is always the predominant vegetation type in QLMA, which could ensure the relative stability of correlations.

4.3. Why Do the Effects of Snow Seasonality Metrics Vary with Vegetation Type?

The response of vegetation to snowpack varies considerably between biomes, and similar phenomena have been observed in the QLMA (Figure 10). We found that the LSP metrics of shrub and grass respond most significantly to LSD compared to other vegetation, namely that later snowmelt extends the growing season of both vegetation species. This is consistent with the findings of many other studies on grasslands [34,36]. It may be because shrub and grass have relatively simple structures and short size, so they are more likely to be completely covered by snow. In addition, due to the strong solar radiation, the snow on the grassland melts more easily and water can be supplied to the soil in a timely manner [73]. The LSP metrics of desert only respond to SCD. Compared to other snow seasonality metrics, SCD reflects not only the timing of snow presence, but also the frequency of snow presence, which is important for drought desert. Conversely, alpine vegetation is more sensitive to FSD and SSL and these may reflect the arrival and duration of cold air during the snow season as temperature is more important for alpine vegetation. We also found that the SOS of the forest responds differently to snow seasonality metrics than other vegetation types. Snow falling on branches does not have a direct and timely effect on the root system [74,75]. Due to the tall structure of the trees, the snow in the canopy and on the ground is exposed to different solar radiation, resulting in differences in snow melt [76,77]. Furthermore, some studies have demonstrated that different vegetation types have different temperature requirements for breaking dormancy [69,78]. Woody plants may require cold conditions to promote germination, while grasses require warmer conditions. Snow protects shallow underground root systems from the cold, resulting in different effects on different vegetation types.

In summary, our study illustrated that different vegetation types have different reflections of LSP metrics on snow seasonality metrics, which explains the spatial variation in the impact of snow seasonality metrics from another perspective than elevation.

4.4. Prediction of Vegetation Phenology from Satellite Data Is Beneficial for Future Research

The concept of using satellite data to estimate vegetation phenology metrics was born long ago. With the continuous improvement of remote sensing image accuracy and cloud computing capability, the estimation of phenology based on satellite data has gradually become reliable. Unlike the phenology metrics obtained from field observations, the remote sensing-based estimation of phenology metrics focuses on the variation of regional greenness. Important time points are calculated from the interannual variation curves of vegetation indices, and thus vegetation phenology is estimated. For stakeholders, satellites can quickly provide long time series and large-scale data, which could also avoid time-consuming field work. Characterizing vegetation phenology at a larger scale is beneficial for studies such as vegetation response in the context of global climate change.

However, the validation for the remote sensing-based phenological indices and in situ phenological indices is also important. Although there may be temporal differences in the phenological indices obtained by the two methods, both can indicate consistent phenological trends. More in situ data is an important way to improve the accuracy of satellite predictions. This is also the objective of our future study.

4.5. Study Limitations and Future Work

The medium-resolution satellite data and complex topographic conditions of the Qilian Mountains make it difficult to estimate land surface phenology. Although we used the NDPI, which is least affected by pre-season snowpack, as a vegetation index, the accuracy of such threshold-based extraction remains uncertain. Land surface phenology is a complex parameter that is influenced by a combination of factors. In addition to snow seasonality, vegetation is influenced by snow depth, pre-growing season temperature and precipitation and light conditions. In addition to elevation, slope and aspect play important roles. Thus, assessing the response of land surface phenology requires more accurate models. As mentioned earlier, the differences in water conditions across the study area may be an important factor influencing the conclusions. The content and depth of groundwater, meltwater from permafrost and glaciers, may also have an impact. These would be good to include in future study. Finally, although the vegetation distribution data we used are very reliable, changes are inevitable over long time series, especially in low-elevation areas that are inherently more susceptible to human activities.

5. Conclusions

The snow season and vegetation phenological indicators in the Qilian Mountains in the northeastern Qinghai–Tibet Plateau were investigated, and their corresponding relationships were analyzed. In this study, we concluded that snow seasonality metrics have distinct spatial distribution characteristics. The snow season started earlier and lasted longer in the central part of the study area. The LSP metrics varied significantly with elevation and most vegetation growing seasons shortened with elevation. The asymmetry of significant correlation between snow seasonality and LSP metrics indicates the main direction of influence. A more snow-prone non-growing season (earlier first snow, later snowmelt, longer snow season and more snow cover days) may trigger a more flourishing vegetation growing season the following year (earlier start and later end of growing season, longer growing season).

The NDPI we used is less affected by spring snowpack. We set thresholds to remove nonseasonal vegetation and delineated more detailed elevation gradients. We described the effect of QLMA snow seasonality as a curve that varies with elevation. Below 3300 m, later first snowfall leads to an earlier growing season and also extends the growing season length, while the effect of first snowfall above 3300 m is reversed. The intensity of the effect of LSD fluctuates with elevation but does not reverse. The effects of SSL and SCD on LSP are small and insignificant below 3500 m, and their increase mainly benefits the extended growing season of high-elevation vegetation. In addition, the sensitivity of LSP metrics to snow seasonality varies among vegetation types. Our research provides more evidence that the impact of snow varies with elevation and underlying vegetation types.

Hydrothermal conditions, changes in temperature and precipitation, extreme weather events and glacial melt are important factors influencing land surface phenology at high elevations and should be investigated in future studies in conjunction with high-resolution data to develop improved models for analyzing them.

Author Contributions: Conceptualization, W.Z., K.Y. and Y.L.; methodology, Y.L. and K.Y.; software, validation, Y.L. and K.Y.; writing—original draft preparation and visualization, Y.L.; writing—review and editing, X.M. and S.G. funding acquisition, W.Z. and K.Y. All authors have read and agreed to the published version of the manuscript.

Funding: This research was funded by the National Natural Science Foundation of China, grant number 41977415, and the Fundamental Research Funds for the Central Universities, grant number 265QZ2022001.

Data Availability Statement: Not applicable.

Acknowledgments: We express our gratitude to anonymous reviewers and editors for their professional comments and suggestions.

Conflicts of Interest: The authors declare no conflict of interest.

References

1. IPCC. *Summary for Policymakers*; IPCC: Geneva, Switzerland, 2018.
2. Kang, W.; Liu, S.; Chen, X.; Feng, K.; Guo, Z.; Wang, T. Evaluation of ecosystem stability against climate changes via satellite data in the eastern sandy area of northern China. *J. Environ. Manag.* **2022**, *308*, 114596. [CrossRef] [PubMed]
3. Wang, Y.; Gu, J. Ecological responses, adaptation and mechanisms of mangrove wetland ecosystem to global climate change and anthropogenic activities. *Int. Biodeterior. Biodegrad.* **2021**, *162*, 105248. [CrossRef]
4. Weiskopf, S.R.; Rubenstein, M.A.; Crozier, L.G.; Gaichas, S.; Griffis, R.; Halofsky, J.E.; Hyde, K.J.W.; Morelli, T.L.; Morisette, J.T.; Muñoz, R.C.; et al. Climate change effects on biodiversity, ecosystems, ecosystem services, and natural resource management in the United States. *Sci. Total Environ.* **2020**, *733*, 137782. [CrossRef] [PubMed]
5. Beniston, M. Climatic Change in Mountain Regions: A Review of Possible Impacts. *Clim. Chang.* **2003**, *59*, 5–31. [CrossRef]
6. Bebi, P.; Krumm, F. *Mountains and Climate Change: A Global Concern*; Centre for Development and Environment (CDE), Swiss Agency for Development and Cooperation (SDC), Geographica Bernensia: Bern, Switzerland, 2014; pp. 8–13.
7. Verrall, B.; Pickering, C.M. Alpine vegetation in the context of climate change: A global review of past research and future directions. *Sci. Total Environ.* **2020**, *748*, 141344. [CrossRef] [PubMed]
8. Gao, Q.; Guo, Y.; Xu, H.; Ganjurjav, H.; Li, Y.; Wan, Y.; Qin, X.; Ma, X.; Liu, S. Climate change and its impacts on vegetation distribution and net primary productivity of the alpine ecosystem in the Qinghai-Tibetan Plateau. *Sci. Total Environ.* **2016**, *554–555*, 34–41. [CrossRef] [PubMed]
9. White, M.A.; Nemani, R.R. Real-time monitoring and short-term forecasting of land surface phenology. *Remote Sens. Environ.* **2006**, *104*, 43–49. [CrossRef]
10. de Beurs, K.M.; Henebry, G.M. Land surface phenology, climatic variation, and institutional change: Analyzing agricultural land cover change in Kazakhstan. *Remote Sens. Environ.* **2004**, *89*, 497–509. [CrossRef]
11. Moody, A.; Johnson, D.M. Land-Surface Phenologies from AVHRR Using the Discrete Fourier Transform. *Remote Sens. Environ.* **2001**, *75*, 305–323. [CrossRef]
12. Zeng, L.; Wardlow, B.D.; Xiang, D.; Hu, S.; Li, D. A review of vegetation phenological metrics extraction using time-series, multispectral satellite data. *Remote Sens. Environ.* **2020**, *237*, 111511. [CrossRef]
13. Piao, S.; Cui, M.; Chen, A.; Wang, X.; Ciais, P.; Liu, J.; Tang, Y. Altitude and temperature dependence of change in the spring vegetation green-up date from 1982 to 2006 in the Qinghai-Xizang Plateau. *Agric. For. Meteorol.* **2011**, *151*, 1599–1608. [CrossRef]
14. Richardson, A.D.; Keenan, T.F.; Migliavacca, M.; Ryu, Y.; Sonnentag, O.; Toomey, M. Climate change, phenology, and phenological control of vegetation feedbacks to the climate system. *Agric. For. Meteorol.* **2013**, *169*, 156–173. [CrossRef]
15. Wang, S.; Zhang, B.; Yang, Q.; Chen, G.; Yang, B.; Lu, L.; Shen, M.; Peng, Y. Responses of net primary productivity to phenological dynamics in the Tibetan Plateau, China. *Agric. For. Meteorol.* **2017**, *232*, 235–246. [CrossRef]
16. Li, F.; Song, G.; Liujun, Z.; Xiuqin, F.; Yanan, Z. Urban vegetation phenology analysis and the response to the temperature change. In Proceedings of the 2017 IEEE International Geoscience and Remote Sensing Symposium (IGARSS), Fort Worth, TX, USA, 23–28 July 2017; pp. 5743–5746.
17. Lang, W.; Chen, X.; Liang, L.; Ren, S.; Qian, S. Geographic and Climatic Attributions of Autumn Land Surface Phenology Spatial Patterns in the Temperate Deciduous Broadleaf Forest of China. *Remote Sens.* **2019**, *11*, 1546. [CrossRef]
18. Clinton, N.; Yu, L.; Fu, H.; He, C.; Gong, P. Global-Scale Associations of Vegetation Phenology with Rainfall and Temperature at a High Spatio-Temporal Resolution. *Remote Sens.* **2014**, *6*, 7320–7338. [CrossRef]
19. Pedersen, S.H.; Liston, G.E.; Tamstorf, M.P.; Abermann, J.; Lund, M.; Schmidt, N.M. Quantifying snow controls on vegetation greenness. *Ecosphere* **2018**, *9*, e02309. [CrossRef]
20. Assmann, J.J.; Myers-Smith, I.H.; Phillimore, A.B.; Bjorkman, A.D.; Ennos, R.E.; Prevéy, J.S.; Henry, G.H.R.; Schmidt, N.M.; Hollister, R.D. Local snow melt and temperature—but not regional sea ice—explain variation in spring phenology in coastal Arctic tundra. *Glob. Chang. Biol.* **2019**, *25*, 2258–2274. [CrossRef]
21. Piao, S.; Liu, Q.; Chen, A.; Janssens, I.A.; Fu, Y.; Dai, J.; Liu, L.; Lian, X.; Shen, M.; Zhu, X. Plant phenology and global climate change: Current progresses and challenges. *Glob. Chang. Biol.* **2019**, *25*, 1922–1940. [CrossRef]
22. Zheng, J.; Jia, G.; Xu, X. Earlier snowmelt predominates advanced spring vegetation greenup in Alaska. *Agric. For. Meteorol.* **2022**, *315*, 108828. [CrossRef]
23. Frei, E.R.; Henry, G.H.R. Long-term effects of snowmelt timing and climate warming on phenology, growth, and reproductive effort of Arctic tundra plant species. *Arct. Sci.* **2021**. *e-First*. [CrossRef]

24. Kumar, M.; Wang, R.; Link, T.E. Effects of more extreme precipitation regimes on maximum seasonal snow water equivalent. *Geophys. Res. Lett.* **2012**, *39*, L20504. [CrossRef]
25. Hanati, G.; Zhang, Y.; Su, L.; Hu, K. Response of water and heat of seasonal frozen soil to snow melting and air temperature. *Arid. Land Geogr.* **2021**, *44*, 889–896. [CrossRef]
26. Bai, J.; Shi, H.; Yu, Q.; Xie, Z.; Li, L.; Luo, G.; Jin, N.; Li, J. Satellite-observed vegetation stability in response to changes in climate and total water storage in Central Asia. *Sci. Total Environ.* **2019**, *659*, 862–871. [CrossRef] [PubMed]
27. Harpold, A.A.; Molotch, N.P. Sensitivity of soil water availability to changing snowmelt timing in the western U.S. *Geophys. Res. Lett.* **2015**, *42*, 8011–8020. [CrossRef]
28. Desai, A.R.; Wohlfahrt, G.; Zeeman, M.J.; Katata, G.; Eugster, W.; Montagnani, L.; Gianelle, D.; Mauder, M.; Schmid, H.P. Montane ecosystem productivity responds more to global circulation patterns than climatic trends. *Environ. Res. Lett.* **2016**, *11*, 024013. [CrossRef] [PubMed]
29. Rixen, C.; Dawes, M.A.; Wipf, S.; Hagedorn, F. Evidence of enhanced freezing damage in treeline plants during six years of CO_2 enrichment and soil warming. *Oikos* **2012**, *121*, 1532–1543. [CrossRef]
30. Peng, S.; Piao, S.; Ciais, P.; Friedlingstein, P.; Zhou, L.; Wang, T. Change in snow phenology and its potential feedback to temperature in the Northern Hemisphere over the last three decades. *Environ. Res. Lett.* **2013**, *8*, 014008. [CrossRef]
31. Sherwood, J.A.; Debinski, D.M.; Caragea, P.C.; Germino, M.J. Effects of experimentally reduced snowpack and passive warming on montane meadow plant phenology and floral resources. *Ecosphere* **2017**, *8*, e01745. [CrossRef]
32. Yang, X.; Henry, H.A.L.; Zhong, S.; Meng, B.; Wang, C.; Gao, Y.; Sun, W. Towards a mechanistic understanding of soil nitrogen availability responses to summer vs. winter drought in a semiarid grassland. *Sci. Total Environ.* **2020**, *741*, 140272. [CrossRef]
33. Tomaszewska, M.A.; Nguyen, L.H.; Henebry, G.M. Land surface phenology in the highland pastures of montane Central Asia: Interactions with snow cover seasonality and terrain characteristics. *Remote Sens. Environ.* **2020**, *240*, 111675. [CrossRef]
34. Qi, Y.; Wang, H.; Ma, X.; Zhang, J.; Yang, R. Relationship between vegetation phenology and snow cover changes during 2001–2018 in the Qilian Mountains. *Ecol. Indic.* **2021**, *133*, 108351. [CrossRef]
35. Wang, S.; Wang, X.; Chen, G.; Yang, Q.; Wang, B.; Ma, Y.; Shen, M. Complex responses of spring alpine vegetation phenology to snow cover dynamics over the Tibetan Plateau, China. *Sci. Total Environ.* **2017**, *593–594*, 449–461. [CrossRef]
36. Wang, X.; Wu, C.; Peng, D.; Gonsamo, A.; Liu, Z. Snow cover phenology affects alpine vegetation growth dynamics on the Tibetan Plateau: Satellite observed evidence, impacts of different biomes, and climate drivers. *Agric. For. Meteorol.* **2018**, *256–257*, 61–74. [CrossRef]
37. Xie, J.; Jonas, T.; Rixen, C.; de Jong, R.; Garonna, I.; Notarnicola, C.; Asam, S.; Schaepman, M.E.; Kneubühler, M. Land surface phenology and greenness in Alpine grasslands driven by seasonal snow and meteorological factors. *Sci. Total Environ.* **2020**, *725*, 138380. [CrossRef] [PubMed]
38. Qiao, D.; Wang, N. Relationship between Winter Snow Cover Dynamics, Climate and Spring Grassland Vegetation Phenology in Inner Mongolia, China. *ISPRS Int. J. Geoinf.* **2019**, *8*, 42. [CrossRef]
39. Choler, P. Growth response of temperate mountain grasslands to inter-annual variations in snow cover duration. *Biogeosciences* **2015**, *12*, 3885–3897. [CrossRef]
40. Delbart, N.; Kergoat, L.; Le Toan, T.; Lhermitte, J.; Picard, G. Determination of phenological dates in boreal regions using normalized difference water index. *Remote Sens. Environ.* **2005**, *97*, 26–38. [CrossRef]
41. Cao, R.; Feng, Y.; Liu, X.; Shen, M.; Zhou, J. Uncertainty of Vegetation Green-Up Date Estimated from Vegetation Indices Due to Snowmelt at Northern Middle and High Latitudes. *Remote Sens.* **2020**, *12*, 190. [CrossRef]
42. Huang, K.; Zhang, Y.; Tagesson, T.; Brandt, M.; Wang, L.; Chen, N.; Zu, J.; Jin, H.; Cai, Z.; Tong, X.; et al. The confounding effect of snow cover on assessing spring phenology from space: A new look at trends on the Tibetan Plateau. *Sci. Total Environ.* **2021**, *756*, 144011. [CrossRef] [PubMed]
43. Li, Z.; Feng, Q.; Li, Z.; Wang, X.; Gui, J.; Zhang, B.; Li, Y.; Deng, X.; Xue, J.; Gao, W.; et al. Reversing conflict between humans and the environment—The experience in the Qilian Mountains. *Renew. Sustain. Energy Rev.* **2021**, *148*, 111333. [CrossRef]
44. Ma, Y.; Guan, Q.; Sun, Y.; Zhang, J.; Yang, L.; Yang, E.; Li, H.; Du, Q. Three-dimensional dynamic characteristics of vegetation and its response to climatic factors in the Qilian Mountains. *Catena* **2022**, *208*, 105694. [CrossRef]
45. Zhang, J.; Zhang, C. *Vegetation Pattern Data (1:100,000) in the Qilian Mountains*; National Cryosphere Desert Data Center: Lanzhou, China, 2020. [CrossRef]
46. Peng, Q.; Wang, R.; Jiang, Y.; Li, C. Contributions of climate change and human activities to vegetation dynamics in Qilian Mountain National Park, northwest China. *Glob. Ecol. Conserv.* **2021**, *32*, e01947. [CrossRef]
47. Yan, K.; Ding, Y. The overview of the progress of Qilian Mountain National Park System Pilot Area. *Int. J. Geoheritage Park* **2020**, *8*, 210–214. [CrossRef]
48. Hall, D.K.; Salomonson, V.V.; Riggs, G.A. *MODIS/Terra Snow Cover Daily L3 Global 500m Grid. Version 6*; NASA National Snow and Ice Data Center Distributed Active Archive Center: Boulder, CO, USA, 2016. [CrossRef]
49. Gafurov, A.; Bárdossy, A. Cloud removal methodology from MODIS snow cover product. *Hydrol. Earth Syst. Sci.* **2009**, *13*, 1361–1373. [CrossRef]
50. Hall, D.K.; Riggs, G.A.; Foster, J.L.; Kumar, S.V. Development and evaluation of a cloud-gap-filled MODIS daily snow-cover product. *Remote Sens. Environ.* **2010**, *114*, 496–503. [CrossRef]

51. Wang, W.; Huang, X.; Deng, J.; Xie, H.; Liang, T. Spatio-Temporal Change of Snow Cover and Its Response to Climate over the Tibetan Plateau Based on an Improved Daily Cloud-Free Snow Cover Product. *Remote Sens.* **2015**, *7*, 169–194. [CrossRef]
52. Liang, T.G.; Huang, X.D.; Wu, C.X.; Liu, X.Y.; Li, W.L.; Guo, Z.G.; Ren, J.Z. An application of MODIS data to snow cover monitoring in a pastoral area: A case study in Northern Xinjiang, China. *Remote Sens. Environ.* **2008**, *112*, 1514–1526. [CrossRef]
53. George, R.; Dorothy, H.; Miguel, R. *USER GUIDE: MODIS/Terra Snow Cover 8-Day L3 Global 500m Grid, Version 61*; NASA Goddard Space Flight Center: Greenbelt, MD, USA, 2021.
54. Parajka, J.; Blöschl, G. Spatio-temporal combination of MODIS images—Potential for snow cover mapping. *Water Resour. Res.* **2008**, *44*, W03406. [CrossRef]
55. Vermote, E. *MOD09A1 MODIS/Terra Surface Reflectance 8-Day L3 Global 500m SIN Grid V006*; NASA EOSDIS Land Processes DAAC: Sioux Falls, SD, USA, 2015. [CrossRef]
56. Cao, M.; Li, X.; Chen, X.; Wang, J.; Che, T. *Remote Sensing of Cryosphere*; Science Press: Beijing, China, 2006; p. 60. (In Chinese)
57. Chu, D.; Hongjie, X.; Pengxiang, W.; Jianping, G.; Jia, L.; Yubao, Q.; Zhaojun, Z. Snow cover variation over the Tibetan Plateau from MODIS and comparison with ground observations. *J. Appl. Remote Sens.* **2014**, *8*, 084690. [CrossRef]
58. Xie, J.; Kneubuhler, M.; Garonna, I.; Notarnicola, C.; De Gregorio, L.; De Jong, R.; Chimani, B.; Schaepman, M.E. Altitude-dependent influence of snow cover on alpine land surface phenology. *J. Geophys. Res. Biogeosci.* **2017**, *122*, 1107–1122. [CrossRef]
59. Wang, C.; Chen, J.; Wu, J.; Tang, Y.; Shi, P.; Black, T.A.; Zhu, K. A snow-free vegetation index for improved monitoring of vegetation spring green-up date in deciduous ecosystems. *Remote Sens. Environ.* **2017**, *196*, 1–12. [CrossRef]
60. White, M.A.; Thornton, P.E.; Running, S.W. A continental phenology model for monitoring vegetation responses to interannual climatic variability. *Glob. Biogeochem. Cycles* **1997**, *11*, 217–234. [CrossRef]
61. Zhang, X.; Friedl, M.A.; Schaaf, C.B.; Strahler, A.H.; Hodges, J.C.F.; Gao, F.; Reed, B.C.; Huete, A. Monitoring vegetation phenology using MODIS. *Remote Sens. Environ.* **2003**, *84*, 471–475. [CrossRef]
62. Reed, B.C.; Brown, J.F.; Van der Zee, D.; Loveland, T.R.; Merchant, J.W.; Ohlen, D.O. Measuring phenological variability from satellite imagery. *J. Veg. Sci.* **1994**, *5*, 703–714. [CrossRef]
63. Salinero-Delgado, M.; Estévez, J.; Pipia, L.; Belda, S.; Berger, K.; Paredes Gómez, V.; Verrelst, J. Monitoring Cropland Phenology on Google Earth Engine Using Gaussian Process Regression. *Remote Sens.* **2022**, *14*, 146. [CrossRef]
64. Broich, M.; Huete, A.; Paget, M.; Ma, X.; Tulbure, M.; Coupe, N.R.; Evans, B.; Beringer, J.; Devadas, R.; Davies, K.; et al. A spatially explicit land surface phenology data product for science, monitoring and natural resources management applications. *Environ. Model. Softw.* **2015**, *64*, 191–204. [CrossRef]
65. Jin, H.; Jönsson, A.M.; Bolmgren, K.; Langvall, O.; Eklundh, L. Disentangling remotely-sensed plant phenology and snow seasonality at northern Europe using MODIS and the plant phenology index. *Remote Sens. Environ.* **2017**, *198*, 203–212. [CrossRef]
66. Paudel, K.P.; Andersen, P. Response of rangeland vegetation to snow cover dynamics in Nepal Trans Himalaya. *Clim. Chang.* **2013**, *117*, 149–162. [CrossRef]
67. Hüsler, F.; Jonas, T.; Riffler, M.; Musial, J.P.; Wunderle, S. A satellite-based snow cover climatology (1985–2011) for the European Alps derived from AVHRR data. *Cryosphere* **2014**, *8*, 73–90. [CrossRef]
68. Beniston, M.; Keller, F.; Goyette, S. Snow pack in the Swiss Alps under changing climatic conditions: An empirical approach for climate impacts studies. *Theor. Appl. Climatol.* **2003**, *74*, 19–31. [CrossRef]
69. Yu, Z.; Liu, S.; Wang, J.; Sun, P.; Liu, W.; Hartley, D.S. Effects of seasonal snow on the growing season of temperate vegetation in China. *Glob. Chang. Biol.* **2013**, *19*, 2182–2195. [CrossRef] [PubMed]
70. Björk, R.G.; Molau, U. Ecology of Alpine Snowbeds and the Impact of Global Change. *Arct. Antarct. Alp. Res.* **2007**, *39*, 34–43. [CrossRef]
71. Wipf, S.; Rixen, C.; Mulder, C.P.H. Advanced snowmelt causes shift towards positive neighbour interactions in a subarctic tundra community. *Glob. Chang. Biol.* **2006**, *12*, 1496–1506. [CrossRef]
72. Freppaz, M.; Celi, L.; Marchelli, M.; Zanini, E. Snow removal and its influence on temperature and N dynamics in alpine soils (Vallée d'Aoste, northwest Italy). *J. Plant Nutr. Soil Sci.* **2008**, *171*, 672–680. [CrossRef]
73. Winkler, R.D.; Moore, R.D. Variability in snow accumulation patterns within forest stands on the interior plateau of British Columbia, Canada. *Hydrol. Process.* **2006**, *20*, 3683–3695. [CrossRef]
74. Pomeroy, J.W.; Parviainen, J.; Hedstrom, N.; Gray, D.M. Coupled modelling of forest snow interception and sublimation. *Hydrol. Process.* **1998**, *12*, 2317–2337. [CrossRef]
75. Pomeroy, J.W.; Gray, D.M.; Hedstrom, N.R.; Janowicz, J.R. Prediction of seasonal snow accumulation in cold climate forests. *Hydrol. Process.* **2002**, *16*, 3543–3558. [CrossRef]
76. Davis, R.E.; Hardy, J.P.; Ni, W.; Woodcock, C.; McKenzie, J.C.; Jordan, R.; Li, X. Variation of snow cover ablation in the boreal forest: A sensitivity study on the effects of conifer canopy. *J. Geophys. Res. Atmos.* **1997**, *102*, 29389–29395. [CrossRef]
77. López-Moreno, J.I.; Stähli, M. Statistical analysis of the snow cover variability in a subalpine watershed: Assessing the role of topography and forest interactions. *J. Hydrol.* **2008**, *348*, 379–394. [CrossRef]
78. Yu, H.; Luedeling, E.; Xu, J. Winter and spring warming result in delayed spring phenology on the Tibetan Plateau. *Proc. Natl. Acad. Sci. USA* **2010**, *107*, 22151. [CrossRef]

remote sensing

Article

Influences of Seasonal Soil Moisture and Temperature on Vegetation Phenology in the Qilian Mountains

Xia Cui [1,*], Gang Xu [2], Xiaofei He [2] and Danqi Luo [2]

[1] Key Laboratory of Western China's Environmental Systems (Ministry of Education), College of Earth and Environmental Sciences, Lanzhou University, Lanzhou 730000, China
[2] State Key Laboratory of Grassland Agro-Ecosystems, College of Pastoral Agriculture Science and Technology, Lanzhou University, Lanzhou 730020, China; xugang@lzu.edu.cn (G.X.); hexf21@lzu.edu.cn (X.H.); luodq20@lzu.edu.cn (D.L.)
* Correspondence: xiacui@lzu.edu.cn

Abstract: Vegetation phenology is a commonly used indicator of ecosystem responses to climate change and plays a vital role in ecosystem carbon and hydrological cycles. Previous studies have mostly focused on the response of vegetation phenology to temperature and precipitation. Soil moisture plays an important role in maintaining vegetation growth. However, our understanding of the influences of soil moisture dynamics on vegetation phenology is sparse. In this study, using a time series of the normalized difference vegetation index (NDVI) from the moderate resolution imaging spectroradiometer (MODIS) dataset (2001–2020), the start of the growing season (SOS), the end of the growing season (EOS), and the length of the growing season (LOS) in the Qilian Mountains (QLMs) were extracted. The spatiotemporal patterns of vegetation phenology (SOS, EOS, and LOS) were explored. The partial coefficient correlations between the SOS, EOS, and seasonal climatic factors (temperature, precipitation, and soil moisture) were analyzed. The results showed that the variation trends of vegetation phenology were not significant ($p > 0.05$) from 2001 to 2020, the SOS was advanced by 0.510 d/year, the EOS was delayed by 0.066 d/year, and the LOS was prolonged by 0.580 d/year. The EOS was significantly advanced and the LOS significantly shortened with increasing altitude. The seasonal temperature, precipitation, and soil moisture had spatiotemporal heterogeneous effects on the vegetation phenology. Overall, compared with temperature and soil moisture, precipitation had a weaker influence on the vegetation phenology in the QLMs. For different elevation zones, the temperature and soil moisture influenced the vegetation phenology in most areas of the QLMs, and spring temperature was the key driving factor influencing SOS; the autumn soil moisture and autumn temperature made the largest contributions to the variations in EOS at lower (<3500 m a.s.l.) and higher elevations (>3500 m a.s.l.), respectively. For different vegetation types, the spring temperature was the main factor influencing the SOS for broadleaf forests, needleleaf forests, shrublands, and meadows because of the relative lower soil moisture stress. The autumn soil moisture was the main factor influencing EOS for deserts because of the strong soil moisture stress. Our results demonstrate that the soil moisture strongly influences vegetation phenology, especially at lower elevations and water-limited areas. This study provides a scientific basis for better understanding the response of vegetation phenology to climate change in arid mountainous areas and suggests that the variation in soil moisture should be considered in future studies on the influence of climate warming and environmental effects on the phenology of water-limited areas.

Keywords: vegetation phenology; Qilian Mountains; soil moisture; remote sensing

Citation: Cui, X.; Xu, G.; He, X.; Luo, D. Influences of Seasonal Soil Moisture and Temperature on Vegetation Phenology in the Qilian Mountains. *Remote Sens.* **2022**, *14*, 3645. https://doi.org/10.3390/rs14153645

Academic Editor: Zhuosen Wang

Received: 2 June 2022
Accepted: 26 July 2022
Published: 29 July 2022

1. Introduction

Vegetation plays an important role in ecosystem carbon and hydrological cycles [1,2] and is a sensitive indicator of ecosystem response to climate change. Vegetation phenology, which is the periodic life activity of plants, provides an independent measure of how ecosystems respond to climate change [3] and varies significantly according to climate zone and vegetation type, especially in temperate and northern regions [4,5]. Affected by climate change and human activity, the start of the growing season (SOS), the end of the growing season (EOS), and the length of the growing season (LOS) show substantial interannual variability [6]. Phenological changes result in small-area changes in plant activity within the community and large-area changes in overall land surface processes, such as the carbon budget, surface energy flux, and regional climate [7]. Therefore, understanding phenological variation and its response to climate change is critical for improving terrestrial biosphere models and climate models [8].

Current research methodologies on vegetation phenology mainly include the traditional ground observation method and satellite remote sensing monitoring method. Traditional ground observations can provide detailed specific plant phenology information at the species-scale or individual plant scale but have limitations in terms of observational stations and spatial coverage [9]. In addition, most ground observation sites only focus on cultivated plants rather than natural vegetation [10]. Remote sensing data from satellites can provide long time series and a high temporal resolution vegetation index (VI) and have been widely applied in large-scale vegetation phenology monitoring [11]. The satellite remote sensing monitoring method primarily uses time series VIs, and the normalized difference vegetation index (NDVI) is the most commonly used VI [12]. The NDVI is simple to calculate and sensitive to plant growth and can track seasonal dynamic changes in vegetation [13].

Climate changes can be directly reflected in vegetation phenology [6]. Temperature can be considered the most important factor affecting vegetation phenology in many regions. Numerous studies have found that advances in spring phenology at middle and high latitudes are primarily controlled by increased global surface mean temperature [14–17]. Additionally, precipitation is a key factor in regulating vegetation phenology, particularly in water-limited arid and semiarid regions [18]. For example, Ren et al. [19] showed that the influence of precipitation on the interannual variation in the SOS and EOS is more important than that of temperature in the Inner Mongolian Autonomous Region. Compared with precipitation, soil moisture is the most direct water supply for vegetation and is susceptible to drought, which can affect vegetation phenology [20]. Some observational studies have suggested that soil water availability is also an important factor that can trigger vegetation growth in water-limited areas [21,22]. An understanding of the impact of soil moisture dynamics on vegetation phenology is very important and can increase our understanding of the influence of climate change on ecosystems. However, there are insufficient studies related to this topic.

The Qilian Mountains (QLMs) are located in the arid/semiarid region of northwestern China, which is a transitional zone between the Qinghai–Tibet Plateau (QTP), Loess Plateau, and Inner Mongolia Plateau. The QLMs have a vulnerable ecosystem and complex climate, and the hydrothermal conditions differ from east to west [23]. In recent years, the QLMs have experienced significant climate changes, which involve a significant trend of warming and wetting, frequent climate anomalies [24], and local vegetation becoming sensitive to climate changes [25]. In addition, the large east–west span and spatial heterogeneity among the vegetation types in the QLMs lead to enormous differences in the response relationship between the vegetation phenology of different vegetation types and climatic factors. Soil moisture plays an essential role in maintaining vegetation growth, especially in arid and semi-arid regions [20]. Soil moisture in the QLMs increases with an increase in altitude and is heterogeneous between different types of land cover [26]. Due to the complex topography and climatic conditions, the ecosystems in the QLMs are fragile and sensitive to climate change, and thus it is necessary to systematically explore the effects of temperature, precipitation, and soil moisture on vegetation phenology.

Based on moderate resolution imaging spectroradiometer (MODIS) NDVI time series products from 2001 to 2020, this study extracted the SOS, EOS, and LOS for the QLMs' vegetation and analyzed the characteristics of the changes in vegetation phenology and the response relationship between vegetation phenology and driving factors, including temperature, precipitation, and soil moisture. The main objectives of this study were to (1) investigate the characteristics of the spatiotemporal patterns of vegetation phenology in the QLMs during the period 2001–2020, (2) evaluate the effects of seasonal temperature, precipitation, and soil moisture on the SOS and the EOS in the study area, and (3) explore the relationship between the phenology of different elevation zones, vegetation type, and climatic factors in the QLMs. This study can contribute to our understanding of the mechanism of the effects of climate change on vegetation phenology in arid mountain areas, and the findings enable the prediction of the future evolution of ecosystems and the implementation of effective ecosystem management.

2. Data and Methods

2.1. Study Area

The QLMs represent the largest mountain system in the marginal northeast of the QTP, which crosses Gansu Province and Qinghai Province. The geographical coordinates lie between 93°25′–103°50′E and 35°52′–39°52′N, with a total area of approximately 184,000 km^2. The terrain gradually rises from northeast to southwest, and the average elevation exceeds 3500 m (Figure 1). The northern slope of the QLMs contain the headwaters of three inland rivers (Heihe, Shulehe, and Shiyanghe rivers) in China [27]. Qinghai Lake is the largest inland saltwater lake in China and is fed from the south slope of the QLMs. The QLMs belong to the midlatitude northern temperate zone, which has a typical continental plateau climate [28]. Due to the obvious vertical zonality and horizontal zonality, the water and heat conditions in the Qilian Mountains dramatically vary spatially. Precipitation mainly occurs in the summer and decreases from east to west, increasing with altitude, but temperature shows the reverse pattern [25]. The main vegetation types in the region include broadleaf forests, needleleaf forests, shrublands, meadows, grasslands, and deserts, and the natural ecosystems are fragile and sensitive to climate change because of complex topographic and climatic conditions.

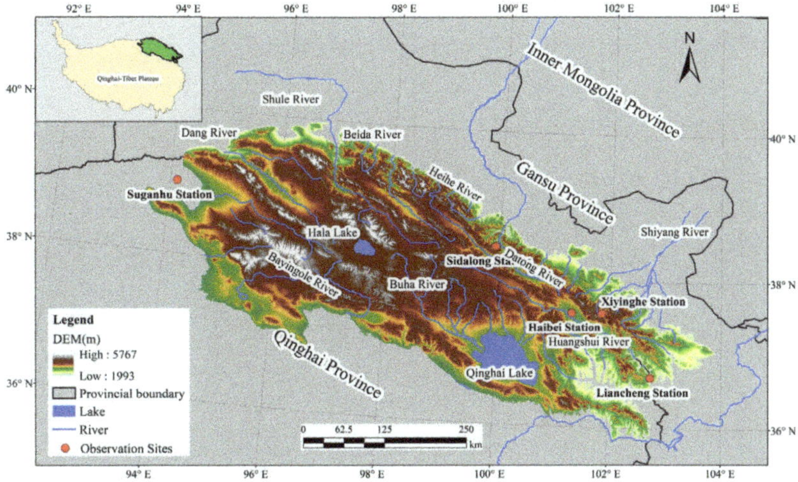

Figure 1. Geographical location of the Qilian Mountains.

2.2. Data Sources

NDVI time series have been widely used as an important proxy for quantifying vegetation photosynthetic activity. The MODIS NDVI products (MOD13A2) from 2001 to 2020 with a 1 km spatial resolution and 16-day time step were used in this study. The NDVI data were obtained from NASA (https://lpdaac.usgs.gov, accessed on 20 November 2021) and were preprocessed using the MODIS reprojection tool (MRT) for band extraction and mosaic, format, and projection conversion. We removed pixels with average annual NDVI values (2001–2020) < 0.1 to prevent the interference of nonvegetation signals [18,29].

The ground-based phenology data were collected from the vegetation phenological observation datasets at Haibei station from 2006 to 2015, which were provided by the Chinese Ecosystem Research Network (CERN) (http://www.cnern.org.cn, accessed on 17 December 2021). In addition, the phenology data of Sidalong station, Liancheng station, Xiyinghe station, and Suganhu station from 2020 were obtained from the National Tibetan Plateau Data Center (https://data.tpdc.ac.cn/en/, accessed on 17 December 2021).

Monthly temperature and precipitation data from 2001 to 2020 were obtained from the National Tibetan Plateau Data Center (http://data.tpdc.ac.cn, accessed on 22 April 2021). These datasets were spatially downscaled from CRU TS v4.02 with WorldClim datasets based on the delta downscaling method and were evaluated using the data of 496 national weather stations across China. The evaluation indicated that the downscaled dataset is reliable for investigations related to climate change across China [30]. The monthly surface soil moisture data (0–7 cm) from 2001 to 2020 were obtained from ERA5-Land and used to represent the water availability indicator to evaluate the water content impacts on vegetation phenology. ERA5-Land provides a soil moisture reanalysis dataset of $0.1^\circ \times 0.1^\circ$ from 1950 to the present. The monthly soil moisture data from ERA5-Land were resampled to the same resolution as the vegetation phenology data using a bilinear interpolation algorithm. The digital elevation model at a spatial resolution of 1 km was obtained from the Resource and Environmental Science and Data Center (http://www.rsdc.cn/, accessed on 17 December 2021).

2.3. Methods

2.3.1. Extraction of Vegetation Phenology

The NDVI time series involves some noise caused by clouds or poor atmospheric conditions and needs to be smoothed using a filter. In this study, NDVI was smoothed using a seven-parameter double logistic function proposed by Gonsamo et al. [31] to reconstruct the NDVI time series at a daily temporal resolution:

$$\mathrm{f}(x) = \alpha_1 + \frac{\alpha_2}{1 + e^{-\partial_1(x-\beta_1)}} - \frac{\alpha_3}{1 + e^{-\partial_2(x-\beta_2)}} \tag{1}$$

where f(x) is the fitted NDVI at day x; x is a specific day of year (DOY); α_1, α_2, α_3, ∂_1, ∂_2, β_1, β_2 are smoothing parameters; α_1 is the background NDVI value; α_2 is the early summer plateau; α_3 is the amplitude of the late summer plateau; ∂_1 and ∂_2 represent the transitions in the slope coefficient; and β_1 and β_2 are the midpoints at the start and end of the growing season transitions, respectively.

For the fitted *NDVI* time series, the dynamic threshold derived from each pixel was used to determine the SOS and EOS. In this method, the SOS and EOS are defined as the DOY when the $NDVI_{ratio}$ reaches a certain threshold during the *NDVI* rising stage in spring and decline stage in autumn. The $NDVI_{ratio}$ is calculated as:

$$NDVI_{ratio} = \frac{NDVI_x - NDVI_{min}}{NDVI_{max} - NDVI_{min}} \tag{2}$$

where $NDVI_x$ represents the *NDVI* value on day x and $NDVI_{max}$ and $NDVI_{min}$ are the maximum and minimum *NDVI* values in the annual NDVI time series, respectively. In this study, the dynamic threshold was defined as $NDVI_{ratio}$ values of 30% and 50% to determine the SOS and EOS, respectively. The LOS was the difference between the SOS and EOS.

2.3.2. Trend Analysis

The temporal trends in the time series of the vegetation phenology were calculated by the Theil–Sen median slope estimator [32] at the pixel level. The Theil–Sen median slope estimator is a nonparametric median-based slope estimator that is less susceptible to noise and outliers [33]. A positive Theil–Sen slope indicates a delayed or extended trend, while a negative value indicates an advanced or shortened trend. The Mann–Kendall (MK) method [34] was used to determine the significance of the long-term advanced/delayed trend in vegetation phenology. In our study, the significance level was based on the MK test value, and $p < 0.05$ was defined as statistically significant. The combined use of the Theil–Sen median slope and MK trend test classified vegetation phenological parameters into five categories, namely, "significant advanced/shortened", "insignificant advanced/shortened", "no change", "insignificant delayed/extended", and "significant delayed/extended".

2.3.3. Partial Correlation Analysis

Partial correlation coefficients were calculated to examine the correlation between vegetation phenology and seasonal driving factors (temperature, precipitation, and soil moisture). In our analysis, the seasons were defined as spring (March–May), summer (June–August), and autumn (September–November). The second-order partial correlation coefficient was calculated as follows:

$$r_{12,34} = \frac{r_{12,3} - r_{14,3} \times r_{24,3}}{\sqrt{(1 - r_{14,3}^2) \times (1 - r_{24,3}^2)}} \tag{3}$$

where $r_{12,34}$ represents the partial correlation coefficient of variables 1 and 2 after controlling for variables 3 and 4. $r_{12,3}$ represents the first order partial correlation coefficient and was computed as follows:

$$r_{12,3} = \frac{r_{12} - r_{13} \times r_{23}}{\sqrt{(1 - r_{13}^2) \times (1 - r_{23}^2)}} \tag{4}$$

where r_{12}, r_{13}, r_{23} represent the Pearson's correlation coefficients between variables 1 and 2, 1 and 3, and 2 and 3, respectively. After we calculated the partial correlation coefficient values, Student's t-test was used to identify the significance of the coefficient, and only the pixels with a significance level of $p < 0.05$ were considered significant. To determine the influence of terrain and vegetation types on the linkage between vegetation phenology and seasonal driving factors, the partial correlation coefficients in different elevation zones and different vegetation types were also analyzed in our study. The elevation was reclassified into four classes (1: <3000 m a.s.l., 2: 3000–3500 m a.s.l., 3: 3500–4000 m a.s.l., and 4: >4000 m a.s.l.).

3. Results

3.1. Temporal and Spatial Variation in Vegetation Phenology

The vegetation phenology derived from the satellite data was consistent with ground observations at the Xiying River, Liancheng, Suganhu, and Sidalong stations in 2020 and the Haibei station from 2006 to 2015. The correlation coefficient (R^2) between the SOS and field data was 0.536 ($p < 0.01$), the mean absolute error (MAE) was 9 d, the root mean square error (RMSE) was 11 d, the R^2 was 0.533 ($p < 0.01$), the MAE was 5 d, and the RMSE was 6 d between the EOS and field data (Figure S1). Based on the validation results described above, the remote sensing monitoring method adopted in this paper can accurately reflect vegetation phenological characteristics in the QLMs.

The interannual changes in vegetation phenology in the QLMs from 2001 to 2020 showed different fluctuation ranges (Figure 2). There was an advanced SOS trend of 0.510 d/year and an extended LOS trend of 0.580 d/year. There was a delayed EOS trend at a rate of 0.066 d/year, which is only a slight change. However, no significant changes were found in these vegetation phenology parameters ($p > 0.05$).

The vegetation phenology parameters varied with altitude (Figure 3). With an increase in altitude, the SOS showed a gentle upward trend, but the correlation between the SOS and altitude was weak ($p > 0.05$). Conversely, with an increase in altitude, the EOS gradually advanced and the LOS gradually shortened. There was a significant negative correlation between the altitude and both EOS and LOS ($p < 0.05$), and the correlation coefficients were high ($R^2 \geq 0.899$). The SOS tended to be delayed by 0.20 d/100 m, while the EOS tended to advance by 0.60 d/100 m, and the LOS tended to extend by 0.80 d/100 m with increasing altitude.

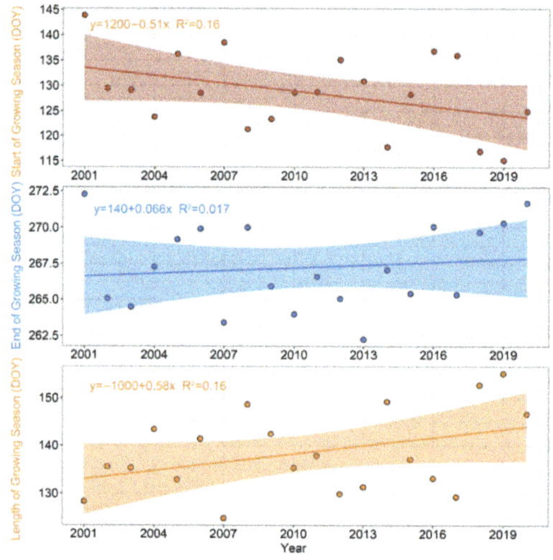

Figure 2. Interannual changes in vegetation phenology in the QLMs.

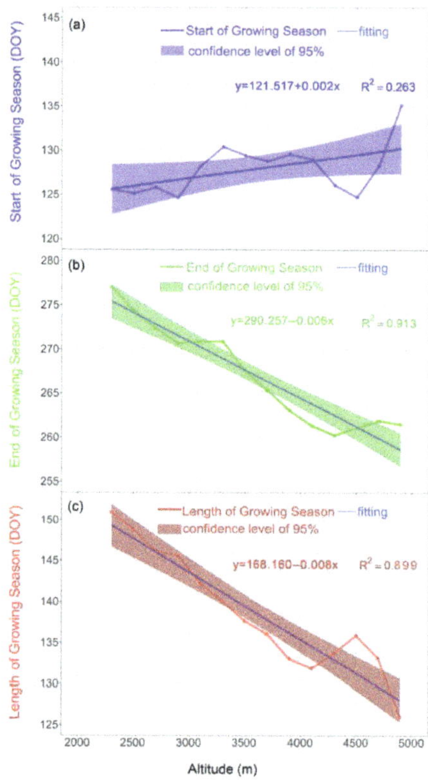

Figure 3. Characteristics of the changes in the SOS (**a**), EOS (**b**), and LOS (**c**) with altitude.

The multiyear average spatial distribution of phenology in QLMs from 2001 to 2020 is shown in Figure 4a,c,e. From east to west, the vegetation phenology showed evident changes. Overall, the SOS in the study area mainly occurred from 115 days to 150 days, which accounted for more than 80% of the vegetation region. The earlier SOS was mainly seen in the eastern and western QLMs, and the later SOS was mainly distributed in the central section. In addition, the multiyear mean EOS of vegetation phenology varied between 255 and 275 d (more than 80% of the overall pixels) from the middle of September to early October. The EOS showed the opposite pattern in terms of spatial distribution compared with the SOS; it was earlier in the central section of the QLMs and later in the western and eastern sections of the QLMs. Due to the combined effects of SOS and EOS, the LOS was mainly between 110 and 160 d. The spatial pattern of LOS was similar to that of EOS, whereby the LOS was shorter in the central section of the QLMs and longer the eastern and western sections of the QLMs.

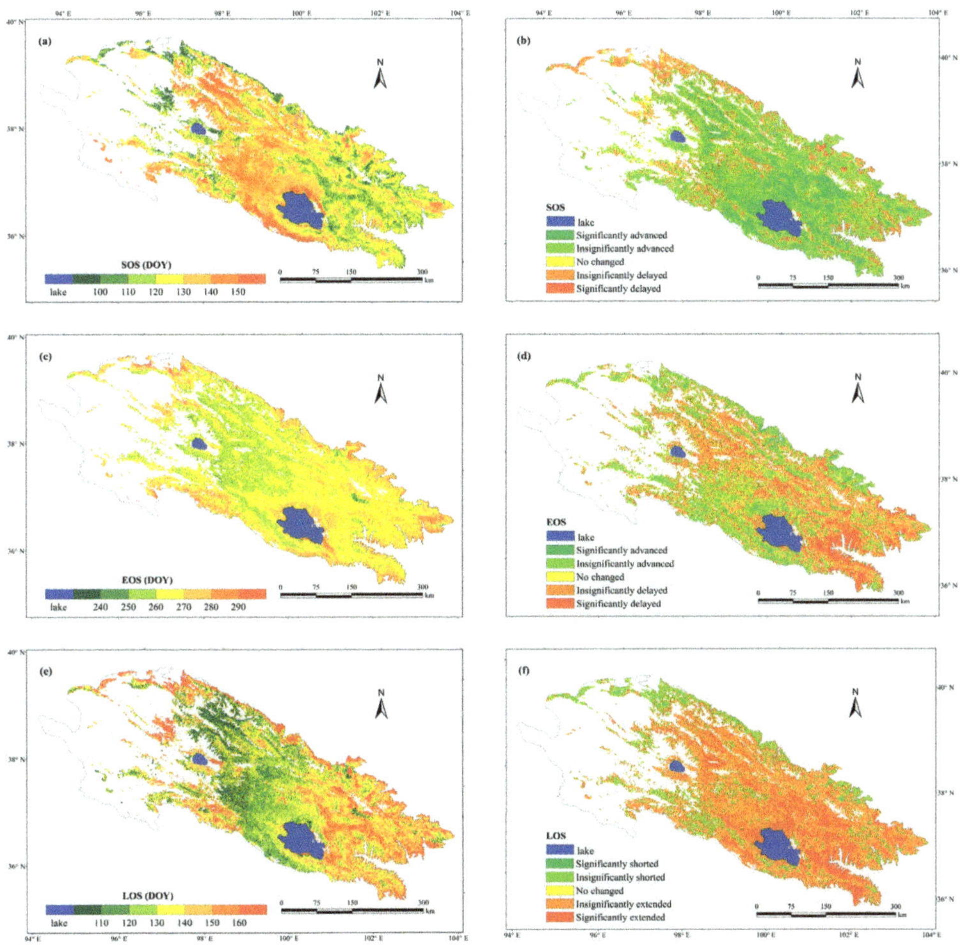

Figure 4. Spatial distribution and the change trend of vegetation phenology parameters. The left side represents the spatial pattern of the average value of the SOS (**a**), EOS (**c**), and LOS (**e**) from 2001 to 2020. The right side represents the spatial distribution of the change trend of the SOS (**b**), EOS (**d**), and LOS (**f**) from 2001 to 2020. The trends are considered significant for pixels according to the MK test ($p < 0.05$).

Figure 4b,d,f and Table 1 show the spatial distribution of the vegetation phenology trend in the QLMs from 2001 to 2020. A total of 72.37% of the vegetation pixels showed an advancing trend of SOS from 2001 to 2020. A total of 13.85% of pixels, which were mainly concentrated in the central and eastern sections of the QLMs, showed a significant advancing trend of SOS. A few areas in the northwest of the QLMs showed a delayed trend of SOS, whereas only 1.44% of the total land area was significantly delayed. Regions with delayed EOS accounted for 47.59% of the vegetation pixels in the study area from 2001 to 2020 and were mainly located in the eastern and central sections of the QLMs. In addition, the areas with advanced EOS were mainly located on the northern margins and at the northwest of Qinghai Lake. Approximately 6.8% of vegetation pixels had a significant delayed trend in terms of the EOS, and 3.9% of vegetation pixels had a significantly advanced trend. There was an overall extended LOS trend for most parts of the vegetation area (71.66% of the vegetation pixels) from 2001 to 2020, with 12.65% being significantly extended and only 1.87% being significantly shortened. The areas with extended LOS trends were mainly distributed in the central and eastern sections of QLMs.

Table 1. The percentage of different trends of vegetation parameters based on MK analysis across the QLMs.

Vegetation Phenology	Insignificantly Advanced/Shortened	Insignificantly Delayed/Prolonged	Significantly Advanced/Shortened	Significantly Delayed/Prolonged
SOS	58.52%	19.62%	13.85%	1.44%
EOS	36.57%	40.79%	3.90%	6.80%
LOS	23.42%	59.01%	1.87%	12.65%

3.2. Response of Vegetation Phenology to Seasonal Driving Factors

The spatial distribution of the partial correlation coefficients between seasonal driving factors and vegetation phenology metrics are displayed in Figure 5. For the QLMs, the SOS was negatively correlated with spring temperature in 73.81% of vegetation pixels, while 21.21% of pixels showed a significant correlation ($p < 0.05$) and were mainly located in the eastern and central parts of the study area (Figure 5a). The percentage of negative and positive correlations between the SOS and spring precipitation was similar (Figure 5c), with a significant negative correlation occurring in the northeast of Hala Lake. More than half of the vegetation pixels (61.34%) of the SOS had a negative correlation with spring soil moisture, of which 9.18% of pixels showed a significant negative correlation (Figure 5e), mainly at the west of Qinghai Lake. The results above indicate that the increases in spring temperature and soil moisture likely cause the SOS to advance in most part of the QLMs.

The partial correlation coefficients between the EOS and autumn temperature showed that the EOS was positively correlated with temperature in 65.20% of vegetation pixels, and 9.89% of the areas passed the significance test (Figure 5b). Approximately 56.19% of the vegetation pixels showed a negative correlation between the EOS and autumn precipitation, and only 4.75% of the areas passed the significance test (Figure 5d). For autumn soil moisture, positive correlations between the EOS and soil moisture occurred in 60.44% of the total vegetation pixels, and approximately 9.50% of the pixels showed a significant positive correlation ($p < 0.05$, Figure 5f), most of which were distributed west of Qinghai Lake and southeast of Hala Lake. In total, the autumn temperature and soil moisture influenced the EOS in most areas, and the increase in autumn temperature and soil moisture likely caused the EOS to be delayed.

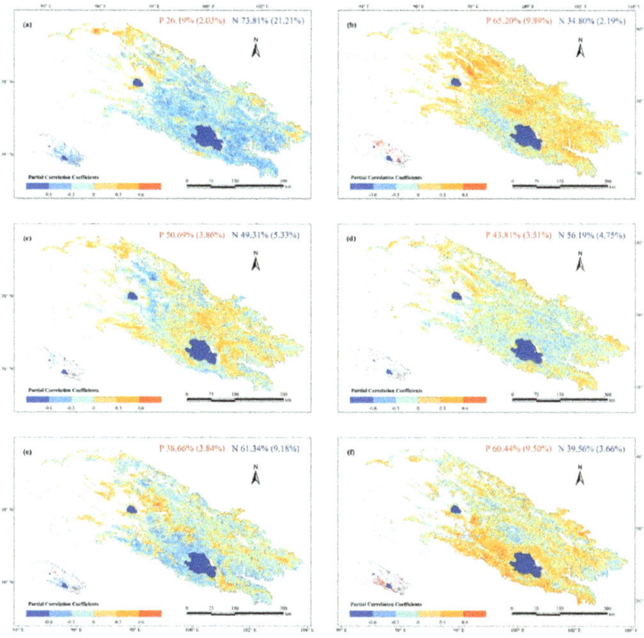

Figure 5. The partial correlation coefficients between vegetation phenology parameters and seasonal driving factors. (**a**,**c**,**e**) represent the partial correlation coefficients between the SOS and spring temperature, precipitation, and soil moisture, respectively. (**b**,**d**,**f**) represent the partial correlation coefficients between the EOS and autumn temperature, precipitation, and soil moisture, respec-tively. The inset panels on the bottom left of each subpicture present pixels with a significantly ($p < 0.05$) negative (blue) and positive (red) correlation. The percentages of positive (P) and negative (N) correlations (the values in brackets indicate the percentage of significant correlations) are shown at the top of each subpicture.

3.3. *Vegetation Phenology Parameters Response to Seasonal Driving Factors Based on Different Elevation Zones*

The driving factors had different effects on vegetation phenology depending on elevation. The results of the partial correlation analysis between the vegetation phenology parameters (SOS, EOS) and seasonal driving factors (temperature, precipitation and soil moisture) for different elevation zones are shown in Figure 6. A mainly negative correlation occurred between the SOS and spring temperature (the percentage of significant negative correlation ranged from 14.59% to 24.87%) at different elevation zones. At middle elevations (3000–4000 m a.s.l.), more than 76% of the vegetation pixels showed a negative correlation between the SOS and spring temperature (Figure 6a), and more than 24% of the areas passed the significance test. The SOS was mainly negatively correlated with spring soil moisture (more than 61%) at the <4000 m elevation zone, and the percentage of areas that passed the significance test at the 95% level ranged from 7.24% to 12.5% (Figure 6c). However, the SOS had the opposite correlation with spring soil moisture in the highest elevation zone (mainly positive), 7.63% of the areas showed a significant positive correlation (Figure 6c). Compared with spring temperature and soil moisture, spring precipitation had a weaker influence on the SOS at the four elevation zones, and the areas of positive and negative correlation between spring precipitation and the SOS were similar, with few pixels showing a significant correlation (Figure 6b).

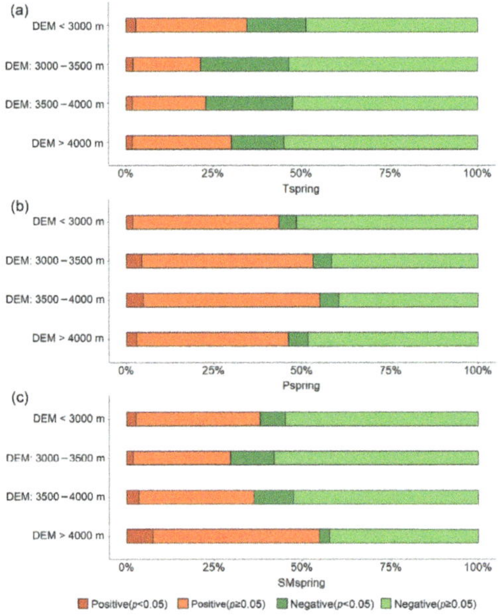

Figure 6. Percentages of correlation between SOS and three driving factors at different elevation zones. The three driving factors were (**a**) spring temperature (Tspring), (**b**) spring precipitation (Pspring), and (**c**) spring soil moisture (SMspring).

At the lowest elevation zone (<3000 m a.s.l.), the EOS was positively correlated with summer temperatures and precipitation in 62.57% and 72.16% of areas, and 8.03% and 18.1% of the areas showed a significant correlation ($p < 0.05$), respectively (Figure 7b,e). Notably, a negative correlation between EOS and summer soil moisture occurred in 80.96% of vegetation pixels at the lowest elevation zone (<3000 m a.s.l.), which was more than four times larger than the positive correlation (19.04%). Approximately 30.06% of the pixels showed a significantly negative correlation between summer soil moisture and EOS ($p < 0.05$) at the lowest elevation zone (<3000 m a.s.l.), while areas with significant positive correlations represented only 0.59% of the total (Figure 7h), indicating that the EOS was advanced in most low elevation regions with an increase in summer soil moisture. A positive correlation between the EOS and autumn temperature covered more than 60% of the area in the different elevation zones, and more than 7.00% of the pixels exhibited a significant correlation (Figure 7c). For the region with elevations less than 4000 m, more than 8.81% of the areas demonstrated a significantly positive correlation between the EOS and autumn soil moisture (Figure 7i). For the regions with elevations of less than 3500 m, the area with a significant positive correlation between the EOS and autumn soil moisture was greater than the EOS and autumn temperature (Figure 7c,i), so at lower elevations (less than 3500 m), soil moisture played a more important role in vegetation growth in autumn. At higher elevations (higher than 3500 m), temperature played a more important role in vegetation growth in autumn (Figure 7c). Overall, compared with autumn temperature and soil moisture, autumn precipitation had a weaker influence on the EOS at the four elevation zones (Figure 7f).

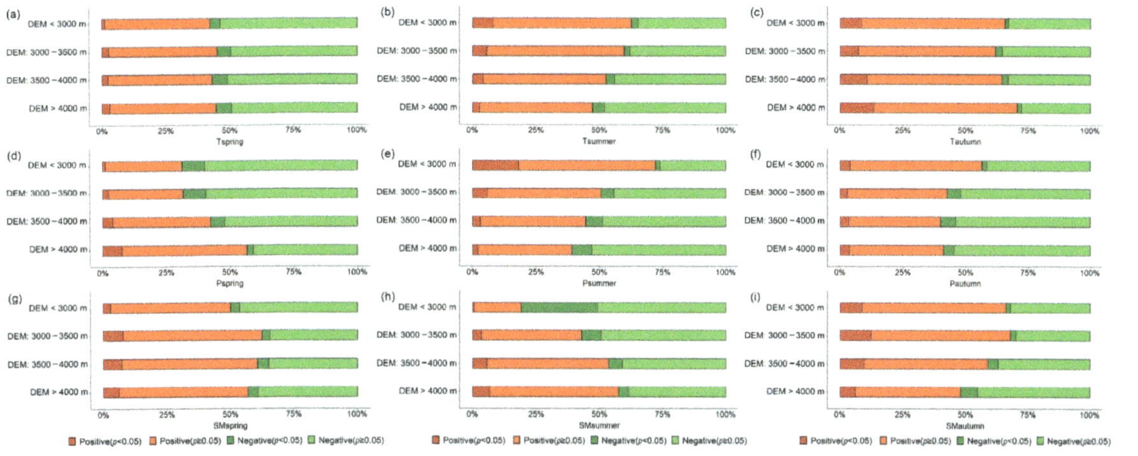

Figure 7. Percentages of correlation between EOS and different driving factors at different elevation zones. The driving factors were (**a**,**b**,**c**) seasonal temperature (Tspring, Tsummer, and Tautumn are spring, summer, and autumn temperature, respectively), (**d**,**e**,**f**) seasonal precipitation (Pspring, Psummer, and Pautumn are spring, summer, and autumn precipitation, respectively), and (**g**,**h**,**i**) seasonal soil moisture (SMspring, SMsummer, and SMautumn are spring, summer, and autumn soil moisture, respectively).

3.4. Vegetation Phenology Response to Seasonal Driving Factors across Vegetation Types

Generally, different vegetation types had different responses to driving factors (Figures 8 and 9). For different vegetation types, the partial correlation coefficients between the SOS and spring temperature were mostly negative (Figure 8a), especially for broadleaf forests, needleleaf forests, shrubland, and meadows (more than 23% of regions had significantly negative correlations). The partial correlation coefficients between the SOS and spring soil moisture were also mostly negative (Figure 8c) for different vegetation types, except alpine vegetation, and more than 7.80% of areas had significantly negative correlations. For alpine vegetation, the SOS was negatively correlated with spring temperature in approximately 65.18% of areas, of which 12.14% of areas showed a significant negative correlation (Figure 8a). However, it was positively correlated with the spring soil moisture in 52.04% of areas, with 7.49% of regions showing a significantly positive correlation (Figure 8c). Compared with spring temperature and soil moisture, spring precipitation had a weaker influence on the SOS in most vegetation types. The areas of positive and negative correlation between spring precipitation and the SOS were similar, with few pixels showing a significant correlation (Figure 8b).

Compared with the correlation between the EOS and autumn temperature and soil moisture, there were limited positive and negative correlations between the EOS and autumn precipitation, and there were relatively few significant pixels for most vegetation types (Figure 9f). Autumn temperature and soil moisture had mainly positive correlations with the EOS for most vegetation types (Figure 9c,i), and the EOS had a more significant relation with soil moisture than temperature for grasslands and deserts. There was a negative correlation between EOS and spring temperature in more than 65% of areas of broadleaf forests, and 13.49% of the areas were significantly correlated (Figure 9a). For needleleaf forests, there was a negative correlation between EOS and spring precipitation in 66.11% of areas, and 10.64% of the areas were significantly correlated (Figure 9d). A significant positive correlation between the EOS and summer precipitation was found in 11.51% of pixels for broadleaf forest and 11.36% of pixels for needleleaf forest (Figure 9e). Summer and autumn soil moisture had opposite correlations with the EOS in QLMs, except meadows. The EOS was mainly negatively correlated with summer soil moisture, especially for broadleaf forests, needleleaf forests, and grasslands (approximately 15.58 %, and 18.78% of the pixels had significant negative correlations, respectively). Th s of most vegetation types were positively correlated with autumn soil moisture, esp ially those of grasslands and deserts (for which approximately 11.35% and 12.97% of the pixels had significant positive correlations, respectively).

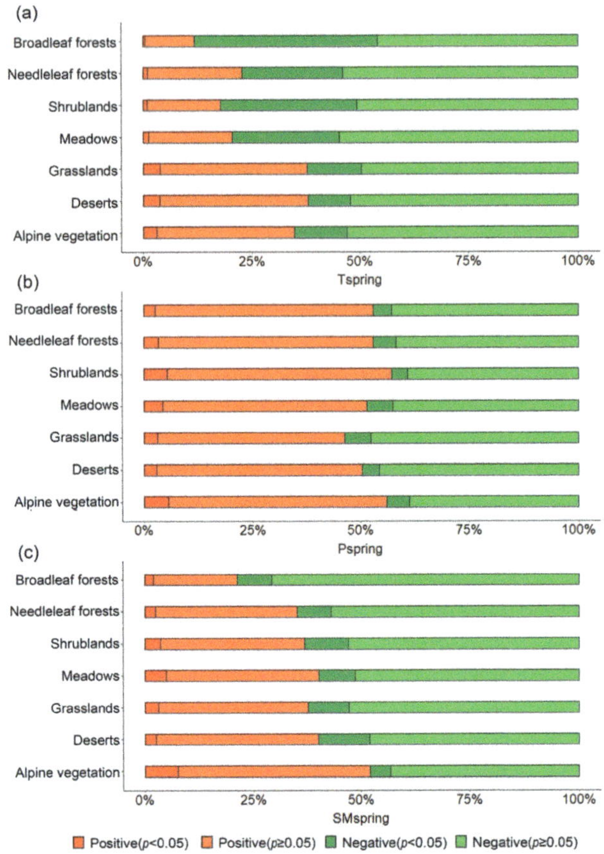

Figure 8. PerPercentages of correlation between the SOS and three driving factors in different vegetation types. The three driving factors were (**a**) spring temperature (Tspring), (**b**) spring precipitation (Pspring), and (**c**) spring soil moisture (SMspring).

Figure 9. Percentages of correlation between EOS and different driving factors in different vegetation types. The driving factors were (**a**,**b**,**c**) seasonal temperature (Tspring, Tsummer, and Tautumn are spring, summer, and autumn temperature, respectively), (**d**,**e**,**f**) seasonal precipitation (Pspring, Psummer, and Pautumn are spring, summer, and autumn precipitation, respectively), and (**g**,**h**,**i**) seasonal soil moisture (SMspring, SMsummer, and SMautumn are spring, summer, and autumn soil moisture, respectively).

4. Discussion

4.1. The Spatial Heterogeneity of Vegetation Phenology in the Qilian Mountains

The vegetation phenology in the QLMs showed significant spatial heterogeneity. In general, the SOS was later in the central region and earlier in the eastern and western regions of the QLMs, and the EOS exhibited the opposite trend in terms of its spatial distribution. These results are consistent with the results reported by Qi et al. [27] and Sun et al. [35] but inconsistent with the results reported by Qiao et al. [36], who observed that the multiyear mean SOS was gradually delayed from southeast to northwest and that the multiyear mean EOS gradually advanced from southeast to northwest in the QLMs. The main reasons for the differences in the results are the different temporal and spatial resolutions of the remote sensing data. Different remote sensing data sources and data time series may obtain different vegetation phenology results [37,38]. The AVHRR and MODIS datasets have a consistently high temporal resolution time series of data and are widely used for phenology studies [39]. MODIS data (1 km) have a higher resolution than AVHRR data (8 km) and can extract more detailed spatial phenological signals for vegetation types, particularly in heterogeneous areas [40]. The vegetation phenology of the QLMs was characterized by an advanced SOS, delayed EOS, and extended LOS during the period from 2001 to 2020, which are consistent with recent results on the QTP [37,38,41] and on the QLMs [27,35].

In our study, the results showed that the SOS gradually delayed, the EOS gradually advanced, and the LOS gradually shortened with increasing altitude. No significant correlation was found between the SOS and altitude, but a significant negative correlation was found between both the EOS and LOS and altitude. These results are consistent with recent findings on the QTP [2]. The SOS change may have almost nothing to do with altitude [2]. At higher altitudes, there is relatively low air temperature, which is not beneficial for delaying leaf senescence, and the EOS advances to avoid harm from frost and has a shorter LOS [42].

4.2. Response of Vegetation Phenology to Different Driving Factors

Vegetation phenology responses to different driving factors are complex and variable. In our study, we found that the SOS was negatively correlated with spring temperature and spring soil moisture in most regions of the study area, implying that the advanced SOS could be associated with a warmer spring air temperature and higher soil moisture. In our study, the spring temperature had a stronger influence on the SOS, and most studies have also reported that higher temperatures were the main factor associated with an earlier SOS around the world over the last several decades [18,43–45].

The impacts of driving factors on vegetation phenology were varied in different elevation zones. For SOS, spring temperature seemed to be the main factor limiting vegetation growth. The QLMs are located in high altitude areas with low temperatures (the annual mean temperature in most areas is below 0 °C); with an increase in altitude, the temperature gradually decreased (Table 2). Vegetation needs a certain amount of cumulative temperature to green up, so the early stages of vegetation growth are more affected by temperature in relatively cold regions [5,46]. The autumn soil moisture was the main limiting factor at lower elevations (<3500 m a.s.l.), and autumn temperature was the main limiting factor at higher elevations (>3500 m a.s.l.). These results are consistent with the research of Peng et al. [47], which demonstrated that soil moisture was the major limiting factor for the radial growth of Qinghai spruce at the lower elevations of the central QLMs and that temperature was the major limiting factor for radial growth of Qinghai spruce at higher elevations. These results also suggest that vegetation management must take elevation differences into account when facing the challenges of climate change. From Table 2, we can see that the annual average soil moisture at lower elevations (0.31 $m^3 \cdot m^{-3}$) was less than that at higher elevations (0.34 $m^3 \cdot m^{-3}$), so this is one possible reason that the autumn soil moisture had a stronger influence on the EOS in lower elevation zones. The EOS showed a significant negative correlation with summer soil moisture in approximately 30.06% of the pixels in the lowest elevation zones (<3000 m a.s.l.). Peng et al. [47] also found that, during the summer at lower elevations, soil moisture is the most important factor limiting xylem cell differentiation based on the Vaganov–Shashkin model.

Table 2. The annual average soil moisture and temperature from 2001 to 2020 in four elevation zones.

	Soil Moisture ($m^3 \cdot m^{-3}$)	Temperature (°C)
DEM < 3000 m	0.31	2.19
DEM: 3000–3500 m	0.31	−0.90
DEM: 3500–4000 m	0.34	−4.67
DEM > 4000 m	0.34	−7.27

At the landscape level, the SOS was negatively correlated with spring temperature in most regions with different vegetation types. More specifically, 42.36%, 23.31%, 31.54%, and 24.83% of the areas of broadleaf forests, needleleaf forests, shrublands, and meadows, respectively, showed a significantly negative correlation between SOS and spring temperature (Figure 8a). This is because the broadleaf forests, needleleaf forests, shrublands, and meadows are mainly located in semi-arid regions (more than 78% of these are located in the semi-arid region) where the climate is relatively humid compared with arid regions; however, the temperature is low in the study area, and higher temperature in spring could decrease the damage from frost and promote spring thawing [5]. The spring soil moisture had a stronger influence on the SOS of deserts (Figure 8c). This is because about 90.37% of deserts are located in arid areas with limited soil water conditions (Table 3). The soil water is an indispensable intermediary used to ensure nutrient substance transport, which is likely to be the main reason for the negative correlation between soil moisture and SOS for deserts. Additionally, there are many shallow-rooted plants in deserts, and these shallow-rooted plants are more sensitive to soil moisture changes than other plants [48,49].

A negative correlation was observed between summer soil moisture and EOS, but a positive correlation was observed between autumn soil moisture and EOS for most vegetation types. This negative correlation shifted to a positive correlation from summer to autumn, indicating that the summer and autumn soil moisture had a great influence on the EOS, but the correlation was the opposite in these two seasons. The main reason for this differential response is that the precipitation in QLMs is mainly concentrated in summer [25], and too much moisture prevents vegetation growth because a high soil moisture can limit the absorption of soil nutrients by vegetation [50]. Ren et al. found that the precipitation played a more important role than temperature in the interannual variation of the SOS and EOS in Inner Mongolia [19]. However, compared with temperature and soil moisture, precipitation had a relatively limited impact on the EOS in the QLMs. This is because precipitation may a have lagged effect on vegetation phenology, meaning that soil moisture is a more straightforward driving factor for vegetation phenology than precipitation and has a number of sources in the QLMs, including precipitation, snowmelt, surface runoff, and groundwater.

Table 3. The annual average soil moisture and temperature from 2001 to 2020 for different vegetation types.

	Soil Moisture ($m^3 \cdot m^{-3}$)	Temperature (°C)
Broadleaf forests	0.36	1.72
Needleleaf forests	0.34	−1.49
Shrublands	0.35	−2.05
Meadows	0.36	−4.87
Grasslands	0.29	−0.75
Deserts	0.20	−2.49
Alpine vegetation	0.33	−7.16

4.3. Limitations and Future Work

It should be noted that there may be some limitations to our current study. The number of ground observations of vegetation phenology is insufficient, especially in the central and western parts of the QLMs because of a lack of phenological observation networks. At the same time, the existing observation stations have a relatively short historical record. Digital cameras have been shown to be valuable tools to validate the phenology derived from satellite imagery at a low cost [40] because of their high temporal and spatial resolutions. In future, automated digital cameras are promising for providing consistent and continuous monitoring of vegetation growth at local and regional scales [51,52].

The vegetation phenology results calculated from remote sensing data may contain some uncertainties that are due to the inaccuracy of satellite data. NDVI data have been widely used for phenology characterization because they are simple to measure for most optical sensors [53,54]. However, because NDVI data are sensitive to the soil background and are easily saturated in high vegetation coverage areas [55], the applications of NDVI data may have some limitations. Considering the sparse vegetation in the western part of the QLMs, the modified vegetation index, such as the soil-adjusted vegetation index (SAVI), may be appropriate for detecting vegetation growth changes because of its ability to minimize the effects of the soil background. In the future, collective analyses of multiple VIs (such as the land phenology index, enhanced vegetation index, and perpendicular vegetation index) may improve the accuracy of phenology estimation [56–58]. The spatial resolution of ERA5-Land soil moisture data is relatively low, which may hide some spatial details of soil moisture parameters. But the high-resolution soil datasets are difficult to obtain for a large study area [59]. Future studies should integrate a series of soil moisture datasets at a higher resolution to further discuss the response relationship between vegetation phenology and soil moisture. Vegetation phenology is also influenced by other factors, such as radiation, soil nutrients, climate extremes, and human activities, so more attentio. should be paid to exploring the phenology variations in response to these driving factors in future work. The vegetation phenology response to driving factors may be nonlinear, and the interactions between climatic factors have critical role in vegetation phenology, so other methods like the GeoDetector model can be used to detect the contribution of driving factors to vegetation phenology and the interactions between driving factors. Our findings suggest that the variation in soil moisture should be considered in future studies on climate warming and the environmental effects of phenology in water-limited areas.

5. Conclusions

Based on the time series MODIS NDVI datasets from 2001 to 2020, we retrieved the vegetation phenological parameters in the QLMs. The spatiotemporal variation in vegetation phenology was analyzed, and divergent correlations between the SOS and EOS and seasonal driving factors were explored. The results demonstrated that vegetation phenology in the QLMs is characteristic of advancing SOS, postponing EOS, and prolonging LOS, but the variation trends of vegetation phenology were not significant ($p > 0.05$) from 2001 to 2020. The seasonal temperature, precipitation, and soil moisture had spatiotemporal heterogeneous effects on the vegetation phenology. Compared with temperature and soil moisture, precipitation had a weaker influence on the vegetation phenology in QLMs. The spring temperature was the key driving factor influencing SOS in the QLMs. The autumn soil moisture and autumn temperature made the largest contributions to the variations in EOS at lower elevations (<3500 m a.s.l.) and higher elevations (>3500 m a.s.l.), respectively. Spring temperature was the key driving factor influencing SOS of most vegetation types. Autumn soil moisture was the main factor influencing EOS in deserts because of the strong soil moisture stress. An increase in summer soil moisture may limit vegetation growth in the QLMs. Under ongoing global change, finding the response of the SOS and EOS to driving factors is beneficial for a better understanding of the interactions between vegetation phenology and future climate change.

Supplementary Materials: The following supporting information can be downloaded at https://www.mdpi.com/article/10.3390/rs14153645/s1: Figure S1: Comparison of satellite-derived phenology data with ground-based phenology data.

Author Contributions: Formal analysis, X.C.; Methodology, X.C.; Software, G.X. and D.L.; Visualization, G.X. and X.H.; Writing—original draft, X.C.; Writing—review & editing, X.C. All authors have read and agreed to the published version of the manuscript.

Funding: This research was supported by the Strategic Priority Research Program of Chinese Academy of Sciences (grant XDA20100102) and Gansu Provincial Science and Technology Major Special Plan (20ZD7FA005).

Conflicts of Interest: The authors declare no conflict of interest.

References

1. Pau, S.; Wolkovich, E.M.; Cook, B.I.; Davies, T.J.; Kraft, N.J.B.; Bolmgren, K.; Betancourt, J.L.; Cleland, E.E. Predicting phenology by integrating ecology, evolution and climate science. *Glob. Change Biol.* **2011**, *17*, 3633–3643. [CrossRef]
2. Liu, X.; Chen, Y.; Li, Z.; Li, Y.; Zhang, Q.; Zan, M. Driving Forces of the Changes in Vegetation Phenology in the Qinghai–Tibet Plateau. *Remote Sens.* **2021**, *13*, 4952. [CrossRef]
3. Parmesan, C. Ecological and evolutionary responses to recent climate change. *Annu. Rev. Ecol. Evol. Syst.* **2006**, *37*, 637–669. [CrossRef]
4. Peñuelas, J.; Rutishauser, T.; Filella, I. Phenology feedbacks on climate change. *Science* **2009**, *324*, 887–888. [CrossRef]
5. Piao, S.L.; Tan, J.G.; Chen, A.P.; Fu, Y.H.; Ciais, P.; Liu, Q.; Janssens, I.A.; Vicca, S.; Zeng, Z.Z.; Jeong, S.J.; et al. Leaf onset in the northern hemisphere triggered by daytime temperature. *Nat. Commun.* **2015**, *6*, 6911. [CrossRef]
6. Richardson, A.D.; Keenan, T.F.; Migliavacca, M.; Ryu, Y.; Sonnentag, O.; Toomey, M. Climate change, phenology, and phenological control of vegetation feedbacks to the climate system. *Agric. For. Meteorol.* **2013**, *169*, 156–173. [CrossRef]
7. Morisette, J.T.; Richardson, A.D.; Knapp, A.K.; Fisher, J.I.; Graham, E.A.; Abatzoglou, J.; Wilson, B.E.; Breshears, D.D.; Henebry, G.M.; Hanes, J.M.; et al. Tracking the rhythm of the seasons in the face of global change: Phenological research in the 21st century. *Front. Ecol. Environ.* **2009**, *7*, 253–260. [CrossRef]

8. Jin, Z.; Zhuang, Q.; He, J.S.; Luo, T.; Shi, Y. Phenology shift from 1989 to 2008 on the Tibetan Plateau: An analysis with a process-based soil physical model and remote sensing data. *Clim. Change* **2013**, *119*, 435–449. [CrossRef]
9. Xu, Q.; Yang, G.; Long, H.; Wang, C.; Li, X.; Huang, D. Crop information identification based on MODIS NDVI time-series data. *Trans. Chin. Soc. Agric. Eng.* **2014**, *30*, 134–144.
10. Fu, Y.; Zhang, H.C.; Dong, W.J.; Yuan, W.P. Comparison of phenology models for predicting the onset of growing season over the Northern Hemisphere. *PLoS ONE* **2014**, *9*, e109544. [CrossRef]
11. Zhao, X.; Zhou, D.; Fang, J. Satellite-based Studies on Large-Scale Vegetation Changes in China. *J. Integr. Plant Biol.* **2012**, *54*, 713–728. [CrossRef]
12. Fitchett, J.M.; Grab, S.W.; Thompson, D.I. Plant phenology and climate change: Progress in methodological approaches and application. *Prog. Phys. Geogr.* **2015**, *39*, 460–482. [CrossRef]
13. Zhang, F.; Wu, B.; Liu, C.; Luo, Z.; Zhang, S.; Zhang, G. A method to extract regional crop growth profile with time series of NDVI data. *Remote Sens.* **2004**, *8*, 515–528.
14. Wolfe, D.W.; Schwartz, M.D.; Lakso, A.N.; Otsuki, Y.; Pool, R.M.; Shaulis, N.J. Climate Change and Shifts in Spring Phenology of Three Horticultural Woody Perennials in Northeastern USA. *Int. J. Biometeorol.* **2005**, *49*, 303–309. [CrossRef]
15. He, Z.B.; Du, J.; Zhao, W.Z.; Yang, J.J.; Chen, L.F.; Zhu, X.; Chang, X.X.; Liu, H. Assessing temperature sensitivity of subalpine shrub phenology in semi-arid mountain regions of China. *Agric. For. Meteorol.* **2015**, *213*, 42–52. [CrossRef]
16. Menzel, A.; Sparks, T.H.; Estrella, N.; Koch, E.; Aasa, A.; Ahas, R.; Alm-Kübler, K.; Bissolli, P.; Braslavska, O.; Briede, A.; et al. European Phenological Response to Climate Change Matches the Warming Pattern. *Glob. Change Biol.* **2006**, *12*, 1969–1976. [CrossRef]
17. Ge, Q.S.; Wang, H.J.; Rutishauser, T.; Dai, J.H. Phenological Response to Climate Change in China: A Meta-Analysis. *Glob. Change Biol.* **2015**, *21*, 265–274. [CrossRef]
18. Liu, Q.; Fu, Y.H.; Zeng, Z.; Huang, M.T.; Li, X.R.; Piao, S.L. Temperature, precipitation, and insolation effects on autumn vegetation phenology in temperate China. *Glob. Change Biol.* **2016**, *22*, 644–655. [CrossRef]
19. Ren, S.L.; Chen, X.Q.; An, S. Assessing plant senescence reflectance index-retrieved vegetation phenology and its spatiotemporal response to climate change in the Inner Mongolian Grassland. *Int. J. Biometeorol.* **2017**, *61*, 601–612. [CrossRef]
20. Luo, M.; Meng, F.H.; Sa, C.L.; Duan, Y.C.; Bao, Y.H.; Liu, T.; De Maeyer, P. Response of vegetation phenology to soil moisture dynamics in the Mongolian Plateau. *CATENA* **2021**, *206*, 105505. [CrossRef]
21. Cleverly, J.; Eamus, D.; Coupe, N.R.; Chen, C.; Maes, W.; Li, L.; Faux, R.; Santini, N.S.; Rumman, R.; Yu, Q.; et al. Soil moisture controls on phenology and productivity in a semi-arid critical zone. *Sci. Total Environ.* **2016**, *568*, 1227–1237. [CrossRef]
22. Garonna, I.; De Jong, R.; De Wit, A.J.W.; Mucher, C.A.; Schmid, B.; Schaepman, M.E. Strong contribution of autumn phenology to changes in satellite-derived growing season length estimates across Europe (1982–2011). *Glob. Change Biol.* **2014**, *20*, 3457–3470. [CrossRef]
23. Song, L.L.; Tian, Q.; Li, Z.J.; Lv, Y.M.; Gui, J.; Zhang, B.J.; Cui, Q. Changes in characteristics of climate extremes from 1961 to 2017 in Qilian Mountain area, northwestern China. *Environ. Earth Sci.* **2022**, *81*, 177. [CrossRef]
24. Lin, P.F.; He, Z.B.; Du, J.; Chen, L.F.; Zhu, X.; Li, J. Recent changes in daily climate extremes in an arid mountain region, a case study in northwestern China's Qilian Mountains. *Sci. Rep.* **2017**, *7*, 2245. [CrossRef]
25. Gou, X.H.; Zhang, F.; Deng, Y.; Ettl, G.J.; Yang, M.X.; Gao, L.L.; Fang, K.Y. Patterns and dynamics of tree-line response to climate change in the eastern Qilian Mountains, northwestern China. *Dendrochronologia* **2012**, *30*, 121–126. [CrossRef]
26. Sun, F.; Lü, Y.; Wang, J.; Hu, J.; Fu, B. Soil moisture dynamics of typical ecosystems in response to precipitation: A monitoring-based analysis of hydrological service in the Qilian Mountains. *CATENA* **2015**, *129*, 63–75. [CrossRef]
27. Qi, Y.; Wang, H.W.; Ma, X.F.; Zhang, J.L.; Yang, R. Relationship between vegetation phenology and snow cover changes during 2001–2018 in the Qilian Mountains. *Ecol. Indic.* **2021**, *133*, 108351. [CrossRef]
28. Zhang, L.F.; Yan, H.W.; Qiu, L.S.; Cao, S.P.; He, Y.; Pang, G.J. Spatial and Temporal Analyses of Vegetation Changes at Multiple Time Scales in the Qilian Mountains. *Remote Sens.* **2021**, *13*, 5046. [CrossRef]
29. Zhou, L.; Tucker, C.J.; Kaufmann, R.K.; Slayback, D.; Shabanov, N.V.; Myneni, R.B. Variations in northern vegetation activity inferred from satellite data of vegetation index during 1981 to 1999. *J. Geophys. Res. Atmos.* **2001**, *106*, 20069–20083. [CrossRef]
30. Peng, S. 1-km monthly precipitation dataset for China (1901–2020). *Natl. Tibet. Plateau Data Center.* **2020**. [CrossRef]
31. Gonsamo, A.; Chen, J.M.; Ooi, Y.W. Peak season plant activity shift towards spring is reflected by increasing carbon uptake by extratropical ecosystems. *Glob. Change Biol.* **2018**, *24*, 2117–2128. [CrossRef] [PubMed]
32. Akritas, M.G.; Murphy, S.A.; Lavalley, M.P. The Theil-Sen estimator with doubly censored data and applications to astronomy. *Am. Stat. Assoc.* **1995**, *90*, 170–177. [CrossRef]
33. Wang, X.Y.; Wu, C.Y. Estimating the peak of growing season (POS) of China's terrestrial ecosystems. *Agric. For. Meteorol.* **2019**, *278*, 107639. [CrossRef]
34. Hamed, K.H. Trend detection in hydrologic data: The Mann–Kendall trend test under the scaling hypothesis. *J. Hydrol.* **2008**, *349*, 350–363. [CrossRef]
35. Sun, Y.F.; Guan, Q.Y.; Wang, Q.Z.; Yang, L.Q.; Pan, N.H.; Ma, Y.R.; Luo, H.P. Quantitative assessment of the impact of climatic factors on phenological changes in the Qilian Mountains, China. *For. Ecol. Manag.* **2021**, *499*, 119594. [CrossRef]
36. Qiao, C.; Shen, S.; Cheng, C.; Wu, J.; Jia, D.; Song, C. Vegetation Phenology in the Qilian Mountains and Its Response to Temperature from 1982 to 2014. *Remote Sens.* **2021**, *13*, 286. [CrossRef]

37. Piao, S.; Cui, M.; Chen, A.; Wang, X.; Ciais, P.; Liu, J.; Tang, Y. Altitude and temperature dependence of change in the spring vegetation green-up date from 1982 to 2006 in the Qinghai-Xizang Plateau. *Agric. For. Meteorol.* **2011**, *151*, 1599–1608. [CrossRef]
38. Shen, M.G.; Piao, S.L.; Dorji, T.; Liu, Q.; Cong, N.; Chen, X.Q.; An, S.; Wang, S.P.; Wang, T.; Zhang, G.X. Plant phenological responses to climate change on the Tibetan Plateau: Research status and challenges. *Natl. Sci. Rev.* **2015**, *2*, 454–467. [CrossRef]
39. Zeng, L.L.; Wardlow, B.D.; Xiang, D.X.; Hu, S.; Li, D.R. A review of vegetation phenological metrics extraction using time-series, multispectral satellite data. *Remote Sens. Environ.* **2020**, *237*, 111511. [CrossRef]
40. Tao, Z.X.; Huang, W.J.; Wang, H.J. Soil moisture outweighs temperature for triggering the green-up date in temperate grasslands. *Theor. Appl. Climatol.* **2020**, *140*, 1093–1105. [CrossRef]
41. Huang, K.; Zu, J.X.; Zhang, Y.J.; Cong, N.; Liu, Y.J.; Chen, N. Impacts of snow cover duration on vegetation spring phenology over the Tibetan Plateau. *J. Plant Ecol.* **2018**, *12*, 583–592. [CrossRef]
42. Chen, L.; Hänninen, H.; Rossi, S.; Smith, N.G.; Pau, S.; Liu, Z.; Feng, G.; Gao, J.; Liu, J. Leaf senescence exhibits stronger climatic responses during warm than during cold autumns. *Nat. Clim. Change* **2020**, *10*, 777–780. [CrossRef]
43. Cleland, E.E.; Chuine, I.; Menzel, A.; Mooney, H.A.; Schwartz, M.D. Shifting plant phenology in response to global change. *Trends. Ecol. Evol.* **2007**, *22*, 357–365. [CrossRef]
44. Che, M.L.; Chen, B.Z.; Innes, J.L.; Wang, G.Y.; Dou, X.M.; Zhou, T.M.; Zhang, H.F.; Yan, J.W.; Xu, G.; Zhao, H.W. Spatial and temporal variations in the end date of the vegetation growing season throughout the Qinghai-Tibetan Plateau from 1982 to 2011. *Agric. For. Meteorol.* **2014**, *189*, 81–90. [CrossRef]
45. Yang, Y.T.; Guan, H.D.; Shen, M.G.; Liang, W.; Jiang, L. Changes in autumn vegetation dormancy onset date and the climate controls across temperate ecosystems in China from 1982 to 2010. *Glob. Change Biol.* **2015**, *21*, 652–665. [CrossRef]
46. Shen, M.G.; Tang, Y.H.; Chen, J.; Zhu, X.L.; Zheng, Y.H. Influences of temperature and precipitation before the growing season on spring phenology in grasslands of the central and eastern Qinghai-Tibetan Plateau. *Agric. For. Meteorol.* **2011**, *151*, 1711–1722. [CrossRef]
47. Peng, X.M.; Du, J.; Yang, B.; Xiao, S.C.; Li, G. Elevation-influenced variation in canopy and stem phenology of Qinghai spruce, central Qilian Mountains, northeastern Tibetan Plateau. *Trees* **2019**, *33*, 707–717. [CrossRef]
48. Derner, J.D.; Hess, B.W.; Olson, R.A.; Schuman, G.E. Functional group and species responses to precipitation in three semi-arid rangeland ecosystems. *Arid Land Res. Manag.* **2008**, *22*, 81–92. [CrossRef]
49. Dorji, T.; Totland, O.; Moe, S.R.; Hopping, K.A.; Pan, J.B.; Klein, J.A. Plant functional traits mediate reproductive phenology and success in response to experimental warming and snow addition in Tibet. *Glob. Chang. Biol.* **2013**, *19*, 459–472. [CrossRef]
50. Zhang, R.R.; Qi, J.Y.; Leng, S.; Wang, Q.F. Long-term vegetation phenology changes and responses to preseason temperature and precipitation in Northern China. *Remote Sens.* **2022**, *14*, 1396. [CrossRef]
51. Zhang, X.Y.; Jayavelu, S.; Liu, L.L.; Friedl, M.A.; Henebry, G.M.; Liu, Y.; Schaaf, C.B.; Richardson, A.D.; Gray, J. Evaluation of land surface phenology from VIIRS data using time series of PhenoCam imagery. *Agric. For. Meteorol.* **2018**, *256*, 137–149. [CrossRef]
52. Sonnentag, O.; Hufkens, K.; Teshera-Sterne, C.; Young, A.M.; Friedl, M.; Braswell, B.H.; Milliman, T.; O'Keefe, J.; Richardson, A.D. Digital repeat photography for phenological research in forest ecosystems. *Agric. For. Meteorol.* **2012**, *152*, 159–177. [CrossRef]
53. Fischer, A. A model for the seasonal variations of vegetation indices in coarse resolution data and its inversion to extract crop parameters. *Remote Sens. Environ.* **1994**, *48*, 220–230. [CrossRef]
54. Moody, A.; Johnson, D.M. Land-surface phenologies from AVHRR using the discrete fourier transform. *Remote Sens. Environ.* **2001**, *75*, 305–323. [CrossRef]
55. Huete, A.; Didan, K.; Miura, T.; Rodriguez, E.P.; Gao, X.; Ferreira, L.G. Overview of the radiometric and biophysical performance of the MODIS vegetation indices. *Remote Sens. Environ.* **2002**, *83*, 195–213. [CrossRef]
56. Walker, J.J.; de Beurs, K.; Wynne, R.H. Dryland vegetation phenology across an elevation gradient in Arizona, USA, investigated with fused MODIS and Landsat data. *Remote Sens. Environ.* **2014**, *144*, 85–97. [CrossRef]
57. Wu, C.Y.; Gonsamo, A.; Gough, C.M.; Chen, J.M.; Xu, S.G. Modeling growing season phenology in North American forests using seasonal mean vegetation indices from MODIS. *Remote Sens. Environ.* **2014**, *147*, 79–88. [CrossRef]
58. Gonsamo, A.; Chen, J.M.; Price, D.T.; Kurz, W.A.; Wu, C.Y. Land surface phenology from optical satellite measurement and CO_2 eddy covariance technique. *J. Geophys. Res. Biogeosci.* **2012**, *117*, 1472. [CrossRef]
59. Kim, J.B.; Kerns, B.K.; Drapek, R.J.; Pitts, G.S.; Halofsky, J.E. Simulating vegetation response to climate change in the Blue Mountains with MC2 dynamic global vegetation model. *Clim. Serv.* **2018**, *10*, 20–32. [CrossRef]